ESO ASTROPHYSICS SYMPOSIA
European Southern Observatory

Series Editor: Bruno Leibundgut

S. Randich   L. Pasquini   (Eds.)

Chemical Abundances and Mixing in Stars in the Milky Way and its Satellites

Proceedings
of the ESO Workshop held
in Castiglione della Pescaia,
Italy, 13-17 September, 2004

 Springer

Volume Editors

Sofia Randich
Osservatorio di Arcetri
Largo E. Fermi, 5
50125 Firenze, Italy

Luca Pasquini
European Southern Observatory
Karl-Schwarzschild-Str. 2
85748 Garching, Germany

Series Editor

Bruno Leibundgut
European Southern Observatory
Karl-Schwarzschild-Str. 2
85748 Garching
Germany

ISBN-10   3-540-34135-8 Springer Berlin Heidelberg New York
ISBN-13   978-3-540-34135-2 Springer Berlin Heidelberg New York

Library of Congress Control Number: 2006925844

Springer is a part of Springer Science+Business Media
springer.com
© Springer-Verlag Berlin Heidelberg 2006
Printed in The Netherlands

Typesetting: by the authors and techbooks using a Springer LaTeX macro package
Cover design: *design & production* GmbH, Heidelberg

Printed on acid-free paper   SPIN: 11741299   55/techbooks   5 4 3 2 1 0

Preface

Ten years after the first ESO Workshop on light elements, and five years after the IAU in conference in Natal (Brazil) on the same subject, we felt that it was time to have a new conference on this topic: a wealth of new observational data have become available from the 8-meter telescopes and in particular from the VLT, and, at the same time, in the last years new theoretical roads have also been inaugurated in the interpretation of the stellar data. It soon became clear, on the other hand, that in order to understand the evolution of the fragile Li and Be in stars and in the Galaxy, the whole problem of internal mixing in stars must be addressed and understood first, and therefore a large number of elements must be investigated, in many different environments.

Adding to this general context the fact that the FLAMES facility at the VLT was offered to the community one year earlier, and that it was starting to produce an impressive wealth of abundance data of stars in the Galaxy and in our neighborhoods, the broad concept of the ESO–Arcetri workshop on "Chemical Abundances and Mixing in Stars in the Milky Way and its Satellites" was built.

The location had to be in our beloved Toscana, and Castiglione della Pescaia was immediately identified as an ideal place.

To achieve from this initial proposal a full conference, which gathered together from September 13 to 17 more than 150 attendees in a smooth organization, was possible only due to the dedicated work of the conference SOC (J. Andersen, G. Gilmore, G. Meynet, P. Molaro, N. Prantzos, R. Rood, C. Sneden, M. Spite, M. Tosi, and A. Weiss), who contributed to the selection of the main topics and speakers and oversaw the scientific organization of the meeting. Special thanks go to E. Masini, and P. Bristow, D. Galli, P. Sestito, C. Stoffer, and C. Travaglio who, as part of the LOC, made our stay productive and pleasant, and to G. Pace and J.D. do Nascimento who trained themselves considerably by walking continuously during the conference days to bring the microphones to all the participants of the lively discussions. A special thank goes to M. Felli, who told us all the secrets of the beautiful region around Castiglione della Pescaia. The workshop would not have been possible without the support of ESO and of the Arcetri Observatory, but a successful workshop needs most of all the enthusiastic contribution of all the participants, to whom we are grateful.

We regret that at least three of our speakers were not able to submit their written contribution, but, due to what is likely a statistical spike, they were quite busy just after the conference in a more fundamental business; we thank them for their participation and we wish all the best to their babies.

Arcetri and Garching,   *Sofia Randich*
December 2005   *Luca Pasquini*

Contents

Part II   Abundances in the Spheroidal Component

Part IV Local Group Galaxies

Part VI   Implications for BBN and Galaxy Formation and Evolution

List of Participants

Abia, Carlos
Universidad de Granada
cabia@ugr.es

Acker, Agnes
Observatoire de Strasbourg
acker@astro.u-strasbg.fr

Affer, Laura
INAF – Osservatorio Astronomico
di Palermo
affer@astropa.unipa.it

Ahumada, Andrea V.
Obs. Astronómico,
Univ. Nacional Córdoba
andrea@mail.oac.uncor.edu

Allen, Dinah M.
University of Sao Paulo, IAG
dinah@astro.iag.usp.br

Aoki, Wako
National Astronomical Observatory
of Japan
aoki.wako@nao.ac.jp

Asplund, Martin
Mt. Stromlo Observatory,
Australian National Univ.
martin@mso.anu.edu.au

Balachandran, Suchitra C.
University of Maryland
suchitra@astro.umd.edu

Balser, Dana S.
National Radio Astronomy Obs.,
Green Bank
dbalser@nrao.edu

Bania, Thomas M.
Inst. for Astrophysical Research,
Boston Univ.
bania@bu.edu

Barbuy, Beatriz
University of Sao Paulo, IAG
barbuy@astro.iag.usp.br

Basu, Sarbani
Yale University
basu@astro.yale.edu

Battaglia, Giuseppina
Kapteyn Institute
gbattagl@astro.rug.nl

Bienaymé, Olivier
Observatoire de Strasbourg
bienayme@astro.u-strasbg.fr

Bonifacio, Piercarlo
INAF – Osservatorio Astronomico
di Trieste
bonifaci@ts.astro.it

Bragaglia, Angela
INAF – Osservatorio Astronomico
di Bologna
angela.bragaglia@bo.astro.it

Bruntt, Hans
Dept. of Physics & Astronomy,
Aarhus University
bruntt@phys.au.dk

Calamida, Annalisa
Univ. di Roma Tor Vergata,
Osservatorio di Roma
calamida@mporzio.astro.it

Canto Martins, Bruno L.
GRAAL –
Université Montpellier II
canto@graal.univ-montp2.fr

Carigi, Leticia
Instituto de Astronomía – UNAM
carigi@astroscu.unam.mx

Carrera, Ricardo
Instituto de Astrofísica de Canarias
rcarrera@ll.iac.es

Carretta, Eugenio
INAF – Osservatorio Astronomico
di Bologna
carretta@pd.astro.it

Chaname, Julio
Ohio State University,
Dept. of Astronomy
jchaname
@astronomy.ohio-state.edu

Charbonnel, Corinne
LAOMP & Observatoire de Genève
Corinne.Charbonnel@obs.unige.ch

Chiappini, Cristina
INAF – Osservatorio Astronomico
di Trieste
chiappini@ts.astro.it

Clariá, Juan
Obs. Astronómico,
Univ. Nacional Córdoba
claria@mail.oac.uncor.edu

Cohen, Judith
California Institute of Technology
jlc@astro.caltech.edu

Collet, Remo
Uppsala Astronomical Observatory
remo@astro.uu.se

Cunha, Katia
University of Texas at El Paso
katia@baade.physics.utep.edu

D'Antona, Francesca
Osservatorio Astronomico di Roma
dantona@mporzio.astro.it

Da Silva, Licio
Observatorio Nacional
and DFTE-UFRN, Brazil
licio@on.br

De La Reza, Ramiro
Observatorio Nacional/MCT – Brazil
delareza@on.br

De Laverny, Patrick
Observatoire de la Côte d'Azur
laverny@obs-nice.fr

De Medeiros, Jose R.
Univ. Federal do Rio Grande
do Norte, Fisica
renan@ufrnet.br

Depagne, Eric
ESO – Chile
edepagne@eso.org

Do Nascimento, José Dias
Univ. Federal do Rio Grande do Norte
dias@dfte.ufrn.br

Ecuvillon, Alexandra
Instituto de Astrofísica de Canarias
aecuvill@ll.iac.es

Fabbian, Damian
RSAA, Mt. Stromlo Obs.,
Australian National Univ.
damian@mso.anu.edu.au

Feltzing, Sofia
Lund Observatory
sofia@astro.lu.se

Fenner, Yeshe
Swinburne University
yfenner@astro.swin.edu.au

Font, Andreea S.
Wesleyan University
afont@astro.wesleyan.edu

Franchini, Mariagrazia
INAF – Osservatorio Astronomico
di Trieste
franchini@ts.astro.it

François, Patrick
Observatoire de Paris
patrick.Francois@obspm.fr

Friel, Eileen D.
US National Science Foundation
efriel@nsf.gov

Frinchaboy, Peter M.
University of Virginia,
Dept. of Astronomy
pmf8b@virginia.edu

Fulbright, Jon
Carnegie Observatories
jfulb@ociw.edu

Galli, Daniele
INAF – Osservatorio Astrofisico
di Arcetri
galli@arcetri.astro.it

Gallino, Roberto
Dip. di Fisica Generale,
Universitá di Torino
gallino@ph.unito.it

Geisler, Doug
Universidad de Concepción,
Depto. de Fisica
doug@kukita.cfm.udec.cl

Gilmore, Gerry
Institute of Astronomy, Cambridge
gil@ast.cam.ac.uk

Girard, Pascal
Observatoire de Bordeaux
girard@obs.bordeaux1.fr

Grebel, Eva K.
Astronomical Institute,
University of Basel
grebel@astro.unibas.ch

Grundahl, Frank
Dept. of Physics & Astronomy,
Aarhus University
fgj@phys.au.dk

Herwig, Falk H.
Los Alamos National Laboratory
fherwig@lanl.gov

Hill, Vanessa
Observatoire de Paris-Meudon
Vanessa.Hill@obspm.fr

Ishimaru, Yuhri
Ochanomizu University, Tokyo
ishimaru@phys.ocha.ac.jp

Israelian, Garik
Instituto de Astrofísica de Canarias
gil@iac.es

Ivans, Inese
California Institute of Technology
iii@astro.caltech.edu

James, Gael
Observatoire de Paris
Gael.James@obspm.fr

Jeffries, Rob
Keele University
rdj@astro.keele.ac.uk

Johnson, Jennifer A.
DAO/HIA/NRC
Jennifer.Johnson@nrc.gc.ca

Käufl, Hans Ulrich
ESO – Garching
hukaufl@eso.org

Kaufer, Andreas
ESO – Chile
akaufer@eso.org

Koch, Andreas
Astronomical Institute,
University of Basel
koch@astro.unibas.ch

Kolaczkowski, Zbigniew
Astronomical Institute,
Wroclaw Univ.
kolaczk@astro.uni.wroc.pl

Korn, Andreas J.
Uppsala Astronomical Observatory
akorn@astro.uu.se

La Cognata, Marco
Università di Catania & INFN-LNS
lacognata@lns.infn.it

Lamia, Livio
Università di Catania & INFN-LNS
llamia@lns.infn.it

Lèbre, Agnes
GRAAL – Université Montpellier
lebre@graal.univ-montp2.fr

Lecureur, Aurelie
Observatoire de Paris-Meudon
aurelie.lecureur@obspm.fr

Lee, Jae-Woo
Dept. Astronomy & Space Science,
Sejong Univ.
jaewoo@arcsec.sejong.ac.kr

Lee, Jung-Kyu
Queen's University, Belfast
j.k.lee@qub.ac.uk

Letarte, Bruno
Kapteyn Institute
bruno@astro.rug.nl

Lucatello, Sara
INAF – Osservatorio Astronomico
di Padova
lucatello@pd.astro.it

Lutz, Julie
University of Washington,
Dept. of Astronomy
jlutz@astro.washington.edu

Maciel, Walter J.
University of Sao Paulo, IAG
maciel@astro.iag.usp.br

Maeder, André W.
Observatoire de Genève
andre.maeder@obs.unige.ch

Malagnini, Maria Lucia
Dip. di Astronomia,
Università di Trieste
malagnini@ts.astro.it

Matteucci, Francesca
Dip. di Astronomia,
Università di Trieste
matteucci@ts.astro.it

McWilliam, Andrew
Carnegie Observatories
andy@ociw.edu

Melo, Claudio
ESO – Chile
cmelo@eso.org

Meynet, Georges
Observatoire de Genève
georges.meynet@obs.unige.ch

Molaro, Paolo
INAF – Osservatorio Astronomico
di Trieste
molaro@ts.astro.it

Momany, Yazan
Dipartimento di Astronomia,
Università di Padova
momany@pd.astro.it

Monaco, Lorenzo
INAF – Osservatorio Astronomico
di Trieste
lmonaco@ts.astro.it

Monelli, Matteo
Osservatorio Astronomico di Roma
monelli@mporzio.astro.it

Morel, Thierry
INAF – Osservatorio Astronomico
di Palermo
morel@astropa.unipa.it

Morossi, Carlo
INAF – Osservatorio Astronomico
di Trieste
morossi@ts.astro.it

Mottini, Marta
ESO – Garching
mmottini@eso.org

North, Pierre
Laboratoire d'Astrophysique
de l'EPFL
pierre.north@epfl.ch

Pace, Giancarlo
ESO – Garching
gpace@eso.org

Palacios, Ana
Inst. d'Astron. & d'Astrophys. –
Univ. Libre Bruxelles
palacios@astro.ulb.ac.be

Pallavicini, Roberto
INAF – Osservatorio Astronomico
di Palermo
pallavic@astropa.unipa.it

Pancino, Elena
INAF – Osservatorio Astronomico
di Bologna
elena.pancino@bo.astro.it

Pasquini, Luca
ESO – Garching
lpasquin@eso.org

Perinotto, Mario
Dipartimento di Astronomia,
Università di Firenze
mario@arcetri.astro.it

Peterson, Ruth C.
UCO/Lick
and Astrophysical Advances
peterson@ucolick.org

Pignatari, Marco
Dip. di Fisica Generale,
Università di Torino
pignatar@ph.unito.it

Pinsonneault, Marc
Ohio State University,
Dept. of Astronomy
pinsonneault.1@osu.edu

Pipino, Antonio
Dipartimento di Astronomia,
Università di Trieste
antonio@ts.astro.it

Pizzone, Rosario G.
INFN LNS Catania
rgpizzone@lns.infn.it

Prantzos, Nikos
Institut d'Astrophysique de Paris
prantzos@iap.fr

Primas, Francesca
ESO – Garching
fprimas@eso.org

Prisinzano, Loredana
INAF – Osservatorio Astronomico
di Palermo
loredana@astropa.unipa.it

Randich, Sofia
INAF – Osservatorio Astrofisico
di Arcetri
randich@arcetri.astro.it

Recio-Blanco, Alejandra
Observatoire de la Côte d'Azur
arecio@obs-nice.fr,
recio@pd.astro.it

Romaniello, Martino
ESO – Garching
mromanie@eso.org

Romano, Donatella
INAF – Osservatorio Astronomico
di Bologna
donatella.romano@bo.astro.it

Rood, Robert T.
University of Virginia
rtr@virginia.edu

Rossi, Silvia
University of Sao Paulo, IAG
rossi@astro.iag.usp.br

Royer, Frédéric
Observatoire de Genève
frederic.royer@obs.unige.ch

Ryan, Sean G.
Open University, Dept. of
Physics & Astronomy (UK)
s.g.ryan@open.ac.uk

Salvati, Marco
INAF – Osservatorio Astrofisico
di Arcetri
salvati@arcetri.astro.it

Santos, Nuno C.
Observatorio Astronomico
de Lisboa/CAAUL
nuno.santos@oal.ul.pt

Sanz Forcada, Jorge
ESA/ESTEC, Astrophysics Division
(SCI-SA)
jsanz@rssd.esa.int

Sbordone, Luca
ESO and
Università di Roma Tor Vergata
lsbordon@eso.org

Searle, Samantha C.
Univ. College London,
Physics & Astronomy
scs@star.ucl.ac.uk

Serenelli, Aldo M.
Institute for Advanced Study,
Princeton
aldos@ias.edu

Sestito, Paola
Dipartimento di Astronomia,
Università di Firenze
sestito@arcetri.astro.it

Shetrone, Matthew D.
Univ. of Texas,
McDonald Observatory
shetrone@astro.as.utexas.edu

Simmerer, Jennifer
Univ. of Texas at Austin,
Dept. of Astronomy
jensim@astro.as.utexas.edu

Smartt, Stephen J.
Queen's University, Belfast
S.Smartt@qub.ac.uk

Soderblom, David R.
Space Telescope Science Institute
drs@stsci.edu

Sollima, Antonio
INAF – Osservatorio Astronomico
di Bologna
s_sollima@astbo4.bo.astro.it

Soubiran, Caroline
Observatoire de Bordeaux
soubiran@obs.u-bordeaux1.fr

Spite, Monique
Observatoire de Paris-Meudon
monique.spite@obspm.fr

Steigman, Gary
Ohio State University
steigman@mps.ohio-state.edu

Szeifert, Thomas
ESO – Chile
tszeifer@eso.org

Tautvaišienė, Gražina
Vilnius Univ.,
Theoretical Phys. & Astron.
taut@itpa.lt

Tolstoy, Eline
Kapteyn Institute
etolstoy@astro.rug.nl

Tosi, Monica
INAF – Osservatorio Astronomico
di Bologna
monica.tosi@bo.astro.it

Travaglio, Claudia
Osservatorio Astronomico di Torino
travaglio@to.astro.it

Tsivilev, Alexander P.
PRAO ASC (Russia) & IRA
(CNR-Bologna)
tsivilev@prao.psn.ru

Tsujimoto, Takuji
National Astronomical Observatory
of Japan
taku.tsujimoto@nao.ac.jp

Umeda, Hideyuki
University of Tokyo,
Dept. of Astronomy
umeda@astron.s.u-tokyo.ac.jp

Valenti, Elena
Università di Bologna
elena.valenti2@studio.unibo.it

Venn, Kim A.
MacAlester College
venn@macalester.edu

Ventura, Paolo
Osservatorio Astronomico di Roma
ventura@mporzio.astro.it

Wallerstein, George
University of Washington
wall@astro.washington.edu

Wanajo, Shinya
Sophia University, Tokyo
wanajo@sophia.ac.jp

Weiss, Achim
MPI für Astrophysik, Garching
weiss@mpa-garching.mpg.de

Zaggia, Simone
INAF – Osservatorio Astronomico
di Trieste
zaggia@ts.astro.it

Part I

Abundances in the Thin and Thick Disks

Metallicities and α-Abundances in Open Clusters

E.D. Friel

National Science Foundation, Arlington, VA, 22230, USA

Abstract. Galactic open clusters exhibit a range of properties that make them valuable probes of stellar and galactic chemical evolution. Metallicities have been determined for open clusters through a variety of photometric and spectroscopic techniques. The status of available data is reviewed and an overview of what these metallicities reveal about Galactic chemical evolution is given. Although many fewer determinations of elemental abundances in open clusters have been made, they serve as useful indicators of nucleosynthesis in previous generations. The available data suggest that the open clusters present a fairly uniform population similar to that of the disk field stars in the solar neighborhood, even to ages of 10 Gyr. There are intriguing indications that the outermost clusters in the disk deviate from the overall abundance gradient and show enhanced α-elements.

1 Introduction and Overview

The properties of clusters reflect those of the interstellar medium at their time and place of formation. As a result, clusters are excellent tracers of galactic chemical evolution. Open clusters span a large range of ages and distances throughout the Galaxy, providing probes over the entire age of the disk and to distances up to 22 kpc from the Galactic center. They allow us to look at how the enrichment of the Galaxy has proceeded, through gradients in the abundance, and their variation with time. They also offer a view that is complementary to that provided by the field stars or other cluster populations in the Galaxy.

In this paper I discuss overall metallicity, a measure of the overall heavy-element abundance in the star, and direct determination of elemental abundances and abundance ratios of Fe, O, the α-elements Mg, Si, Ca, and Ti, and also the light elements Na and Al.

2 Open Cluster Metallicities

2.1 Techniques for Abundance Determinations

Photometric methods are widely used to determine overall metallicity, and measure the blanketing in either broad or narrow bands due primarily to Fe and Fe-peak blends, or common molecular features, such as CN. These measures can be calibrated to a true [Fe/H] or can be interpreted through analysis of color-magnitude diagrams with the help of theoretical isochrones through their sensitivity to metallicity. Photometric measures have the advantages of reaching large samples and faint, distant objects, but suffer from contamination by non-members and reddening.

Low resolution spectroscopy, by comparison, has the advantage of providing discrimination against cluster non-members through the use of radial velocities, and can still reach large samples of reasonably faint objects. Common metallicity indicators from low resolution work are Fe and Fe-peak blends, CN, and the Ca II infrared triplet, which are calibrated against high-resolution abundance analyses.

Ultimately, however, one seeks abundances from high-resolution spectroscopy, and a full elemental abundance analysis. Data on a broad range of elemental abundances in open clusters have been limited until recently. Fortunately, this situation is beginning to change.

2.2   Available Metallicity Data

The current most complete catalog of available data on cluster abundances is [1], which includes approximately 120 clusters with metallicity determinations selected from the literature. These abundances are drawn from many disparate sources which use a variety of techniques and systematic differences between different studies may be significant. Several other large samples on uniform systems, that were the basis for much of the Dias catalog entries are found in [2] which in turn, merges results from spectroscopic samples of [3] and photometric metallicities of [4]. Large samples of clusters analyzed in a uniform and homogeneous way are critical for measuring some of the finer, and most interesting, features of chemical evolution of the disk.

3   Galactic Trends in Metallicity

3.1   Abundance Gradients

The first use of open clusters to trace the Galactic abundance gradient was made by [5] using DDO photometry for a sample of 41 clusters of all ages over a distance range of 8 to 14 kpc and found a gradient of -0.05 dex/kpc. Later work based on spectroscopy [3,10] and photometry [4] and the combined literature samples of the Dias catalog [6] has led to similar results, using expanded samples to yield tighter relationships. For the purposes of comparison to observations and theory, and as the simplest description of the overall decrease of metallicity with distance, these studies have fit the gradient with linear functions. Reference [2], however, combining results from [3] and [4] with new DDO photometry, and with an independent assessment of cluster reddenings, distances, and membership determinations, concluded that the abundance gradient is discontinuous, with a break at 10 kpc, where the metallicity drops from [Fe/H] of roughly solar to $\sim$ -0.3 dex.

The most distant clusters in these samples reach to Galactocentric distances of 16 kpc, where [Fe/H] $\sim$ -0.5 to -0.6. Recently two open clusters have been discovered at Galactocentric distances of 19 and 22 kpc, and their abundances ([7,8,9]) do not continue this downward trend, suggesting either a flattening of the gradient, or an origin distinct from the rest of the Galactic disk.

There has been mixed or inconclusive evidence for time variations in the gradient from open cluster data. Most studies that comment on the question ([5,10,6,11]) have shown slight evolution from steeper gradients in the past to more shallow ones, though the effect is at the level of only a few sigma. However, [12] find significantly different results, with shallower slopes for older clusters, and very steep slopes for younger clusters. Small sample sizes, and the differing distance coverage of clusters at different ages, as well as possible systematic differences between merged data sets and the adoption of significantly different cluster parameters between studies may strongly affect these results.

Theoretical models of galactic chemical evolution offer varying predictions of how the gradient should change with time, so using open clusters to measure that time-dependence can provide important constraints on model parameters.

3.2   Age-Metallicity Relationships

For years, it has been known that some of the oldest open clusters, such as NGC 188, had roughly solar metallicity, but increasing samples have shown clearly that there is no relationship between overall cluster metallicity and age ([3,10,6,11]) in the solar neighborhood. The existence of clusters like NGC 6791, a 9 Gyr old cluster with metallicity twice solar, requires prompt initial enrichment of the Galactic disk, at least in the solar neighborhood. The metallicity of open clusters depends much more strongly on their position in the disk than on their age.

4   Elemental Abundance Ratios

Only recently have large numbers of open clusters become the subject of elemental abundance determinations other than those of the lightest elements (Li, C). A survey of the literature (not guaranteed to be complete), including some unpublished data kindly made available for this contribution, revealed [Fe/H] determinations for 45 open clusters. Of these, 33 have determinations of at least some of the α-elements, and these are collected in Table 1. There are very few clusters in common between studies, so it is not possible to investigate systematic differences between studies, nor to bring these measures to a common system.

4.1   Oxygen and α-Elements

Oxygen abundances in clusters show solar [O/Fe] ratios overall, and fall within the envelope of the distribution with [Fe/H] displayed by the disk field stars [13]. The α-elements Mg, Ca, and Ti generally show solar ratios as well, while the [Si/Fe] value is enhanced slightly in the average (~ 0.1) in the clusters. Compared to the behavior for field stars, the clusters do not show the slightly enhanced [Mg/Fe] or [Ti/Fe] values seen in the fields stars at the same [Fe/H].

[Ca/Fe] agrees very closely with the field stars, while [Si/Fe] is somewhat more enhanced in the clusters stars compared to the field stars at the same metallicity.

Working with clusters, though, allows us to look at trends of these abundances with age and with Galactocentric distance (Figure 1). There is no evidence for enhanced values of [O/Fe] in the oldest of the clusters (Be 17, Cr 261, and NGC 6791 at 7 -10 Gyr). There is a suggestion that the most distant clusters (Saurer 1 and Be 29) have enhanced [O/Fe], but these values come from the OI infrared triplet, which has been known to yield values systematically higher than values from the forbidden [OI] lines, which are used for the other clusters in the sample.

The α-elements show no strong trends with age either; Figure 1 plots the average of the α-elements for clusters in Table 1. The clusters older than about 1 Gyr may have [α/Fe] up by about 0.1 dex, and show larger scatter relative to those clusters younger than 1 Gyr, particularly if values from [17], which deviate strongly from other clusters in some elements are omitted. The enhancement in [α/Fe] is largest for clusters in the outer disk ($R_{gc} > 12$ kpc), where mean [α/Fe] values reach an average of 0.15 dex compared to an average of 0 for those in the solar neighborhood.

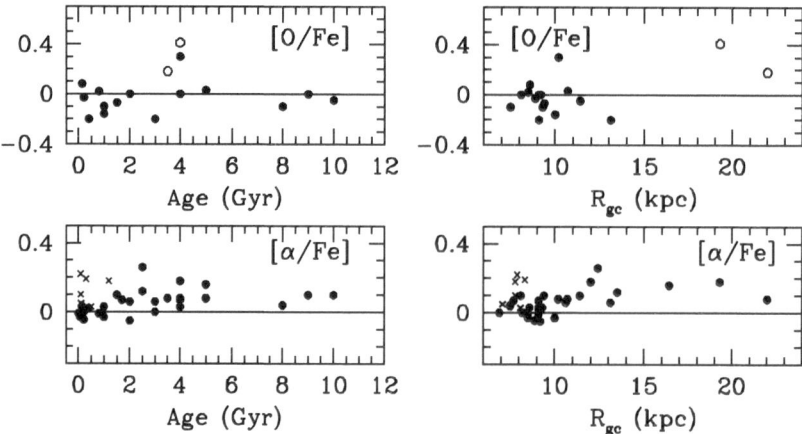

Fig. 1. Cluster mean [α/Fe] and [O/Fe] versus cluster age (left) and Galactocentric distance (right) for clusters in Table 1. The open circles refer to cluster [O/Fe] determinations made from the OI triplet; all other [O/Fe] determinations are from the [OI] forbidden lines. Crosses indicate data from [17] which deviate systematically from the cluster trends in some elements, particularly Mg.

4.2   Sodium and Aluminum

Unlike the field stars, the open clusters show enhanced values of [Na/Fe] (~ 0.2) and [Al/Fe] (~ 0.1). Sodium and aluminum seem to be high and show a large dispersion at all ages and Galactocentric distances. There is recent evidence that

Table 1. Abundances in Open Clusters

Cluster	R_gc (kpc)	Age (Gyr)	[Fe/H]	[O/Fe]	[Mg/Fe]	[Si/Fe]	[Ca/Fe]	[Ti/Fe]	[Na/Fe]	[Al/Fe]	# Stars
Blanco 1[1]	8.5	0.06	+0.23			−0.15	−0.10	−0.04	+0.18	−0.04	4
NGC 1039[2]	8.9	0.2	+0.07	−0.03		−0.10	−0.05	+0.03	−0.06	−0.05	9
Pleiades[3]	8.6	0.14	−0.03	+0.08		+0.07	+0.05	+0.02	−0.02		16
Hyades[4]	8.5	0.8	+0.13	+0.02	−0.06	+0.04	+0.06	−0.06	+0.01	−0.05	55
Be 17[5]	11.4	10	−0.11	−0.05	+0.12	+0.30	−0.02	+0.02	+0.32	+0.25	3
Be 20[6]	16.4	5	−0.72			+0.23	+0.08	+0.23			2
Be 21[6]	13.5	2.5	−0.97			+0.12					1
Be 21[7]			−0.53			+0.32	−0.10	+0.13	+0.45	+0.15	3
NGC 2112[8]	9.2	2	−0.1	0.0	+0.06	−0.34	+0.02	+0.07	+0.08	+0.29	2
NGC 2141[6]	12.4	2.5	−0.47			+0.28	+0.23				1
NGC 2243[9]	10.7	5	−0.48	+0.03	+0.02		+0.18	+0.05	+0.08	+0.08	2
NGC 2264[10]	9.1	0.01	−0.18			+0.01	−0.01	−0.02	−0.02	0.0	3
Be 29[11]	22.0	3.5	−0.44	+0.18	−0.01	+0.22	+0.11	+0.02	+0.39	+0.20	2
Be 29[6]			−0.43			+0.00	+0.03	+0.21			2
Be 31[6]	12.0	4	−0.84			+0.23	+0.14	+0.46			1
To 2[8]	13.1	3	−0.4	−0.2	+0.13	+0.20	−0.15	+0.04	−0.25	+0.17	3
NGC 2360[12]	9.3	1	+0.07	−0.1	0.0	−0.04	+0.08	+0.06	+0.10		4
Saurer 1[11]	19.3	4	−0.38	+0.41	+0.03	+0.38	+0.19	+0.12	+0.44	+0.33	2
Melotte 66[9]	10.2	4	−0.38	+0.3	+0.04		+0.08	+0.13	+0.12	+0.25	2
Melotte 71[8]	10.0	1	−0.32	−0.16	+0.01	+0.14	−0.13	−0.15	+0.19	+0.23	2
NGC 2420[13]	10.6	2	−0.47				+0.04	+0.09			4
NGC 2447[12]	9.1	0.4	+0.03	−0.2	−0.02	0.00	+0.02	+0.09	+0.18		3
M 67[14]	9.1	4	−0.03	0.0	+0.10	+0.10	+0.05	+0.04	+0.19	+0.14	9
M 67[6]			+0.03			+0.10	−0.03	+0.20			3
NGC 3532[15]	8.3	0.3	+0.06		+0.53	+0.30	−0.09	+0.03	+0.31	+0.03	7
Cr 261[16]	7.5	8	−0.22	−0.1	+0.07	+0.21	−0.04	−0.09	+0.48	+0.39	4
NGC 5822[15]	7.8	1.2	+0.10		+0.56	+0.15	−0.15	+0.16	+0.18	+0.02	1
NGC 6067[15]	7.1	0.12	−0.05		+0.27	+0.25	−0.25	−0.07	+0.40	+0.28	5
NGC 6087[15]	7.8	0.1	−0.01		+0.44	+0.21	+0.03	−0.06	+0.46	+0.33	1
IC 4651[17]	7.7	1.7	+0.10		+0.09	+0.08	0.00	+0.12	+0.19	+0.07	6
IC 4725[15]	7.9	0.09	+0.15		+0.55	+0.38	−0.03	−0.02	+0.37	+0.09	2
IC 4756[15]	8.1	0.5	−0.03		+0.13	+0.17	−0.14	−0.06	+0.11	+0.02	4
NGC 6705[18]	6.9	0.2	+0.10			+0.14	−0.19	+0.05	+0.21	+0.20	10
NGC 6791[19]	8.1	9	+0.4	0.0	+0.2	+0.2	0.0	0.0	+0.4	0.0	1
NGC 6819[20]	8.2	3	+0.09		−0.12	+0.18	−0.04	-0.01	+0.47	−0.07	3
NGC 7789[21]	9.4	1.5	−0.04	−0.07	+0.18	+0.14	+0.14	−0.03	+0.28	+0.18	9

(1) Edvardsson et al, A&A, 293, 75, (1995); (2) Schuler et al, AJ, 125, 2085, (2003); (3) Wilden et al, AJ, 124, 2799, (2002); Schuler et al, ApJL, 602, L117, (2004); (4) Cayrel et al, A&A, 146, 249, (1985); Paulson et al, AJ, 125, 3185, (2003); (5) Friel et al, in prep, (2004); (6) de Almeida & Carney, priv. comm, (2004); (7) Hill & Pasquini, A&A, 348, L21, (1999); (8) Brown et al, AJ, 112, 1551, (1996); (9) Gratton & Contarini, A&A, 283, 911, (1994); (10) King et al, AJ, 533, 944, (2000); (11) Carraro et al, astro-ph/0406679, (2004); (12) Hamdani et al, A&A, 360, 509, (2000); (13) Smith & Suntzeff, AJ, 92, 359, (1987); (14) Tautvaisiene et al, A&A, 360, 499, (2000); (15) Luck, ApJS, 91, 309, (1994); (16) Friel et al, AJ, 126, 2372, (2003); (17) Pasquini et al, A&A, 424, 951, (2004); (18) Gonzalez & Wallerstein, PASP, 112, 1081, (2000); (19) Peterson & Green, ApJ, 502, L39, (1998); (20) Bragaglia et al, AJ, 121, 329, (2001); (21) Tautvaisiene et al, priv. comm., (2004)

Na is higher in giants than in dwarfs in IC 4561 [14], which suggests that we may be seeing signs of dredge-up of processed material, although that is not expected on theoretical grounds for clusters of these metallicities and ages and there are other inconsistencies with this picture.

5   Conclusions

The open clusters are key probes of the chemical evolution of the Galactic disk. Although the abundances of open clusters indicate a fairly uniform population, sharing many characteristics with the field stars of the disk, there are some interesting possible differences, such as indications that oxygen and the α-elements may be enhanced in clusters of the outer disk ($R_{gc} > 12$-15 kpc). This is in contrast to the behavior of these elements in the oldest of the clusters, which show solar abundance ratios even to ages of 10 Gyr. Are the higher α-abundances in the outer disk due to localized contributions from Type II SN to the natal clouds of the clusters, perhaps triggered by infall? Or are the most distant clusters not part of the 'normal' disk at all, but instead associated with the stellar streams of a disrupted satellite galaxy identified as the Monoceros Ring or the Galactic Anticenter Stellar System ([15,16])?

Only additional abundance studies, of both clusters and the field, and the inclusion of additional elements, such as the r- and s-process elements will tell. Fortunately, there are efforts underway, many of them described at this conference, that promise some answers to these questions.

I would like to thank those colleagues who provided results in advance of publication. Any opinions, findings, and conclusions expressed in this material are those of the author and do not necessarily reflect the views of the NSF.

References

1. W. S. Dias et al: A&A, **389**, 871 (2002)
2. B.A. Twarog et al: AJ, **114**, 2556, (1999)
3. E.D. Friel, J.A. Janes: A&A, **267**, 75, (1993)
4. J. Claria et al: AJ, **110**, 2813, (1995)
5. K.A. Janes: AJ, **39**, 135, (1979)
6. L. Chen et al: AJ, **125**, 1397, (2003)
7. G. Carraro et al: astro-ph/0406679, (2004)
8. P. Frinchaboy et al: this meeting (2004)
9. L. de Almeida, B.W. Carney: private comm. (2004)
10. E.D. Friel et al: AJ, **124**, 2693, (2002)
11. G. Carraro et al: MNRAS, **296**, 1045, (1998)
12. M. Salaris et al: A&A, **414**, 163, (2004)
13. B. Edvardsson et al: A&A, **275**, 101, (1993)
14. L. Pasquini et al: A&A, **424**, 951, (2004)
15. H. Newberg et al: ApJ, **569**, 245, (2002)
16. P. Frinchaboy et al: ApJL, **602**, L21, (2004)
17. E. Luck: ApJS, **91**, 309, (1994)

Old Open Clusters as Tracers of Galactic Chemical Evolution: the BOCCE Project

A. Bragaglia and M. Tosi

INAF - Osservatorio Astronomico di Bologna, via Ranzani 1, I-40127 Bologna, Italy

Our Sample of Open Clusters

Open clusters (OCs) are important tools both for stellar and for galactic astrophysics, as tests of stellar evolution theory for low and intermediate mass stars and as tracers of the Galactic disk properties. Since old OCs allow us to probe the lifetime of the Milky Way disk, up to about 10 Gyr ago, they can be used to study the disk evolution with time, and in particular its chemical history.

One question that still needs to be answered concerns the evolution of the chemical abundance gradients in the Galactic disk. Many indicators concur on the existence (and slope) of a radial negative abundance gradient at the present time, but the slope at epochs earlier than a few Gyr ago is still a matter of debate, due to the lack of robust observational constraints. The time behaviour of the gradient is strictly related to the disk formation scenarios, as shown by a wealth of chemical evolution models which predict very different gradient evolutions, in spite of being all able to reproduce the observational data relative to recent epochs and to the solar neighborhood [1].

One problem encountered so far in using OCs to derive the abundance gradient at different epochs, is the inhomogeneity in OCs parameters (distance, age, and metal abundance) found in literature; not only absolute values, but also the ranking can be uncertain. For this reason, we are building a large sample of clusters (at least 30) for which age, distance and metallicity are all derived in a homogeneous way. To this end, we use accurate photometry coupled with the synthetic colour-magnitude diagrams (CMD) technique [2], and high resolution spectroscopy with fine abundance analysis. Detailed descriptions of our procedures are presented e.g. in [3] and [4], or [5] and [6], respectively.

We have collected photometric data for about 30 OCs, and have determined their parameters through synthetic CMDs for about half of them. A paper presenting homogeneous results and discussing what is possible to learn from our sample is in preparation. An example is shown in Fig. 1: from our database it is possible e.g., to calibrate the difference in magnitude between clump and turn off as a function of age (Fig. 1a), to notice the absence of relation between age and Galactocentric distance or metallicity (Fig. 1b,c), to study the distribution of metallicity with Galactocentric distance (i.e. the abundance gradient: Fig. 1d). In the latter panel we have also used information coming in part from our spectroscopic work ([5], [6]), and in part from literature ([7], [8], [9]). While photometric and spectroscopic metallicities do reasonably agree (the most discrepant cases being NGC6253 for which the spectroscopic value is only preliminary, and

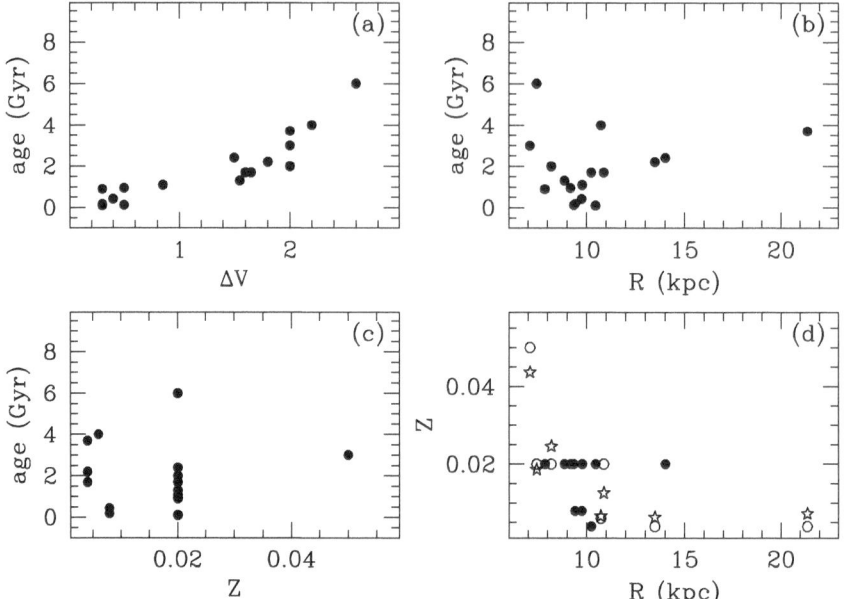

Fig. 1. Results based on 17 OC's: (a) Difference in magnitude between the red clump and the turn-off as a function of age; (b) and (c) age distribution with Galactocentric distance or metallicity; (d) radial distribution of metallicity (filled dots refer to photometric metallicity; when metallicity from high resolution spectroscopy is also available it is represented by asterisks and photometric metallicity by open dots).

NGC2506, for which new and better spectra have now been acquired), we will be able to correctly interpret the run of abundances with radius only when all measures will come from a precise and homogeneous spectroscopic analysis. Spectra of several OCs have already been collected using SARG, FEROS, UVES and FLAMES, but this part of the program is in a less advanced status. since abundance analysis for these metal rich stars is very time consuming.

References

1. M. Tosi: ASP Conference Series **98**, 299 (1996)
2. M. Tosi, L. Greggio, G. Marconi, P. Focardi: AJ **102**, 951 (1991)
3. M. Tosi, L. Di Fabrizio, A. Bragaglia, P.A. Carusillo, G. Marconi: MNRAS **354**, 225 (2004)
4. J.S. Kalirai, M. Tosi: MNRAS **351**, 649 (2004)
5. A. Bragaglia, *et al.*: AJ **121**, 327 (2001)
6. E. Carretta, A. Bragaglia, R.G. Gratton, M. Tosi: A&A **422**, 951 (2004)
7. R.G. Gratton, G. Contarini: A&A **283**, 911 (1994)
8. V. Hill, L. Pasquini: A&A **348**, L21 (1999)
9. G. Carraro, F. Bresolin, S. Villanova, F. Matteucci, F. Patat, M. Romaniello: AJ, **128**, 1676 (2004)

Chemical Abundances and Mixing in Red Clump Stars of the Galaxy

G. Tautvaišienė[1], B. Edvardsson[2], E. Puzeras[1], E. Stasiukaitis[1], and I. Ilyin[3]

[1] Vilnius University Institute of Theoretical Physics and Astronomy, Gostauto 12, Vilnius 01108, Lithuania
[2] Department of Astronomy and Space Physics, Uppsala Astronomical Observatory, Box 515, S-751 20 Uppsala, Sweden
[3] Astrophysikalisches Institut Potsdam, An der Sternwarte 16, D-14482 Potsdam, Germany

Abstract. In this contribution we present the results based on high-resolution spectra of 45 clump stars of the Galactic field. The main atmospheric parameters and abundances of ^{12}C, ^{13}C, N, O and other mixing sensitive chemical elements were investigated. Elemental ratios in the sample of field stars are compared to the results available for evolved stars in open clusters and to the theoretical prediction of extra mixing in stellar interiors.

1 Introduction

More than thirty years have passed since the first discovery of helium-core-burning giants in the Galactic field (Sturch & Helfer 1971). However, because of difficulties of identification (see Tautvaišienė 1996 for a review), the progress in high-resolution spectral analysis of these evolved stars was rather slow. The *Hipparcos* orbiting observatory revealed a large population of helium-core-burning "clump" giants in the Galactic field (Perryman et al. 1997). These stars are bright enough for a detailed spectroscopic investigation, thus are excellent targets for the analysis of mixing processes in evolved low-mass stars.

2 Observations and Method of Analysis

The red clump stars were selected from the Hipparcos Catalog (Perryman 1997). The spectra for 17 stars were obtained at the Nordic Optical Telescope (La Palma) with the SOFIN échelle spectrograph in 2001. The 2nd optical camera ($R \approx 60,000$) was used to observe simultaneously 13 spectral orders, each of $40 - 60$ Å in length, located between 5600 Å and 8130 Å. The spectra for 18 stars were obtained at the Elginfield Observatory (Canada) with the 1.2 m telescope and the high-resolution spectrograph ($R \approx 100,000$). The spectral interval from 6220 Å to 6270 Å was observed. The sample of spectra was supplemented by spectroscopic observations ($R \approx 37,000$) of 15 red clump stars obtained with the 2.16 m telescope of Beijing Astronomical Observatory (China) taken from the literature (Zhao et al. 2001). The spectra were analyzed using a differential model atmosphere technique. The program packages, developed at the Uppsala

Astronomical Observatory, were used to carry out the calculations of theoretical equivalent widths of lines, synthetic spectra and a set of plane parallel, line-blanketed, flux constant LTE model atmospheres. The effective temperatures of the stars were determined from photometry, the infrared flux method and corrected, if needed, in order to achieve the LTE excitation balance in the iron abundance results. The gravities were found by forcing Fe I and Fe II to yield the same iron abundances. The microturbulent velocities were determined by forcing Fe I line abundances to be independent of the equivalent width. For more details on the method of analysis and atomic data see Tautvaišienė et al. (2001).

3   Results

The sample of field clump stars investigated form quite a homogeneous sample. The ratios of [C/Fe] in the stars lie below the trend obtained for the dwarf stars of the Galactic field (Gustafsson et al. 1999) which is in agreement with results of other red helium-core burning stars (Tautvaišienė et al. 2001, Gratton et al. 2000, Charbonnel et al. 1998). The carbon isotope ratios $^{12}C/^{13}C$ and C/N ratios are lowered relative to unevolved field stars and are in agreement with the theoretical predictions of the 1st dredge-up and extra-mixing by Boothroyd & Sackmann (1999). Most of the stars show enhanced abundances of sodium and a slight enhancement of aluminium and silicon can be suspected. Abundances of oxygen, magnesium, calcium, titanium and iron group elements are consistent with results obtained in the disk dwarfs (Edvardsson et al. 1993). Our field clump stars show elemental abundance patterns which are very similar to those of open-cluster evolved stars (c.f. Friel et al. 2003, Bragaglia et al. 2001, Hamdani et al. 2000, Tautvaišienė et al. 2000, Gratton & Contarini 1994).

References

1. Boothroyd A.I., Sackmann I.J. 1999, ApJ **510**, 232
2. Bragaglia A., Carretta E., Gratton R.G., et al. 2001, AJ **121**, 327
3. Charbonnel C., Brown J.A., Wallerstein G. 1998, A&A bf 332, 204
4. Edvardsson B., Andersen J., Gustafsson B., Lambert D.L., Nissen P.E., Tomkin J. 1993, A&A **275**, 101
5. Friel E.D., Jacobson H.R., Barrett E., Fullton L., Balachanrdan S.C., Pilachowski C.A. 2003, ApJ bf 126, 2372
6. Gratton R.G., Contarini G. 1994, A&A **283**, 911
7. Gratton R.G., Sneden C., Carreta E., Bragaglia A. 2000, A&A **354**, 169
8. Hamdani S., North P., Mowlavi N., Raboud D., Mermilliod J.-C. 2000, A&A **360**, 509
9. Perryman M.A.C. 1997, ESASP **404**, 231
10. Sturch C., Helfer H.L. 1971, AJ **76**, 334
11. Tautvaišienė G. 1996, Baltic Astronomy **5**, 503
12. Tautvaišienė G., Edvardsson B., Tuominen I., Ilyin I. 2000, A&A **360**, 499
13. Tautvaišienė G., Edvardsson B., Tuominen I., Ilyin I. 2001, A&A **380**, 578
14. Zhao G., Qiu H.M., Mao S. 2001, ApJ **551**, L85

Abundance Trends in the Thin and Thick Disks

S. Feltzing

Lund Observatory, Box 43, SE-221 00 Lund, Sweden

Abstract. The Milky Way harbours two disks that appear distinct concerning scale-heights, kinematics, and elemental abundance patterns. Recent years have seen a surge of studies of the elemental abundance trends in the disks using high resolution spectroscopy. Here I will review and discuss the currently available data. Special focus will also be put on how we define stars to be members of either disk, and how current models of galaxy formation favour that thick disks are formed from several accreted bodies. The ability for the stellar abundance trends to test such predictions are discussed.

1 Current Observational Knowledge About the Thick Disk

During the last decade there has been an increasing interest in trying to establish the chemical abundance trends for the thick disk in the Milky Way. Some of these have included stars from both the thin and the thick disk. These are the most useful ones as they perform differential studies between the two disks. This means that shortcomings in the abundance analyses are, to first order, canceled. This is especially true if the stars are selected to span a small range in effective temperature and surface gravity. Then any resulting difference, at a given [Fe/H], between the two disks may be regarded as real.

Recent important observational studies include the following (sorted according to the stellar samples): *Differential studies*: Fuhrmann, 1998, [10] and 2004, [11], Chen et al., 2000, [8], Mashonkina et al., 2003, [14], Gratton et al., 2003, [13], Bensby et al., 2003, [3] and 2004, [4], Mishenina et al., 2004,[15] *Thick disk only*: Prochaska et al., 2000, [18] *Thin disk only*: Reddy et al., 2003, [20], Allende Prieto et al., 2004, [2]

Although the various studies cited above take different approaches to defining their samples and though some of them are only concerned with one of the disks, there is agreement on the following:

- The thick disk is, at a given [Fe/H], more enhanced in the α-elements than the thin disk
- The abundance trends in the thin disk is a gentle slope
- The solar neighbourhood thick disk stars that have been studied are all old

Other issues are less clear cut. There is some agreement on the following issues (e.g. [14], [3], and Fig. 1 & 2).

- The thick disk shows evidence of extensive star formation
- The thick disk shows evidence of pollution from SNe Ia
- The thick disk shows evidence of pollution from AGB stars

In addition to the indication of pollution from SNe Ia (i.e. the "knee") found in the [Mg/Fe] and [O/Fe] trends,for kinematically selected thick disk samples, there is also evidence that thick disk stars well above the galactic plane show the same trend. In Fig. 2 we show the first results of a study of "in situ" metal-rich thick disk stars. It appears that there might be a downward trend in [Mg/Fe] as a function of [Fe/H] also well above the thin disk. This must, however, be further established before any firm conclusions can be drawn.

There is no agreement on the following two issues ([13], [11], [3], and [5]):

- The thick disk extends to [Fe/H] = 0
- A true age-gap between the end of star formation in the thick disk and the onset of star formation in the thin disk

- Finally: Age-metallicity relation in the thick disk – this is a very tentative statement and, to our knowledge, there is only one study that claims the *possibility* of such an age-metallicity relation ([5]). This would suggest an extended period of star formation in the thick disk

2   Ways To Define a Thick Disk Star

Kinematics Selection is clean, even though not trivial, as the velocities of the thick disk overlaps those of the thin. It is easy to compare selection criteria between different studies.

Position Sufficiently high above the galactic plane the thick disk dominates over both the thin disk and the halo. Thus if we can pick stars 1000-1500 pc above the plane we have a good chance of picking a thick disk star. But, these stars are faint due to their large distance (typically V~ 16) and hence difficult to study.

[Fe/H] and [α/Fe] The metallicity distribution functions of the various stellar populations present in the solar neighbourhood overlap greatly and hence metallicity on its own is not useful for classifying stars. The enhancements of α-elements (e.g. O and Si) are sometimes used to identify the various populations. There is currently general agreement that the thick disk is enhanced in these elements relative to the thin disk. This should be regarded as something we are looking for rather than prior knowledge that can be used to select stars for further study.

Age Stellar ages can be derived using isochrones. However, such data can not be used to select stars for studies of the thick disk as this must imply a prior knowledge of the age of the stellar population in this disk. This is knowledge that we seek, but do not have.

Given that we in principle can never select a sample of local thick disk stars that is guaranteed to not contain any intervening thin disk stars, I would argue that we should keep the selection schemes as simple and as transparent as possible. In this sense the simplest and most robust selection is based on the kinematics of the stars. This is also the least model dependent method. Of course, should positions be available, i.e. height over the galactic plane, these could, and should, be used.

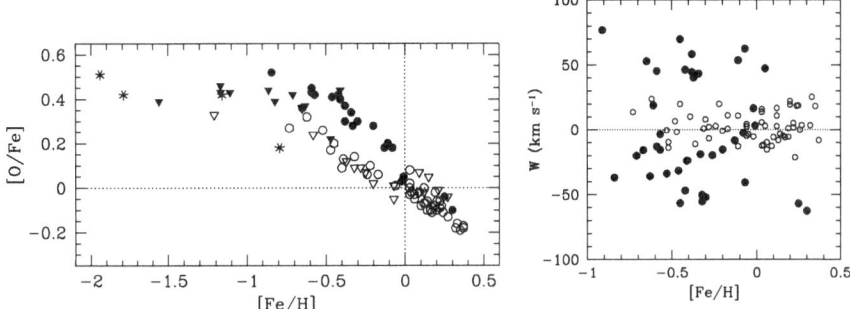

Fig. 1. Left panel: Demonstration that SNe Ia have contributed to the chemical enrichment for stars that are kinematically selected to trace the thick disk (i.e. "knee" in the trend for [O/Fe]). Figure taken from [4]. Thick disk stars are represented by filled symbols (circles from [4] and triangles from [16]), thin disk by open symbols (circles from [4] and triangles from [16]), and halo stars by ∗ (from [16]). **Right panel:** Shows the W-velocity as a function of [Fe/H] for stars kinematically selected to be thin (open circles) or thick (filled circles) disk stars. Data taken from [3] and Bensby et al. submitted. This plot establishes that metal-rich stars can also have high velocities perpendicular to the plane (sometimes taken as an evidence for the star to belong to the thick disk)

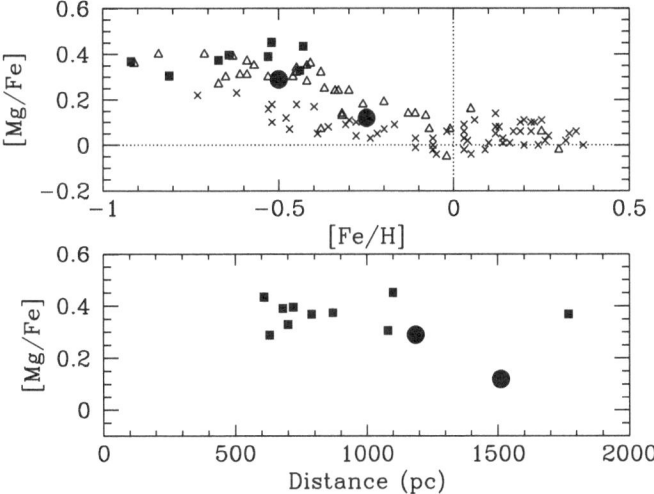

Fig. 2. Upper panel: First results from a study of five "in situ" thick disk stars (Feltzing et al. in prep.). The Mg abundance for two of them are indicated by large • while data from [3] and Bensby et al. in prep. are shown with △ for thick disks stars and × for thin disk stars. The thick disk stars from [18] are shown as filled squares. **Lower panel:** [Mg/Fe] vs the distance above the galactic plane for the two stars from Feltzing et al. in prep. and the thick disk stars from [18]

3   Thick Disks in Other Galaxies and How To Form Them

3.1   Thick Disks in Other Galaxies and Disks and Old Stars at High z

Thick disks are not unique to the Milky Way. Thick disks are seen in many spiral and lenticular galaxies, see e.g. [17], and in galaxies in merging environments, see e.g. [21]. Some, [9], even suggest that all spiral galaxies have thick disks. It is an important observational task to verify and extend these findings.

It is also interesting to note that solar metallicities are reported for z=2, e.g. [22], and that disks of old stars have been found at redshifts as high as z=2.5, see e.g. [23]. These types of findings indicate that indeed the formation of the thick disk in our galaxy might have happened well in the past.

3.2   Ways To Form Thick Disks

Meanwhile, the quest to find formation scenarios for the thick disks is ongoing. Earlier studies focused on fast and slow monolithic collapse and various scenarios for increasing the velocity dispersion in a pre-existing thin disk. The latter was envisaged to be able to happen in two distinct ways; either through the general diffusion of orbits and stars being scattered by molecular clouds and other stars or by a violent encounter and merging with a smaller galaxy. A good summary and discussion of why some of these scenarios do not work (e.g. diffusion of orbits and slow collapse) while others (such as fast collapse) is still viable can be found in [12]. Current models of galaxy formation envisages that a galaxy formed from many building blocks. These types of models have only recently become detailed enough that we can attempt to compare them in greater detail with the data for the Milky Way, e.g. [1] and [6].

I'll review here three different scenarios that still appear viable (based on the currently available data for the Milky Way, see summary in Fig. 3). For each example I have chosen one or two references that have done detailed models as illustrations to compare the observational data to. This is not an exhaustive account for all the possibilities within each scenario but it gives a flavour of the types of comparisons we ought to make and, hopefully, it also illustrates the shortcomings both of the observed as well as simulated data.

In an early phase with enhanced star formation ([7])

Predictions: no abundance gradients the thick disk; thin disk stars will all have [Fe/H] larger than found in the thick disk; period of star formation in thick disk was short < 1 billion years; thin disk always younger than thick disk

Observations: no vertical gradients; metallicity distributions for the disks overlap; star formation in thick disk includes SNe Ia and AGB; star formation in thick disk probably > 1 billion years

Conclusions: probably not a viable scenario (if an age-metallicity relation in thick disk is established this formation scenario is in serious trouble)

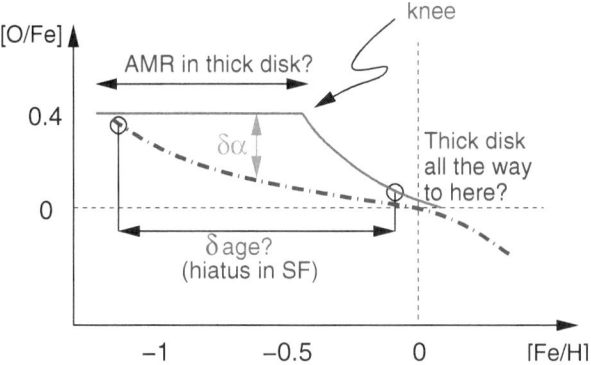

Fig. 3. Summary of current observational knowledge about the thin and the thick disks in the Milky Way. The two trends for [O/Fe] vs [Fe/H] are depicted in blue, dashed line (thin disk) and red (thick disk). The most debated issues are marked in blue (i.e. thick disk all the way to [Fe/H]=0, hiatus in star formation, SF, and AMR in thick disk) and topics of some debate in purple ("knee" and δage between various sub-populations, such as δage between the youngest thin disk and the oldest thick disk). The one issue all agree upon, the α-enhancement, is indicated in green (light grey)

As the result of violent heating of a pre-existing thin disk ([19])

Predictions: infalling satellite probably has to be large; thin disk younger than thick disk and an age gap between the disks; likely that the abundance trends differ; gradients will be preserved (if they exist in pre-existing thin disk); star formation in original thin disk could be as long as needed (e.g. to create the "knee" in the α-element trends)

Observations: no vertical gradients; abundance trends differ; star formation in thick disk includes SNe Ia and AGB

Conclusions: quite possible, no obvious problems with currently available observations

Are made from (largish) satellites that accrete over time ([1], [6])

Predictions: we will only see the abundance trend in the one satellite that finally settles in a torus at our R; the abundance trends will differ depending on the potential well of the satellite, it must be largish and not a dSph; no age gap between thin and thick disk as thin disk is accreted also at early times

Observations: tight trends in kinematically selected samples; so far all studied thick disk stars are older than thin disk stars

Conclusions: satellite must be fairly large; only one satellite contributes at the solar radius; the age structure remains a problem (i.e. that very old thin disk stars are predicted and not observed, so far?)

4   Summary

There are a number of important observational facts that we now have established for the stellar population of solar neighbourhood stars that have kinematics that are typical of the galactic thick disk. Most importantly it is now convincingly shown by independent studies that the abundance trends for kinematically selected samples differ; e.g. the thick disk is more enhanced in α-elements than the thin disk. This is also true for other elements as well, e.g. Eu and Ba. From the, currently few, studies of the r- and the s-process elements it is clear that the thick disk has experienced a contribution from AGB stars.

Some studies find that the thick disk has also experienced a contribution from SNe Ia. This is true in purely kinematically defined samples (e.g. [3]), but is not present, obviously, if the more metal-rich stars, > -0.5 dex, are excluded as in [10].

Based on currently available elemental abundance data and age determinations, the thick disk could have formed either through a violent, heating merger or through accretion of (substantial) satellites in a hierarchical galaxy formation scenario. The fast monolithic-like collapse is getting more and more problematic as data are gathered. It would be especially crucial to establish if there is an age-metallicity relation in the thick disk or not as in that case the thick disk could not have formed in that way (since the models indicate that the formation time-scale for the stars in the thick disk would be very short, see [7]).

Thick disks are common in other galaxies, especially in merger environments, hence perhaps we should prefer the merger scenario for the Milky Way?

References

1. M.G. Abadi, J.F. Navarro, M. Steinmetz, V.R. Eke: ApJ **597**, 21 (2003)
2. C. Allende Prieto, P.S. Barklem, D.L. Lambert, K. Cunha: A&A **420**, 183 (2004)
3. T. Bensby, S. Feltzing, I. Lundström: A&A **410**, 527 (2003)
4. T. Bensby, S. Feltzing, I. Lundström: A&A **415**, 155 (2004)
5. T. Bensby, S. Feltzing, I. Lundström: A&A **421**, 969 (2004)
6. C.B. Brook, D. Kawata, B.K. Gibson, K.C. Freeman: astro-ph/0405306 (2004)
7. A. Burkert, J.W. Truran, G. Hensler: ApJ **391**, 651 (1992)
8. Y.Q. Chen, P.E. Nissen, G. Zhao, H.W. Zhang, T. Benoni: A&AS **141**, 491 (2000)
9. J.J. Dalcanton, R.A. Bernstein: AJ **124**,1328 (2002)
10. K. Fuhrmann: A&A **338**, 161 (1998)
11. K. Fuhrmann: AN **325**, 3 (2004)
12. G. Gilmore, R.F.G: Wyse, K. Kuijken ARA&A, **27**, 555 (1989)
13. R.G. Gratton, E. Carretta, S. Desidera, S. Lucatello, P. Mazzei, M. Barbieri: A&A **406**, 131 (2003)
14. L. Mashonkina, T. Gehren, C. Travaglio, T. Borkova: A&A **397**, 275 (2003)
15. T.V. Mishenina, C. Soubiran, V.V. Kovtyukh, S.A. Korotin: A&A **418**, 551 (2004)
16. P.E. Nissen, F. Primas, M. Asplund, D.L. Lambert: A&A, **390**, 235 (2002)
17. M. Pohlen, M. Balcells, R. Lütticke, R.-J. Dettmar:. A&A **422**, 465 (2004)
18. J.X. Prochaska, S.Q. Naumov, B.W. Carney, A. McWilliam, A.M. Wolfe: ApJ **120**, 2513 (2000)
19. P.J. Quinn, L. Hernquist, D.P. Fullagar: ApJ **403**, 74 (1993)
20. B.E. Reddy, J. Tomkin, D.L. Lambert, C. Allende Prieto: MNRAS **340**, 304 (2003)
21. U. Schwarzkopf, R.-J. Dettmar: A&A **361**, 451 (2000)
22. A.E. Shapley, D.K. Erb, M. Pettini, C.C. Steidel, K.L. Adelberger: ApJ **612**, 108 (2004)
23. A. Stockton, G. Canalizo, T. Maihara: AJ **605**, 37 (2004)

On the Chemical Abundances
of Stars with Giant Planets

N.C. Santos

Centro de Astronomia e Astrofísica da Universidade de Lisboa,
Observatório Astronómico de Lisboa, Tapada da Ajuda, 1349-018 Lisboa, Portugal

Abstract. One particular fact that is helping us to understand the mechanisms of
planetary formation has to do with the planet host stars themselves. In fact, these
were found to have, on average, a metal content higher than the one found in stars
without detected planetary companions. In this contribution we will mainly focus on
the most recent results on the chemical abundances of planet-host stars, and what kind
of constraints they are bringing to the theories of planet formation.

1 Introduction

Following the discovery of the first extra-solar planet orbiting a solar type star
[13], a multitude of other planetary systems has been announced (for a review
of their main characteristics see e.g. [14])[1]. At first, only very short period giant
companions were found, something that was quite unexpected by the theories of
planet formation. However, the increase in precision of the current radial-velocity
"machines" and the long baseline of the measurements, have brought to light
a variety of worlds, whose existence defies current models. Interestingly, these
discoveries now include what may be the first rocky worlds [18,15,5].

With the number of known exoplanets increasing very fast, current results are
giving us the chance to undertake the first statistical studies of the properties
of the exoplanets, as well of their host stars [24,21,8]. This is bringing new
interesting constraints for the models of planet formation and evolution.

In this paper we will review the current situation regarding the study of the
chemical abundances of stars with giant planets, and discuss the implications
these results have on the theories of planetary formation.

2 The Metallicity of Stars with Planets

Soon after the discovery of the first extra-solar planets, it has been noticed that
planet-host stars these were particularly metal-rich when compared with "single"
field dwarfs [12], i.e., on average they present a metal-content that is above the
one found in stars now known to have any planetary-mass companion. This
result, clearly confirmed by an uniform spectroscopic analysis of large samples
of stars with and without detected giant planets [22] is obtained by using different

---

[1] For an updated list of the known extra-solar planetary systems see table at
http://obswww.unige.ch/Exoplanets

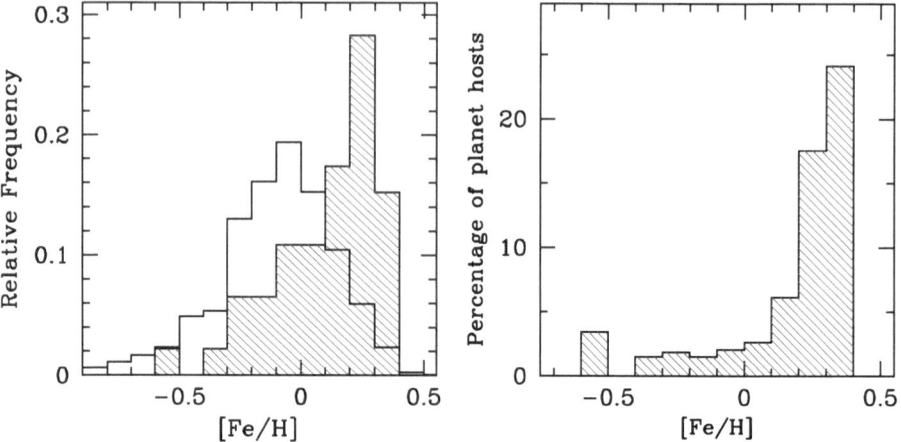

Fig. 1. Left panel: [Fe/H] distributions for planet host stars (hashed histogram) included in the CORALIE planet-search sample, when compared with the same distribution for ∼900 stars in the whole CORALIE program (solid-line open histogram). *Right panel*: percentage of planet hosts found amid the stars in the CORALIE sample as a function of stellar metallicity. Taken from [19]

kinds of techniques to derive the stellar metallicity (see e.g. [17]). Planet host stars are indeed significantly more metal-rich that stars without known giant planets (see e.g. [19]). The average metallicity difference between the two samples is ∼0.25 dex. This result is found for all the metals studied so far (e.g. [2,6,7])[2].

In the left panel of Fig. 1 we present the [Fe/H] distributions for planet host stars (hashed histogram) included in the CORALIE planet-search sample [25], when compared with the same distribution for ∼900 stars in the whole CORALIE program (open histogram). The knowledge of the uniformly determined metallicity distribution for stars in the solar neighborhood (and included in the CORALIE sample) permits us to determine the percentage of planet host stars per metallicity bin.

The result is shown in the right panel of Fig. 1, and shows that the probability of finding a planet is a strong function of the stellar metallicity. About 25% of the stars in the CORALIE sample having metallicity between 0.3 and 0.4 dex have been discovered to harbor a planet. On the other hand, only about 3% of the stars having solar metallicity seem to have a planetary-mass companion [22,17,19]. This result is thus probably telling us that the probability of forming a giant planet, or at least a planet of the kind we are finding now, depends strongly on the metallicity of the could of gas and dust that gave origin to the star and planetary system. Although it is unwise to draw any strong conclusions based on only one point, it is also worth noticing that our own Sun is in the "metal-poor" tail of the planet host [Fe/H] distribution.

---

[2] The studies of other elements in metal-rich planet-host stars is also giving important information about the chemical evolution of the Galaxy.

It is important to remember at this point that the metallicities for the two samples of stars plotted in Fig. 1 were derived using exactly the same techniques, and are thus both in the same scale (see [19]). Also, in the CORALIE planet search sample we have never used the stellar [Fe/H] as a criterion to chose a star. The comparison shown in Fig. 1 is thus not sample-biased. Finally, and as shown in [21], the precision in the derived radial-velocities is not a strong function of the stellar metallicity. The observed increasing frequency of planets with increasing [Fe/H] is thus also not due to any bias in the planet searches.

3 The Source of the [Fe/H] "Excess"

All the conclusions discussed above are true in the assumption that the metallicity excess observed is original to the cloud of gas and dust that gave origin to the star and planetary system. In other words, we are supposing that the higher frequency of planets around metal-rich stars is reflecting an higher probability of forming a planet around such a star before the disk dissipates.

However, one other interpretation has been discussed in the literature to explain the [Fe/H] excess observed for stars with planets. In fact, it has been suggested that the high metal content is the result of the accretion of planets and/or planetary material into the star (e.g. [12]). In such a case, the observed metallicity excess would itself be a by-product of the planetary formation process.

There are multiple ways of deciding between the two former scenarios, and in particular to try to see if "pollution" might indeed have played an important role in increasing the metal content of the planet host stars relative to their non-planet host counterparts. Probably the most clear argument is based on stellar internal structure. Material falling into a star's surface would induce a different increase in [Fe/H] depending on the depth of its convective envelope, where mixing can occur. However, no correlation is found between the metallicity of the planet host stars and their convective envelope mass (e.g. [16,21]).

Some doubts have recently been put for this lack of correlation being a good reason to exclude that stellar pollution could have caused the observed [Fe/H] "excess" ([23]). Furthermore, it has been shown that in a few cases stellar pollution may have played some role [11], although not strong enough to be responsible for a large [Fe/H] variation (e.g. [10,20]). However, the evidence for an "original" source is further supported by the huge quantities of "pollution" by hydrogen poor (planetary) material needed to explain the metallicity excess observed in a few late-type very metal-rich dwarfs known to harbor giant planets, as well as for a few sub-giant planet-host stars (e.g. [21]). In other words, the bulk metallicity "excess" observed has most probably a "primordial" origin.

4 Implications for the Models of Planetary Formation

These conclusions have many important implications for the theories of planetary formation. In this respect, two main cases are now debated in the literature.

On the one side, the traditional core accretion scenario (e.g. [1]) tells us that giant planets are formed as the result of the runaway accretion of gas around a previously formed icy core with about 10-20 times the mass of the Earth. Opposite to this idea, some authors have proposed that giant planets may form by a disk instability process [4].

According to the instability model, the efficiency of planetary formation should not be dependent on the metallicity of the star/disk [3]. This is opposite to the expected from the traditional core-accretion scenario [9], since the higher the grain content of the disk, the easier the "metal" cores that will later-on accrete gas should be formed before the gas disk dissipates.

The results presented above, showing that the probability of finding a planet is a strong function of the stellar metallicity, thus favor the core-accretion model as the main mechanisms responsible for the formation of giant planets (although they do not completely exclude the disk instability model – see e.g. [19]). Indeed, it has even be shown that according to the core-accretion model it is possible to predict the observed [Fe/H] distribution of planet-host stars [9].

References

1. Y. Alibert, C. Mordasini, W. Benz: A&A **417**, L25 (2004)
2. A. Bodaghee, N.C. Santos, G. Israelian, M. Mayor: A&A **404**, 715 (2003)
3. A.P. Boss: ApJ **567**, L149 (2002)
4. A.P. Boss: Science **276**, 1836 (1997)
5. R.P. Butler, S.S. Vogt, G.W. Marcy, et al.: ApJ Letters, in press (2004)
6. A. Ecuvillon, G. Israelian, N.C. Santos, et al.: A&A, in press – astro-ph/0406584 (2004a)
7. A. Ecuvillon, G. Israelian, N.C. Santos, et al.: A&A **418**, 703 (2004b)
8. A. Eggenberger, S. Udry, M. Mayor: A&A **417**, 353 (2004)
9. S. Ida, D.N.C. Lin: ApJ, in press – astro-ph/0408019 (2004)
10. G. Israelian, N.C. Santos, M. Mayor, R. Rebolo: A&A **414**, 601 (2004)
11. G. Israelian, N.C. Santos, M. Mayor, R. Rebolo: A&A **405**, 753 (2003)
12. G. Gonzalez: A&A **334**, 221 (1998)
13. M. Mayor, D. Queloz: Nature **378**, 355 (1995)
14. M. Mayor, N.C. Santos, in: Proc. of the ESO-CERN-ESA Symposium on "Astronomy, Cosmology and Fundamental Physics", p.359 (2003)
15. B.E. McArthur, M. Endl, W.D. Cochran, et al.: ApJ Letters, in press – astro-ph/0408585 (2004)
16. M.H. Pinsonneault, D.L. DePoy, M. Coffee: ApJ **556**, L59 (2001)
17. I.N. Reid: PASP **114**, 306 (2002)
18. N.C. Santos, F. Bouchy, M. Mayor, et al.: A&A Letters in press – astro-ph/0408471 (2004a)
19. N.C. Santos, G. Israelian, M. Mayor: A&A **415**, 1153 (2004b)
20. N.C. Santos, G. Israelian, R.J. García López, et al.: A&A, in press – astro-ph/0408108 (2004c)
21. N.C. Santos, G. Israelian, M. Mayor, R. Rebolo, S. Udry: A&A **398**, 363 (2003)
22. N.C. Santos, G. Israelian, M. Mayor: A&A **373**, 1019 (2001)
23. S. Vauclair: ApJ **605**, 874 (2004)
24. S. Udry, M. Mayor, N.C. Santos: A&A **407**, 369 (2003)
25. S. Udry, M. Mayor, D. Naef, et al.: A&A **356**, 590 (2000)

Observational Constraints on Nucleosynthesis in AGB Stars

C. Abia[1], I. Domínguez[1], O. Straniero[2], and O. Zamora[1]

[1] Dpto. Física Teórica y del Cosmos, Universidad de Granada, 18071 Granada, Spain
[2] INAF – Osservatorio Astronomico di Collurania, 64100 Teramo, Italy

Abstract. AGB stars, in particular those of *carbon* types, are excellent laboratories to constraint the theory of stellar structure, evolution and nucleosynthesis. Despite the uncertainties still existing in the chemical analysis of these stars, the determination of the abundances of several key species in their atmospheres (lithium, s-elements, carbon and magnesium isotopic ratios etc.) is an useful tool to test these theories and the mixing processes during the AGB phase. This contribution briefly review some recent advances on this subject.

1   Observational Constraints

The chemical analysis of AGB stars is a quite difficult task. Their spectra look *horrible*: crowded and plenty of strong molecular absorptions; for many of these molecules accurate spectroscopic information does not exist. AGB stars are variable, have huge and thin atmospheres so the common hypotheses of planoparallel, static and LTE atmosphere strictly cannot be applied. Analyses based on these hypotheses can lead to systematic errors. This can be reduced, however, using the differential method of chemical analysis *line-by-line* respect to a comparison star of similar and well known stellar parameters. In any case, large abundance (non-systematic) errors ($\pm 0.3 - 0.5$ dex) are expected because of the uncertainty in the derivation of the basic stellar parameters ($T_{\rm eff}$, gravity etc.). This is an important problem which has to be reminded when comparing abundance determinations with theoretical predictions. This fact still strongly limits the observational constraints upon the modelling of the AGB phase. Some of these constraints are, nevertheless, described next.

1.1   s-Process Elements

Nowadays it is widely accepted that the $^{13}\mathrm{C}(\alpha, n)^{16}\mathrm{O}$ reaction is the main source or neutrons of the *s*-process in AGB stars. Comparison between the *s*-element abundance patterns found in AGB stars of different classes and metallicity with theoretical predictions show a nice agreement (see e.g. Busso et al. 2001 and references therein). This comparison would indicate also that, at a given stellar metallicity, a dispersion in the quantity of $^{13}\mathrm{C}$ burnt may exists as one would expect, on the other hand. In fact, *s*-element patterns for individual stars can be fitted assuming that the amount of $^{13}\mathrm{C}$ burnt ranges from 10^{-7} to almost 10^{-5} $M_\odot$. However, the large error bar in the abundances precludes to put more stringent limitations. This is unfortunate since the amount of protons that may mix in the He-intershell determining the total amount of ^{13}C available for the s-process, cannot be derived from first principles and has to be parametrised.

On the other hand, the simple comparison of theoretical predictions and observations cannot either stringently limit the role played by the ^{22}Ne neutron source. Although, the predicted s-element pattern is quite different depending on which neutron source plays the main role, again the large observational errors avoids a clear discern. The only way to address this problem is to study the abundance ratios between elements placed at the branchings of the s-process. This has been done in near solar metallicity AGB stars (Lambert et al. 1995; Abia et al. 2001) at the ^{85}Kr-branching by comparing the abundance ratios between Rb and its neighbours Zr,Y and Sr. The conclusion is that the ^{13}C source plays the main role, however similar studies in low metallicity (intrinsic) AGB stars are necessary. In any case, the ^{22}Ne source has to play some role since the observed evolution with [Fe/H] of the ^{24}Mg/^{25}Mg/^{26}Mg ratios in field stars (Yong et al. 2002) is fitted better if AGB stars are significant producers of ^{25,26}Mg (Fenner et al. 2003). Only in a few O-rich AGB stars the magnesium isotopic ratios have been derived showing, however, solar ratios. Recently, Zamora et al.(2004) attempt to derive these isotopic ratios in C-rich AGB-stars with s-element enhancements. Unfortunately, the very crowded spectral regions available for that, prevented them to derive even upper or lower limits. Therefore, the issue of the ^{22}Ne neutron source role still remains open from the observational point of view.

1.2 CNO Isotopes

One of the most surprising results concerning the chemical composition of AGB stars is the fact that the overwhelming majority of the galactic C-rich AGBs show a C/O ratio only slightly exceeding the unity. This figure has been found by many authors (see Abia et al. 2001 and references therein) using different methods. The same figure seems to hold in the AGB C-rich stars so far analysed in the Magellanic Clouds and other satellite galaxies, although, the statistics (still scarce) show that the *alien* carbon stars have larger C/O ratios that their galactic counterparts, probably due to their lower metallicity. The explanation to this figure may be just an observational bias: as the C/O ratio increases in the envelope, C-stars become extremely red and cannot be *seem* at visual wavelengths. However, according with the lifetimes predicted by theoretical models, it would be statistically improbable to catch all the observed C-stars within a very short range of C/O ratios around one; this difficulty increases as the metallicity of the star decreases. May be the observed C/O$\sim$ 1 is telling us something more about the internal structure and nucleosynthetic processes along the AGB phase. Exceptions exist: D461 is an intrinsic carbon star in the dwarf spheroidal galaxy Draco showing C/O$\sim$ 3 − 5 (Domínguez et al. 2004).

An other important constraint comes from the ^{12}C/^{13}C ratios derived in AGB C-stars. Some discrepancy still exists among different authors concerning the typical ^{12}C/^{13}C ratio in these stars (see e.g. de Laverny & Gustafsson 1998). However, all the authors agree in the fact that there are a significant number of

C-stars showing isotopic ratios (< 15) that cannot be explained by the standard evolutionary models in the AGB phase, on the basis that most of these stars have low mass (< 3 M$_\odot$, Abia et al. 2002). Even assuming an extra-mixing process during the previous RGB phase (see e.g. contributions by A. Weiss and A. Palacios in this issue), AGB models fail to reproduce these low values. Similarly to RGB stars, Wassenburg et al. (1995) suggested that existence of a non-standard mixing process coupled with some kind of *cool bottom burning* at the base of the convective envelope in low mass AGB stars. This extra mixing is described as the circulation of material from the base of the convective envelope into the thin radiative region located on top of the H-burning shell. Here the material is processed by proton captures and then carried back to the envelope, thus producing the signature of CNO processing at the stellar surface. Although the specific mechanism that trigger this extra-mixing and burning process is not known, recent modelling of this mechanism (Nollett et al. 2003) can marginally account for the low $^{12}C/^{13}C$, $^{17,18}O/^{16}O$ ratios and other isotopic anomalies observed in many AGB stars and stellar dust grains with an AGB origin (Harris et al. 1987; Zinner 1998 and references therein).

1.3   Lithium

Probably the lithium abundances found in AGB carbon stars constitutes the most compelling evidence that an extra-mixing process occurs in the AGB phase. Lithium can be produced by the Cameron & Fowler mechanism in intermediate mass AGB stars (e.g. Sackmann & Boothroyd 1992). Although observations and theoretical modelling of Li production widely agree, still the peak Li abundances ($12+ \log (\text{Li/H}) \sim 5$) derived in some AGB stars cannot reproduced by the models. Furthermore, and more important, there are a significant number of galactic low luminosity AGB *carbon stars* which do show some Li enhancements ($12+ \log (\text{Li/H}) > 1$). Many of them present low $^{12}C/^{13}C$ ratios (< 15, named J-type carbon stars) and in some cases, also *s*-element enhancements. These figures are very difficult to reconcile with the standard models of Li production in AGB stars. Note, that the number of carbon stars with these characteristics among the AGB galactic population is quite large (10-15%; Abia & Isern 2000), figure which has also been found in AGB stars of the Magellanic Clouds (Hatzidimitriou et al. 2003), although with a lower statistics. Therefore, one has necessarily to conclude that this phenomenon may constitute a frequent stage during the evolution on the AGB phase. Some attempts have been done to model these Li-enhancements. Again, a non-standard mixing below the base of the convective envelope seems to be necessary. However, a fine-tunning is required in the models concerning the maximum extend of this mixing (i.e., maximum temperature reached by the extra-mixing) and duration of this phenomenon (see e.g. Domínguez et al. 2004). Although, it is not expected these stars would contribute significantly to the galactic Li abundance, it is obvious that we cannot evaluate the real contribution of AGB (carbon) stars to the galactic Li until its production in AGB stars is fully understood.

1.4   Fluorine

Fluorine abundances in stars are of considerable importance to constraint AGB models. It seems that AGB stars are the main producers of this element although some contribution from Wolf Rayet stars and type II supernovae is also expected (Reda et al. 2004). Spectroscopic observations of AGB stars of different classes (Jorissen et al. 1992) show that the ^{19}F abundance is enhanced by factors of 2 to 20 with respect to the solar one. The observed abundances show also a positive correlation with the carbon enhancement which suggests that it is produced in the He intershell and dredged-up to the surface after the TDU together with carbon and s-elements. The nucleosynthesis path for the production of ^{19}F is quite complex involving many nuclear reactions with uncertain rates for some of them. Although, the observed correlation of ^{19}F and carbon is reproduced for moderate enhancements by theoretical models, these cannot fit the larger ([F/O]> 1) enhancements, as models predict larger C/O ratios than observed. Recently, Lugaro et al. (2004) perform a detailed analysis of the reaction rates involved in the ^{19}F production and find that the major uncertainties are related with the reactions ^{14}C$(\alpha, \gamma)^{18}$O and ^{19}F$(\alpha, p)^{22}$Ne. They also evaluate the effect of the existence of a partial mixing zone below of the convective envelope (similar to this necessary for the s-elements production) during the interpulse period, and conclude that the highest ^{19}F enhancements observed in AGB stars are not matched by the models. Again, it is suggested that the existence of a *cool bottom processing* may help matching the observed ^{19}F abundances by decreasing the C/O ratio at the surface of the star, while leaving the fluorine abundance unchanged. This kind of simulations have not been done yet.

References

1. C. Abia, J. Isern: ApJ, **536**, 438 (2000)
2. C. Abia, M. Busso, R. Gallino, I. Domínguez, et al. : ApJ, **559**, 1117 (2001)
3. C. Abia, I. Domínguez, R. Gallino, M. Busso, et al.: ApJ, **579**, 817 (2002)
4. M. Busso, D.L. Lambert, R. Gallino, et al.: ApJ, **557**, 802 (2001)
5. P. de Laverny, B. Gustafsson: A&A, **332**, 661 (1998)
6. I. Domínguez, C. Abia, O. Straniero, et al.: A&A, **422**, 1045 (2004)
7. Y. Fenner, B.K. Gibson, H.c.-Lee, A.I. Karakas et al.: PASA, **20**, 340 (2003)
8. M.J. Harris, D.L. Lambert, K.H. Hinkle, et al.: ApJ, **316**, 294 (1987)
9. D. Hatzidimitriou, D. H. Morgan, R.D. Cannon, B.F.W. Croke: MNRAS, **341**, 1290 (2003)
10. A. Jorissen, V.V. Smith, D.L. Lambert: A&A, **261**, 164 (1992)
11. D.L. Lambert, V.V. Smith, M. Busso, et al.: ApJ, **450**, 302 (1995)
12. M. Lugaro, C. Ugalde, A.I. Karakas, J. Görres et al.: ApJ, **in press**, (2004)
13. K.M. Nollett, M. Busso, G.J. Wasserburg: ApJ, **582**, 1036 (2003)
14. F. M. Reda et al: MNRAS, in press (2004)
15. I.J. Sackmann, A.I. Boothroyd: ApJ, **393**, L21 (1992)
16. G.J. Wasserburg, A. I. Boothroyd, I.-J. Sackmann: ApJ, **440**, L101 (1995)
17 D. Yong, D.L. Lambert, I.I. Ivans: ApJ, **500**, 1357 (2002)
18. O. Zamora, C. Abia, I. Domínguez, P. de Laverny: Memorie SAIt, in press (2004)
19. E. Zinner: ARE&PS, **26**, 147 (1998)

Advances in Chemical Abundances in Planetary Nebulae

M. Perinotto[1], A. Scatarzi[1], and L. Morbidelli[2]

[1] Dipartimento di Astronomia, Universitá di Firenze, Largo E. Fermi 2,
I-50125 Firenze
[2] IRA/CNR, Sez. Firenze, Largo E. Fermi 5, I-50125 Firenze

Abstract. We present recent advances in the determination of chemical abundances of galactic Planetary Nebulae and discuss implications resulting from the comparison with theoretical predictions. From the analysis of diagrams of abundances of N/O vs He/H, N/O vs N/H and N/O vs O/H we argue that very likely the often used solar photospheric abundance of oxygen of 8.9, in usual units, is overestimated by a factor of 2-3, as suggested by very recent work in the Sun. This would solve an astrophysical problem with the measured abundances in planetaries.

1 Introduction

Planetary nebulae (PNe) offer the opportunities : 1) to study stellar nucleosynthesis in the advanced phases of stellar evolution of stars in the wide mass range ~ 0.8 to $\sim 8 M_\odot$ and 2) to probe radial and as well horizontal/vertical chemical gradients in spiral galaxies by the time of formation of their progenitors.

Linked to 1) is of course the enrichment of the interstellar medium, to which they are important contributors in nuclearly "processed elements" as He, C, N, s-elements (Ba etc). Goal 2) can be pursued with nuclearly "unprocessed elements", the best accessible of them being O, Ne, Ar and S.

As a consequence chemical abundances in PNe are of primary importance for the chemical evolution of spiral galaxies, including our own and related topics.

We will summarize some aspects of the status of art on chemical abundances of galactic PNe, focussing on a comprehensive study we have recently completed (Perinotto, Morbidelli & Scatarzi, 2004; hereafter PMS04) and comparing with theoretical predictions.

2 Observations of Chemical Abundances in PNe

Basic aim of PMS04 was not to waste valuable sets of measurements acquired during the years by several authors in galactic PNe, whose resulting chemical abundances are however not easy to handle, because of the inhomogeneity of the analyses due to different atomic data and different procedures used by the various authors to interpret the data. We decided to go back to the original measurements and interpret all of them with an unique best assessed procedure.

Essential steps have been : 1) Accept only observational papers, whose line lists included with reasonably accuracy the critical lines to determine the nebular

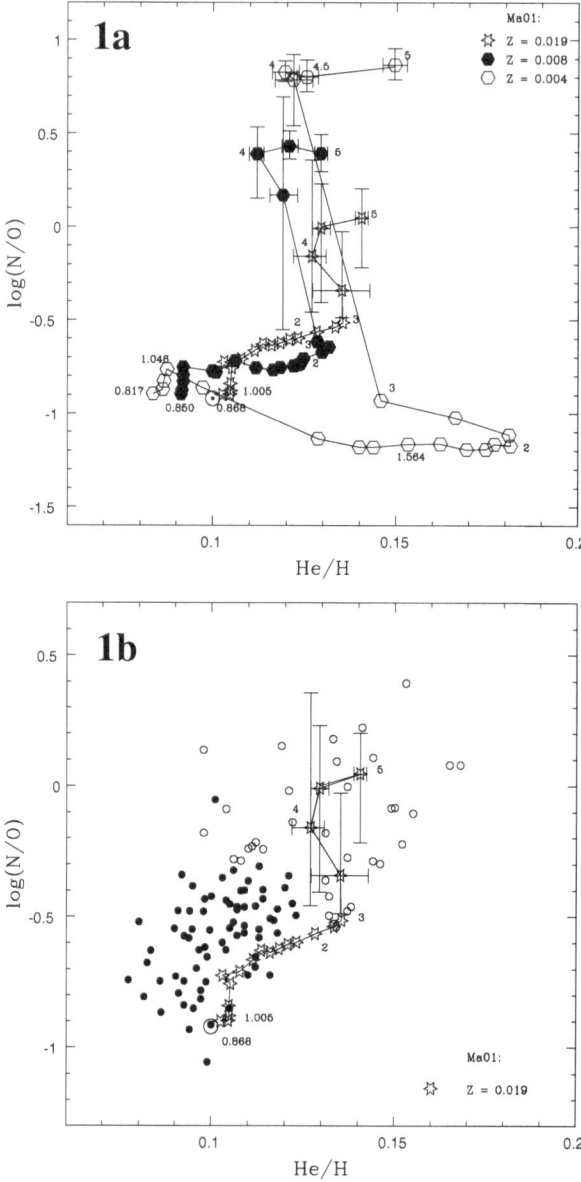

Fig. 1. a) Theoretical predicted abundances of planetaries coming from progenitors of different initial metallicity and mass (in solar units) in the MS (Marigo, 2001). b) Observed abundances from PMS04 are overimposed on the theoretical predictions for solar initial metallicity. Open circles are PNe of type I of Peimbert (1978) , while full dots refer to PNe of Peimbert's types II-III. Type I PNe are more nearly concentrated closer to the galactic plane and so should refer to more massive progenitors.

electron temperature for a basically two–zones model nebula, i.e. λ 4363 Å [OIII] and λ 5755 Å [NII], and at least the density sensitive doublet ratio λ 6731/6720 [SII]; 2) Use the same reddening function (Mathis, 1990) and then, when the case, reconstruct the original observational data, and deredden them; 3) Not mix up measurements by different authors which practically always pertain to different portions of the nebulae, but re-derive chemical abundances separately for each set of original data ; 4) Use always the same model representation, consisting basically of two T_e zones plus adjustments for zones appropriate to higher ionization potential ions; 5) Adopt a single set of ionization correction factors (ICFs) for unseen ions (that of Kingsbourgh and Barlow, 1994, with minor adjustments) ; 6) Perform on objective calculation of errors of the final individual abundances of each ion & atom by propagating the observational error of each line flux across the whole procedure; 7) Derive final abundances in each nebula with their errors by weighted averages of single results. This would be not so appropriate if chemical gradients in the nebula are present. But the whole literature and our own work devoted in the past to detect such "chemical differences" within PNe, convince us that such differences are not revealed at the present level of accuracy, except for about 3 objects where differences have been recognized between the very inner and the outer parts of the nebula, but were the inner parts show a special "knotty" morphology (c.f. Perinotto, 1991).

We mention that recently (c.f. Liu et al., 2004) the dichotomy between chemical abundances of heavy elements derived using: a) the classical intense forbidden lines, as in the "traditional" method or b) the much fainter recombination lines, appears solved with both methods correct: a) probes the main body of the nebula, ionized and warm (10.000 K or so), while b) refers to a minor fraction in mass (few per cent) of the nebula (ionized but cold : 1000-2000 K).

Because of that, we claim that results by PMS04, obtained with method a), can be considered to represent the best up-to-date single collection of chemical abundances in galactic PNe, with reference to essentially all of their mass.

The final number of galactic PNe passing the selection procedures, for which it was possible to calculate "high quality" abundances resulted of 131 coming from some 400 papers.

3   The N/O Abundance Ratio

We touch now essentially a single chemical aspect using results by PMS04. This refers to the correlation N/O vs He/H (Fig. 3 upper left, PMS04).

This relationship has been predicted by the stellar evolution theory, as resulting essentially from the so called first and second dredge-up episodes. In Fig. 1a we see the corresponding predictions by the theory (Marigo, 2001). The sensitivity of the predicted abundances both to the metallicity at the epoch of formation of the progenitors and to their stellar mass in the Main Sequence is evident, making such diagrams powerful tools to trace back the evolutionary history of individual PNe. Uncertainties prevent however us to pursue now this ultimate goal. In Fig. 1b we overimpose the theoretical predictions (for solar initial metallicity) to our data points (avoiding some discrepant points).

4   Possible Interpretations of the N/O Behaviour

Considering the type II-III PNe in Fig. 1b coming from the more common less massive progenitors, we have two evident "naive" interpretations: either 1) the initial metallicity of the progenitors was lower than the solar one or 2) the solar photospheric abundance used here has been overestimated.

The first possibility is clear in Fig. 4 (upper right) of the PMS04 paper, where N/O is plotted vs N/H. Looking at that Figure we can clearly exclude any role from progenitors with an initial metallicity of the order of Z = 0.004, while progenitors with a metallicity 2-3 times less than the value classically attributed to the Sun (Z= 0.019) would be consistent with the observations.

The second possibility is evident from Fig. 5 (upper right) of PMS04, where N/O is plotted vs O/H. In that diagram the theoretical prediction used for the Sun is the "classical" abundance of oxygen equal to 8.9 (in usual units) (Grevesse & Anders, 1989; Grevesse & Sauval 1998). If we instead accept the recent oxygen solar photospheric abundance of 8.6 (Asplund et al., 2004; Melendez, 2004, ApJ in press), the theoretical predictions would move in that diagram to the left, going closer to the observations. We favour this latter possibility, even if the line fluxes on which this lower solar oxygen abundance is based are extremely small and the abundances from the single lines depend in some cases on details of the photosphere modelling. We prefer the latter, because with the first possibility one does not see why all PNe, from those originated from more massive progenitors to those coming from about solar mass stars, should have started with an original oxygen abundance quite less than the solar one, as it's outstandingly evident in the mentioned Fig. 5 (upper right) of PMS04.

In any case the above one seems to us an important case which requires to be fully settled.

References

1. M. Asplund, N. Grevesse, A.J. Sauval, C. Allende Prieto, D. Kiselman: A&A **417**, 751 (2004)
2. N. Grevesse, E. Anders: AIP Conf. Proc. **183**, Cosmic Abundances of Matter, p. 1, Ed. C.J. Waddington (1989)
3. N. Grevesse, A.J. Sauval, SSRv, **85**, 161 (1998)
4. R.L. Kingsbourgh, M.J. Barlow: MNRAS **271**, 257 (1994)
5. Y. Liu, X.-W. Liu, M.J. Barlow, S.-G. Luo: MNRAS **353**, 1251 (2004)
6. P. Marigo: A&A **370**, 194 (2001)
7. J.S. Mathis: ARA&A **28**, 37 (1990)
8. M. Peimbert: IAU Symp. 76, Planetary Nebulae, Ed. Y. Terzian, p.215 p. 233 (1983)
9. M. Perinotto: ApJS **76**, 687 (1991)
10. M. Perinotto, L. Morbidelli, A. Scatarzi: MNRAS **349**, 793 (2004)

A Comparison of Methods for Photospheric Abundance Determinations in K-Type Stars

L. Affer[1], G. Micela[1], T. Morel[1], J.S. Forcada[2], and F. Favata[2]

[1] INAF-Osservatorio Astr. di Palermo, Piazza del Parlamento 1, Palermo, Italy
[2] Astrophysics Division - Research and Science Support Department of ESA,
ESTEC, Postbus 299, NL-2200 AG Noordwijk, The Netherlands

Abstract. We have performed a detailed abundance analysis of six inactive K-type stars using high-resolution optical spectra. We have used three different techniques and compared the results obtained in order to establish their respective merits and faults. The two spectroscopic methods give consistent results suggesting that non-LTE effects are small, whereas the 'mixed' spectroscopic-photometric method leads to photospheric parameters and abundances systematically lower than those obtained with the other two. We have also determined the stars' positions in H-R diagrams and made a comparison between the gravities derived from the ionization equilibrium of the iron lines and from the evolutionary tracks: the agreement is reasonably good.

1 Observations

We have observed three subgiants HD 23249 (K0), HD 198149 (K0), HD 222404 (K1) and three dwarfs HD 10780 (K0), HD 4628 (K2), HD 201091 (K5), on 2002 November 28 and 29, with the high-resolution cross-dispersed echelle spectrograph SOFIN, mounted on the Nordic Optical Telescope (NOT). They are in the solar neighbourhood (≤ 15 pc), are very bright ($V \leq 6$) and have modest projected rotational velocities ($v \sin i \leq 4$ km s^{-1}) to limit blends between spectral lines. They also do not present any evidence for emission (or a moderate one, as in the case of the three dwarfs) in the core of Ca II H and K lines.

2 Methods of Analysis

The atmospheric parameters and metal abundances were determined using the measured equivalent widths (EWs) and a standard local thermodynamic equilibrium (LTE) analysis with the most recent version of the line code MOOG. We used a grid of Kurucz ATLAS9 atmospheres computed without the overshooting option and with a mixing length to the pressure scale height ratio $\alpha = 0.5$. We have determined the abundance ratios with respect to iron of 12 elements. The reader is referred to [1] for further details.

Three different iterative methods were used for the analysis.

Method 1: The effective temperature was derived from the excitation equilibrium of the Fe I lines and the surface gravity from the ionization equilibrium of the iron lines.

Method 2: It follows a similar approach as for Method 1, but discards the Fe I

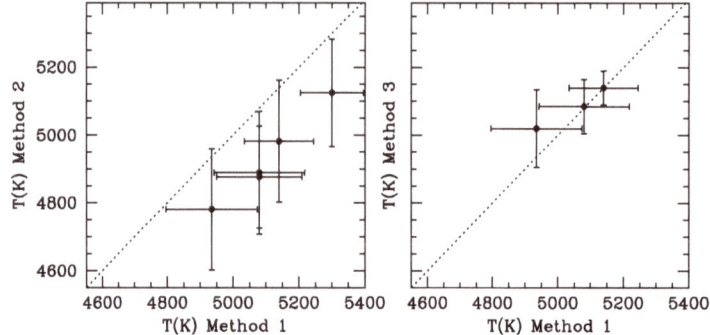

Fig. 1. Comparison between the temperatures obtained by methods 1 and 2 (*left-hand panel*) and by methods 1 and 3 (*right-hand panel*).

low excitation potential transitions (which are potentially affected by non-LTE effects) and relies on the $B - V$ colour index to determine the temperature [2,3]. *Method 3:* It relies on the detailed fitting of the 6162 Å Ca I line to derive the surface gravity, using the same restricted line list as for Method 2. We have used the excitation equilibrium to determine the effective temperature.

2.1   Comparison of the Methods

Methods 1 and 3 give consistent results for the program stars. The good agreement between the atmospheric parameters and chemical abundances derived suggests that the Fe I low excitation potential transitions are not significantly affected by non-LTE effects (at least for the subgiant stars, for which Method 3 has led to convergent solutions). The second method leads to systematically lower T_{eff} (Fig. 1) and $\log g$ values with respect to the first one, and a similar trend is shown by the chemical abundances (with the exception of the oxygen abundance).

3   Evolutionary Status

We have determined the positions of our stars in H-R diagrams for the appropriate Fe and $[\alpha/\text{Fe}]$ abundances, using T_{eff} obtained by Method 1 and the absolute magnitudes, M_{v}, derived from Hipparcos parallaxes. The good agreement between the gravities obtained from the evolutionary tracks ([4]) and those from Method 1 suggests that non-LTE effects are unlikely in Method 1.

References

1. L. Affer, G. Micela, T. Morel, J. Sanz Forcada,A&A, in press (2004)
2. A. Alonso, S. Arribas, C. Martínez-Roger, A&A, **313**, 873 (1996)
3. A. Alonso, S. Arribas, C. Martínez-Roger, A&A, **140**, 261 (1999)
4. S. K Yi, Y. Kim, P. Demarque, A&A, **144**, 259 (2003)

S/R Ratio in Barium Stars

D.M. Allen

Universidade de São Paulo, Rua do Matão 1226
05508-900 São Paulo - Brazil

Abstract. [La/Eu] and [Ba/Eu] for a sample of Barium stars were determined in order to evaluate the ratio of abundances of s- and r-elements. The results have been compared to previous work dealing with normal red giants and dwarfs with metallicities in the range -3 < [Fe/H] < +0.3.

1 Introduction

Barium stars were recognized as a distinct group of peculiar stars by [1]. The objects initially included in this group were red giants of spectral type G and K, which showed strong lines of s-process elements, particularly Ba II and Sr II, as well as enhanced CH, CN and C_2 bands. The discovery that HR 107, a dwarf star, shows a composition similar to that of a mild Barium giant by [6] has pushed the search for new Barium dwarfs.

The relation between s- and r-process for a sample of Barium stars are showed, using europium as representative of r-process, since it is a nearly pure r-process element, and La and Ba as representatives of s-process.

2 Atmospheric Parameters and Abundances

The temperatures of all stars were determined from colour indices, including our own observations and data from the literature. The log g values were derived by fundamental relations using distances from [5], and the metallicities were derived from equivalent widths of Fe II lines.

The atmospheric parameters of the Barium stars sample were found to be in the range 4300 K $\leq$ T$_{eff}$ $\leq$ 6500 K, -1.0 $\leq$ [Fe/H] $\leq$ +0.2, and 1.4 $\leq$ log g $\leq$ 4.5, suggesting that there are giants, subgiants and dwarfs.

Elemental abundances were determined by spectral synthesis of individual lines, compared to observed spectra obtained with the FEROS spectrograph at the ESO-1.5m telescope.

At the early stages of the galaxy, the r-process dominated, being responsible for the abundances of the otherwise s-production dominated elements. When stars reached the AGB phase, the s-process was turned on, causing a significant increase in the s-elements abundances. The Fig. 1 shows [Ba/Eu] and [La/Eu] relations, for Barium stars (full circles) and normal stars from [2] (full squares), [4] (full triangles), [7] and [3] (opened squares). These are normal red giants and dwarfs with metallicities in the range -3 < [Fe/H] < +0.3.

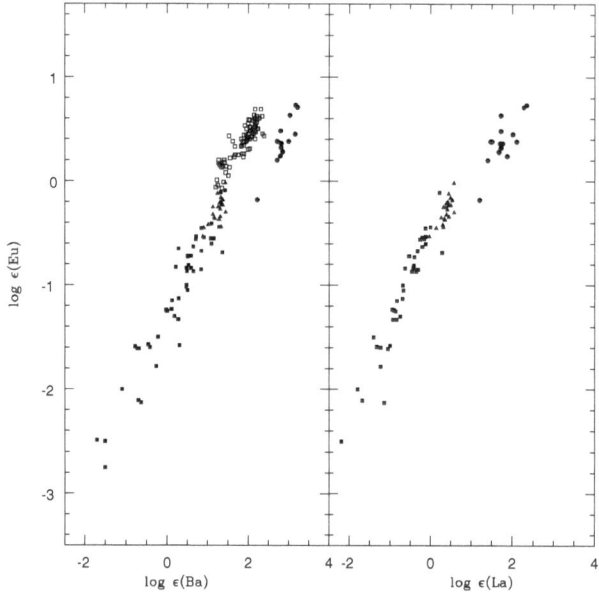

Fig. 1. Relation between log ϵ(Eu) and log ϵ(Ba,La)

Ba (La) and Eu abundances have a linear relation for stars whose metallicities are below -2.75. At higher metallicities the relation is still linear, but the fit shows different slope. Since Barium stars are rich in s-elements, that relation is different relative to normal stars with similar metallicities.

According to the Fig. 1, Ba is more overabundant than La in Barium stars relative to normal stars with similar metallicities. No dependence on luminosity classes was found in the s/r behavior among those Barium stars.

References

1. W.P. Bidelman, P.C. Keenan: ApJ **114**. 473 (1951)
2. D.L. Burris, C.A. Pilachowski, T.E. Armandroff, C. Sneden, J.J. Cowan, H. Roe, ApJ **544**, 302, (2000)
3. B. Edvardsson, J. Andersen, B. Gustafsson, D.L. Lambert, P.E. Nissen, J. Tomkin, A&A **275**, 101, (1993)
4. E. Jehin, P. Magain, C. Neuforge, A. Noels, G. Parmentier, A.A. Thoul, A&A **341**, 241, (1999)
5. M.O. Mennessier, X. Luri, F. Figueras, A.E. Gómez, S. Grenier, J. Torra, P. North, A&A **326**, 722, (1997)
6. J. Tomkin, D.L. Lambert, B. Edvardsson, B. Gustafsson, P.E. Nissen, A&A **210**, L15, (1989)
7. V.M. Woolf, J. Tomkin, D.L. Lambert, ApJ 453, 660, (1995)

Observations of ^{3}He in Planetary Nebulae

D.S. Balser[1], T.M. Bania[2], R.T. Rood[3], and W.M. Goss[1]

[1] National Radio Astronomy Observatory, USA
[2] Institute for Astrophysical Research, Boston University, USA
[3] Astronomy Department, University of Virginia, USA

Abstract. We discuss new observations of ^{3}He towards planetary nebulae (PNe) using the Very Large Array (VLA), the 305 m Arecibo telescope, which is now capable of observing the ^{3}He$^+$ spectral transition, and the recently commissioned 100 m Green Bank Telescope (GBT).

The light element ^{3}He is produced in significant amounts during the era of primordial nucleosynthesis. This abundance is then altered by nuclear processing in stars. Our current hypothesis is that there is approximately a net zero gain in the production of ^{3}He over the lifetime of the Galaxy. This conclusion is based on: (1) measurements of Galactic H II regions that reveal no ^{3}He gradient with metallicity (Figure 1); (2) ^{3}He/H abundances that are consistent with primordial values determined using other methods (e.g., CMB experiments such as WMAP that measure the baryon-to-photon ratio); and (3) evidence that mixing in low-mass stars inhibits the production of ^{3}He as predicted by standard stel-

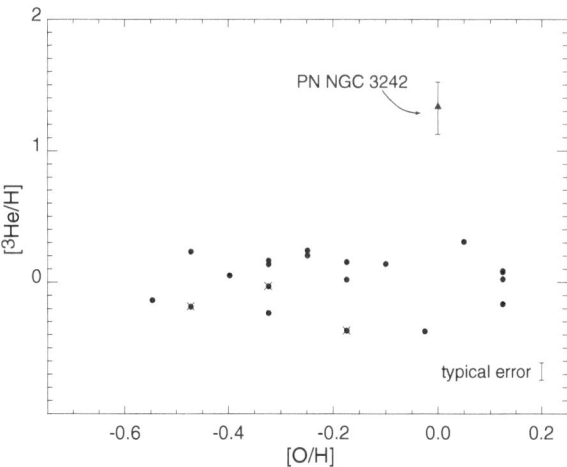

Fig. 1. Interstellar ^{3}He/H abundances as a function of source metallicity [2]. The [^{3}He/H] abundances by number derived for the H II region sample are given with respect to the solar ratio. Also shown is the abundance derived for the planetary nebula NGC 3242 (triangle). We note that there is no trend in the ^{3}He/H abundance with source metallicity

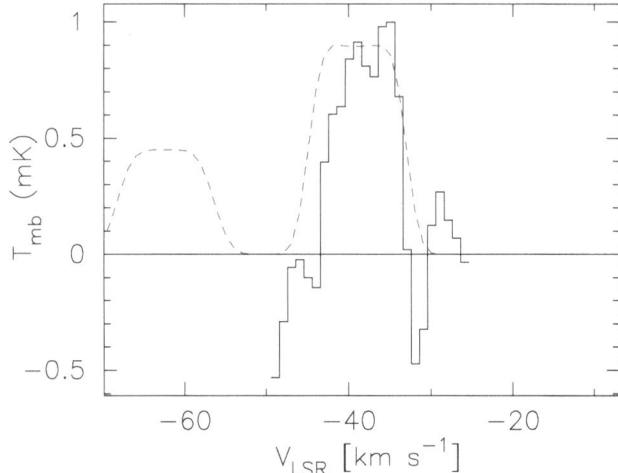

Fig. 2. VLA detection of ^{3}He$^+$ in the PN J 320. We have modeled the radio continuum and line emission using the radiative transfer code NEBULA [1], assuming an expanding shell of ionized gas. The dashed line is the model including the H171η and ^{3}He$^+$ transitions. The solid line shows the observed spectrum and only includes the ^{3}He$^+$ transition. The model fits the data reasonably well even though the morphology is bipolar as indicated by the HST image [6]

lar evolution models [5,4,2,7]. Nevertheless, measurements of ^{3}He in PNe that directly probe ^{3}He production in stars are required to confirm this hypothesis.

We have observed ^{3}He towards several PNe that have been selected to maximize the likelihood of ^{3}He$^+$ detections. First epoch observations with the GBT are discussed in [3]. Figure 2 shows a $4\,\sigma$ detection towards the PNe J 320 with the VLA. Both of these results are consistent with ^{3}He/H abundances between $10^{-4} - 10^{-3}$ by number and standard stellar evolution models. Observations with Arecibo are planned for winter 2005. Our goal is to be able to make a connection between some of the selection criteria and a high ^{3}He abundance. In this way we can use subsidiary measures, e.g., the N abundance, to estimate what fraction of PNe have preserved their ^{3}He.

References

1. D. Balser, T. Bania, R. Rood, & T. Wilson: Ap. J. **510**, 759 (1999)
2. T. Bania. R. Rood, & D. Balser: Nature **415**, 54 (2002)
3. T. Bania et al.: 'First Epoch Observations of ^{3}He with the Green Bank Telescope'. In: this volume
4. C. Charbonnel & J. do Nascimento Jr: Astron. & Astrophys. **336**, 915 (1998)
5. D. Galli, L. Stanghellini, M. Tosi, & F. Palla: Ap. J. **477**, 218 (1997)
6. D. Harman, M. Bryce, J. López, J. Meaburn, & A. Holloway: MNRAS **348**, 1047 (2004)
7. D. Romano, M. Tosi, F. Matteucci, & C. Chiappini: MNRAS **346**, 295 (2003)

Chemical Composition and Kinematics of Disk Stars

O. Bienaymé[1], T. Mishenina[2], C. Soubiran[3], V. Kovtyukh[2], and A. Siebert[4]

[1] Strasbourg Observatory (France)
[2] Odessa Observatory (Ukraine)
[3] Bordeaux Observatory (France)
[4] Steward Observatory, Tucson (USA)

Abstract. High resolution spectral data of red clump stars towards the NGP have been obtained with the spectrograph Elodie at OHP stars. Nearby Hipparcos red clump stars were also observed. We determine the thin and thick properties: kinematics and chemical abundances in the solar neighbourhood. We estimate the surface mass density of the galactic disk, we also determine the thin and thick disk chemical properties.

Nearly 700 Tycho-2 stars have been observed in the solar neighbourhood at distances smaller than 100 pc or in a 720 square degree field in the direction of the North Galactic Pole with the high resolution echelle spectrograph ELODIE at the OHP (Observatoire de Haute Provence, France). This work is an extension of the Soubiran et al (2003) and Siebert et al. (2003) studies: we present a new data set of stars with higher distances above the galactic plane, and we determine detailed abundances for a large set of Hipparcos red clump stars.

Absolute magnitudes, effective temperatures, gravities and metallicities have been estimated, as well as distances and 3D velocities. Abundances of Fe, Si and Ni have been determined from equivalent widths under LTE approximation, whereas abundances of Mg have been determined under NLTE approximation using equivalent widths of 4 lines and profiles of 5 lines.

Most of these stars are clump giants and span typical distances from 0 pc to 800 pc to the galactic mid-plane. This new sample, free of any kinematical and metallicity bias, is used to investigate the vertical distribution of disk stars.

The old thin disk and thick disk populations are deconvolved from the velocity-metallicity distribution of the sample and their parameters are determined. The thick disk is found to have a moderate rotational lag with respect to the Sun with a mean metallicity of $[Fe/H] = -0.48 \pm 0.05$ and a high local normalization of $15 \pm 7\%$.

We also determine both the gravitational force law perpendicular to the Galactic plane and the total surface mass density and thickness of the Galactic disk. The surface mass density of the Galactic disk within 800 pc derived from this analysis is Σ ($|z| < 800$ pc) $= 76$ Msun pc^{-2}.

The thickness of the total disk mass distribution is dynamically measured for the first time and is found to be 390 pc in relative agreement with the old stellar disk scaleheight. All the dynamical evidences concerning the structure of the disk (its local volume density - i.e. the Oort limit-, its surface density and its thickness) are compatible with our current knowledge of the corresponding stellar disk properties. This result implies that the dark matter component of our Galaxy cannot be distributed in a flat or disk-like component but must be distributed in a round halo.

References

1. C. Soubiran, O. Bienaymé, A. Siebert: Astron. Astrophys. **398**, 141 (2003)
2. A. Siebert, O. Bienaymé, C. Soubiran: Astron. Astrophys. **399**, 531 (2003)

Cu and Zn in Thick-Disk and Thin-Disk Stars

S. Bisterzo, R. Gallino, and M. Pignatari

Dipartimento di Fisica Generale, Universitá di Torino, Via P. Giuria 1, 10125 Torino, Italy

Previous works [6,13,5,8,14] already outlined the existence of two distinct stellar populations in the Galactic disk, the thick-disk and the thin-disk stars, based on both kinematical properties, as deduced from Hipparcos measurements, and spectroscopic abundance determinations. The metallicity range of the thick-disk stars extends from [Fe/H] ≤ -1.2 up to solar, the major fraction of them being concentrated in the range $-1 <$ [Fe/H] < -0.4. This is particularly evident for the α-elements, like Mg, Ti and Si: the [α/Fe] ratio for thick-disk stars is flat until [Fe/H] ~ -0.4, continuing the trend of halo stars, and then decreases linearly towards 0 at solar metallicity. The thin-disk stars instead show a continuous decreasing trend, from the halo typical enhancement at [Fe/H] ~ -1 down to solar at [Fe/H] $= 0$. For higher than solar metallicities, the [α/Fe] distribution flattens or even increases ([2], [1]). The knee shown by the thick-disk stars does not appear in the thin-disk population. The two distributions are sufficiently well separated. The age of the thick-disk stars is mainly greater than 8 Gyr, with an average age of 11.2 ± 4.3 Gyr [2]. This suggests a faster stellar formation rate for the thick-disk population, and consequently a delay in the appearance of the Fe contribution by Type Ia Supernovae. We have collected in Fig. 1 recent high-resolution spectroscopy data of [Cu/Fe] (left panel) and [Zn/Fe] (right panel) in the metallicity range $-1.2 <$ [Fe/H] $< +0.5$ [11,14,13,1]. Further data for [Zn/Fe] are from [2,9,7]. As in the case of the α-rich elements, both [Cu/Fe] and [Zn/Fe] show distinct trends versus metallicity, according if they belong to the thick-disk population (open symbols) or to the thin-disk population (full symbols). The [Cu/Fe] data for the thick-disk stars show a linear increase versus

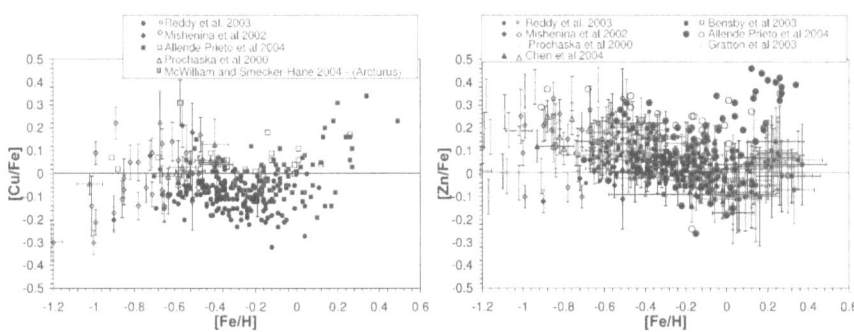

Fig. 1. *Left panel.* Spectroscopic observations of [Cu/Fe] versus [Fe/H] for thick-disk (open symbols) and thin-disk stars (full symbols). *Right panel.* The same for [Zn/Fe].

metallicity, with a bending towards a flat distribution for $[Fe/H] > -0.5$, continuing and overlapping with the trend of halo stars [11]. The thin-disk stars instead show a flat behaviour, with an average $[Cu/Fe] = -0.1$ [14]. For higher than solar metallicities a linear increase of $[Cu/Fe]$ is apparent [1]. The zinc abundances for thick-disk stars show a flat $[Zn/Fe] \sim +0.2$ up to $[Fe/H] \sim -0.5$, and a possible lower average value for higher metallicities. For the thin-disk stars $[Zn/Fe] \sim 0$ [14,2], with again an increasing trend for higher than solar metallicities. Cu and Zn have a different nucleosynthesis origin ([3], [4]): the complex trend in $[Cu/Fe]$ is due to competition among the Galactic production of copper, which is mostly originated in massive stars by the weak $sr(p)$ component as a secondary-like element (see Pignatari et al., these Proceedings), and Fe, which receives a large contribution by the long-lived Type Ia Supernovae. The more flat behaviour of $[Zn/Fe]$ has to be ascribed to the fact that about 50% of solar Zn by made of ^{64}Zn, which is most likely produced in a primary way during the α-rich free-zout of exploding massive stars. The residual 50% of solar Zn derives from the secondary-like weak-sr(p) component in massive stars SNIa produce negligible Cu and Zn. The contribution by AGB stars is also marginal, amounting to 5% and 3% of solar Cu and Zn, respectively. In conclusion, distinct trends versus metallicity for the two populations of thick-disk and thin-disk stars are apparent for both $[Cu/Fe]$ and $[Zn/Fe]$. It is tempting to ascribe the increasing trend of $[Cu/Fe]$ at higher than solar metallicities to the secondary-like nucleosynthesis origin of copper, once a kind of general Galactic equilibrium between SNII and SNIa ejecta is established.

References

1. Allende Prieto, C., Barklem, P.S., Lambert, D.L., Cunha, K.: A&A 420, 183 (2004)
2. Bensby, T., Feltzing, S., Lundstrom, I.: A&A 410, 527 (2003)
3. Bisterzo, S., et al.: in Nuclei in the Cosmos VIII, Nucl. Phys. A, in press
4. Bisterzo, S., et al. : Mem. Soc. Astron. It., in press
5. Carretta, E., Gratton, R.G., Sneden, C.: A&A 356, 238 (2000)
6. Chen, Y.Q., Nissen, P.E., Zhao, G., Zhang, H.W., Benoni, T.: A&ASS 141, 491 (2000)
7. Chen, Y.Q., Nissen, P.E., Zhao, G.: A&A 425, 697 (2004)
8. Edvardsson, B., et al.: A&A 275, 101 (1993)
9. Gratton, R.G., Carretta, E., Claudi, R., Lucatello, S., Barbieri, M.: A&A 404, 187 (2003)
10. McWilliam, A., Smecker-Hane, T.A.: *The Composition of Sagittarius Dwarf Spheroidal Galaxy* astroph/0409083 (2004)
11. Mishenina, T.V., Kovtyukh, V.V., Soubiran, C., Travaglio, C., Busso, M.: A&A 396, 189 (2002)
12. Mishenina, T.V., Soubiran, C., Kovtyukh, V.V., Korotin, S.A.: A&A 418, 551 (2004)
13. Prochaska, J.X., Naumov, S.O., Carney, B.W., McWilliam, A., Wolfe, A.M.: AJ 120, 2513 (2000)
14. Reddy, B.E., Tomkin, J., Lambert, D.L., Allende Prieto, C.: MNRAS 340, 304 (2003)

NGC 2324: A Relatively Young, Metal-Poor Open Cluster Located Beyond the Perseus Spiral Arm

J.J. Clariá[1], A.E. Piatti[2], and A.V. Ahumada[1]

[1] Observatorio Astronómico de Córdoba, Laprida 854, Córdoba, Argentina
[2] Instituto de Astronomía y Física del Espacio, CC 67, Suc. 28, 1428, Buenos Aires, Argentina

We present CCD photometry in the Johnson V, Kron-Cousins I and Washington system CT_1 passbands for NGC 2324, a rich open cluster located near the Galactic anticentre direction. We believe that the high discrepancy in the basic cluster parameters derived in previous studies, particularly in the cluster metal content, warrants their redetermination on the basis of more reliable data.

We built four CMDs extracted from different circular regions and we used the innermost one as representative of the cluster. Fig. 1 shows that the isochrone of the Geneva group corresponding to log t = 8.65 and Z = 0.008 (solid line) is the one which most accurately reproduces the cluster features. To match this isochrone, we used the reddening (E(V-I) = 0.33) and apparent distance modulus (V-Mv = 13.70) derived from the fit of the ZAMS for the same metallicity level. Open circles are red giant members confirmed from Coravel radial velocities. In the Washington (T_1,C-T_1) CMD, the same isochrone for the same metallicity level (Z = 0.008) is clearly the best representative of the cluster features. In this CMD the red giant clump appears better defined by the isochrones in comparison with those of the former figure.

To derive the cluster metal content, we used the reddening here derived (E(B-V) = 0.25) and the original Washington photometric data obtained by [2] at Cerro Tololo Inter-American Observatory for the red cluster giants confirmed from Coravel radial velocities. We applied an iterative method described by [1] to derive this cluster metal content ([Fe/H] = -0.31 ± 0.04), which is in good agreement with the best fits of isochrones. Fig. 2 shows the five Washington colour-colour relations for 4 of the 5 Coravel cluster giants, since one of them falls outside the range of the calibrations. In this figure, we have drawn three different isoabundance lines. Therefore, NGC 2324 is found to be a relatively young (440 Myr), metal-poor and distant open cluster (3.8 kpc) located beyond the Perseus spiral arm. A comparison with 10 well-known open clusters of nearly the same age shows that the cluster metal abundance and its position in the Galaxy are both consistent with the existence of a radial abundance gradient in the disk. We also present arguments which demonstrate that the current abundance gradient is very close to the paleogradient one, namely, the gradient in metallicity of the gas out of which the clusters have been formed.

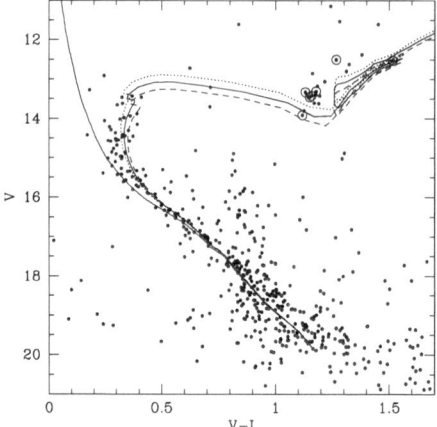

Fig. 1. (V,V-I) CMD for stars in NGC 2324. Isochrones from [3] of log t = 8.60, 8.65 and 8.70 are overplotted. Open circles are red giant members confirmed from Coravel radial velocities.

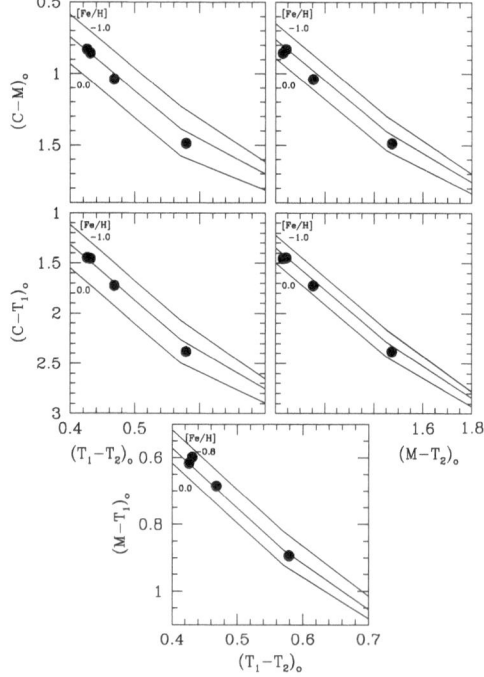

Fig. 2. Unreddened Washington colour-colour diagram for 4 red giants confirmed from Coravel radial velocities. Isoabundance relations from [1] for 0.5 dex intervals from [Fe/H] = -1.0 to 0.0 are shown, except for the $(M - T_1)_0/(T_1 - T_2)_0$ diagram wherein isoabundance relations for 0.4 dex intervals from -0.8 to 0.0 are given.

References

1. D. Geisler, J. Clariá, D. Minniti: AJ **102**, 1836 (1991)
2. D. Geisler, J. Clariá, D. Minniti: AJ **104**, 1892 (1992)
3. T. Lejeune, D. Schaerer: A&A **366**, 538 (2001)

Fluorine Abundances in the Galaxy

K. Cunha[1] and V.V. Smith[2]

[1] Observatório Nacional - MCT, Rio de Janeiro, Brazil
[2] National Optical Astronomy Observatory, Tucson, Arizona, USA

Abstract. The astrophysical origins of the element fluorine remain uncertain due in part to the availability of just a small number of abundance results for this element, that has readily observable transitions only in the infrared via vibration-rotation lines of HF. In this paper, we discuss all the available Galactic fluorine abundances to date, and add results for field stars with metallicities between [Fe/H] = -0.5 and -1.0, plus two stars that are members of the Orion association. The fluorine abundances obtained for the young Orion members are found to be in agreement with the trend of [F/O] versus O observed for the disk and they are a good representation of the present day value in the Galactic disk.

1 Fluorine Sources and Observations

Recent observations of the HF (1–0) R9 line at $\sim$ 2.3 μm with the Phoenix spectrograph on the Gemini-South telescope has opened a new window that sheds light on understanding the chemical evolution of fluorine and the nuclear processes that produce this element. Until recently, only a small number of observations of fluorine were available and the trend of fluorine abundances with metallicity had yet to be probed in the Galaxy.

Three sources have been proposed to produce fluorine in the Galaxy. The first was suggested by Forestini et al. (1992) and refers to production in low-mass stars during the AGB phase; while two others are related to massive stars: production in Wolf-Rayet stars (Meynet & Arnould 2000) and in type II Supernovae, via the neutrino-induced nucleosynthesis (Woosley et al. 1990).

The current status of HF abundances from infrared spectroscopy in samples of red-giants from different Galactic stellar populations are summarized in Figure 1. The abundance results displayed in this figure are from Cunha et al. (2003), plus new results for stars at the lowest metallicities, as well as two Orion pre-main-sequence stars. The run of fluorine with metallicity is now probed between oxygen abundances from roughly 7.7 to 8.7.

2 Discussion

A few comments can be drawn from the abundance results presented in Figure 1. The abundances obtained for the red giants in the globular cluster ω Centauri seem to indicate the existence of a sharp decline in the [F/O] ratios as the metallicity approaches the lowest observed oxygen abundance in this globular

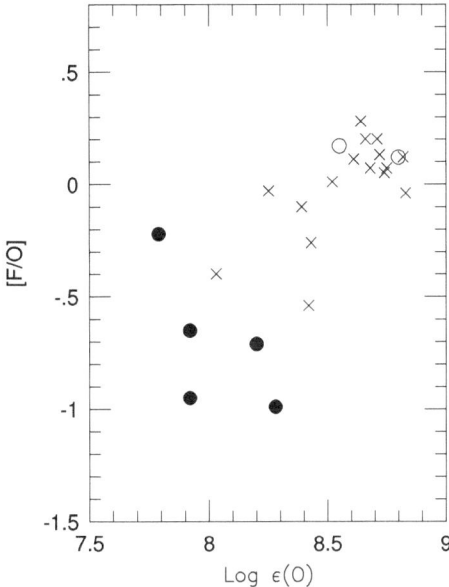

Fig. 1. The Galactic fluorine abundances obtained to date. Three samples are represented: the disk of the Milky Way (crosses), including the two young Orion pre-main-sequence stars (open circles), and ω Centauri giants (filled circles).

cluster. There is also considerable decline in the trend observed for the Milky Way disk stars. However, the [F/O] ratios in the Milky Way seem to be consistently higher than in ω Centauri. The new fluorine results for the young stars in the Orion association agree well with the trend observed for the Milky Way disk. These fluorine abundances all together can be used to constrain chemical evolution models including all possible sources for fluorine production, as discussed in Renda et al. (2004).

References

1. K. Cunha, V. V. Smith, D. L. Lambert, K. Hinkle: AJ 126, 1305 (2003)
2. M. Forestini, S. Goriely, A. Jorissen, M. Arnould: A&A 261, 157 (1992)
3. G. Meynet, M. Arnould: A&A 355, 176 (2000)
4. A. Renda, Y. Fenner, B. Gibson, et al.: MNRAS, 354, 575 (2004)
5. S.E. Woosley, D. H. Hartmann, R.D. Hoffman, W.C. Haxton: ApJ, 356, 272 (2000)

Determination of the Age of the Galactic Thin Disk from Th/Eu Stellar Abundance Ratios

E.F. del Peloso[1], L. da Silva[1], L.I. Arany-Prado[2], and G.F. Porto de Mello[2]

[1] Observatório Nacional/MCT, Rua General José Cristino, 77,
    20921-400 Rio de Janeiro, Brazil
[2] Observatório do Valongo/UFRJ, Ladeira do Pedro Antônio, 43,
    20080-090 Rio de Janeiro, Brazil

1 Introduction

Current Galactic disk age determinations are carried out by dating either the oldest open clusters with theoretical isochrones, or the oldest white dwarfs with cooling sequences. Nucleocosmochronology is the determination of time scales using abundances of radioactive nuclides. It can be used to date the Galactic disk with low dependence on stellar evolution calculations. The nuclides used by us to accomplish this are ^{232}Th, with a 14 Gyr half-life (i.e., of the order of magnitude of the age being assessed), and ^{151,153}Eu, the two stable isotopes of the comparison element (97% produced by the same nucleosynthesis process as all Th, the r-process). The sample is composed of 20 thin disk dwarfs/subgiants with F5–G9 spectral types, $5600 < T_{\text{eff}}(\text{K}) < 6300$ and $-0.84 \leq [\text{Fe/H}] \leq +0.28$. A complete description of this work can be found in [1] and [2].

2 Determination of Th and Eu Abundances

For our stars, only one adequately strong and uncontaminated line is available for Eu abundance determination, at 4129.72 Å. It has a significantly non-gaussian profile, due to its hyperfine splitting and isotope shift. Thus, spectral synthesis must be employed. High resolution, high S/N spectra were obtained for Eu with ESO's 1.52 m telescope and FEROS, and with ESO's CAT and CES. Th also has only one adequate line available, at 4019.13 Å. It is highly contaminated, also requiring spectral synthesis. High resolution, high S/N spectra were obtained for Th with ESO's CES fed by CAT and by the 3.60 m telescope. Atmospheric parameters were determined by photometric and spectroscopic differential analysis. Abundances of elements contaminating the Eu and Th spectral regions were determined using equivalent widths of lines measured in 10 orders of the same FEROS spectra used for Eu. Comparison with the literature ([3], [4], and [5]) shows that our results present the same behaviour, but with lower scatter.

3 Galactic Chemical Evolution (GCE) Model

We have developed a GCE model based on [6], with the inclusion of refuses, following an improved version of the formulation of [7]. Refuses are composed of stellar remnants (white dwarfs, neutron stars and black holes) and low-mass stellar formation residues (terrestrial planets, comets, etc.). Residues evaporate a considerable amount of H and He, retaining metals and diluting the interstellar medium. The effect of this dilution is equivalent, mathematically, to a second source of metal-poor infall, and contributes to a better fit to observational constraints. We adopted the delayed production approximation developed by [8], which takes into account the delay in production of elements that are synthesised mainly in stars of slow evolution, like Fe, which is predominantly generated in Type Ia supernovae (with a typical timescale of 1 Gyr).

4   The Age of the Galactic Disk

We used two different methods to determine the age:

1. Comparison of GCE model parameters with production ratio data from the literature: $T_G = 9.9 \pm 3.5$ Gyr. The high uncertainty reflects the difficulties of estimating theoretically the production ratio of r-process elements, whose production sites are not well known.
2. Comparison of our stellar abundance data with [Th/Eu] vs. [Fe/H] curves obtained from the GCE models, calculated for four different Galactic disk ages – 6, 9, 12, and 15 Gyr: $T_G = 8.2 \pm 1.9$ Gyr. The uncertainty is relative only to the abundance ratio uncertainties, not considering the uncertainties intrinsic to the GCE model itself, which are difficult to evaluate.

Taking the weighted mean of these results, using the reciprocal of the square of the uncertainties as weights, we arrive at the final result:

$$\text{Age of the Galactic thin disk} = 8.4 \pm 1.7\,\text{Gyr}$$

This agrees well with the latest white dwarf cooling estimations, which favour a low disk age (≤ 10 Gyr).

References

1. E.F. del Peloso, L. da Silva, G.F. Porto de Mello: A&A, accepted for publication (2004)
2. E.F. del Peloso, L. da Silva, L.I. Arany-Prado: A&A, accepted for publication (2004)
3. O. Morell, D. Källander, H.R. Butcher: A&A **259**, 543 (1992)
4. V.M. Woolf, J. Tomkin, D.L. Lambert: ApJ **453**, 660 (1995)
5. A. Koch, B. Edvardsson: A&A **381**, 500 (2002)
6. B.E.J. Pagel, G. Tautvaišienė: MNRAS **276**, 505 (1995)
7. H.J. Rocha-Pinto, L.I. Arany-Prado, W.J. Maciel: Ap&SS **211**, 241 (1994)
8. B.E.J. Pagel: 'The G-Dwarf Problem and Radio-Active Cosmochronology'. In: *Evolutionary Phenomena in Galaxies, Summer School at Puerto de la Cruz, Spain, July 4–15, 1988*, ed. by J.E. Beckman, B.E.J. Pagel (Cambridge University Press, Cambridge, New York 1989), pp. 201–223

CNO Abundances in Giants
with Precise Radial Velocity Measurements

J.D.Jr. do Nascimento[1], L. da Silva[1,2], J.R. De Medeiros[1], and J. Setiawan[3], and L. Pasquini[4]

[1] Departamento de Física, UFRN, Natal, RN., Brazil
[2] Observatorio Nacional, RJ., Brazil
[3] Max-Planck-Institut für Astronomie, Königstuhl, Germany
[4] European Southern Observatory, Germany

1 Introduction

We present oxygen abundances derived from the forbidden oxygen line for G and K giants with precise radial velocity (RV) measurements. Our goal is determine the evolutionary stage for 75 G and K giant stars using CNO abundances. The abundance analysis was done in standard Local Thermodynamic Equilibrium (LTE) with appropriate NLTE corrections and high precision spectra obtained with the FEROS (Setiawan et al. 2003) spectrograph. The fundamental parameters $T_{\rm eff}$, $\log g$ and [Fe/H] were determined from a detailed analysis to obtain the position of stars on the HR diagram as precise as possible. The abundances were derived from a synthetic analysis using plane-parallel, line-blanketed atmospheric models (Kurucz et al. 1993) and an updated version of the spectral analysis code MOOG (Sneden 1973). These abundances offer a unique possibility to establish the evolutionary stage of the objects, specially for those stars presenting a misclassification from the evolutionary tracks. Our stars sample presents three giants with extra solar planets detection HD 47536, HD 122430 and HD 11977 (Figure 1). In addition, derived CNO abundances permit to study the behavior of these elements in low mass ($M \leq 1 - 3.0\ M_\odot$) evolved stars.

2 Observations, Atmospheric Parameters and Chemical Analysis

The observations were performed at ESO using the 1.52m telescope and FEROS. The obtained spectra have high nominal resolving power ($R \sim 48000$), and S/N ~ 500 at maximum and a coverage from 4000 Å to 9200 Å. Many spectra were acquired for all sample stars. The atmospheric parameters ($T_{\rm eff}$, $\log g$, [Fe/H] and microturbulence velocities) have been obtained through an iterative and totally self-consistent procedure from Fe lines of the observed spectrum. The initial values of $T_{\rm eff}$ were obtained from a (B-V) vs $T_{\rm eff}$ calibration and $\log g$ were determined from Hipparcos parallaxes and evolutionary tracks. The [O/Fe] abundances were derived by fitting synthetic spectra to the observed one.

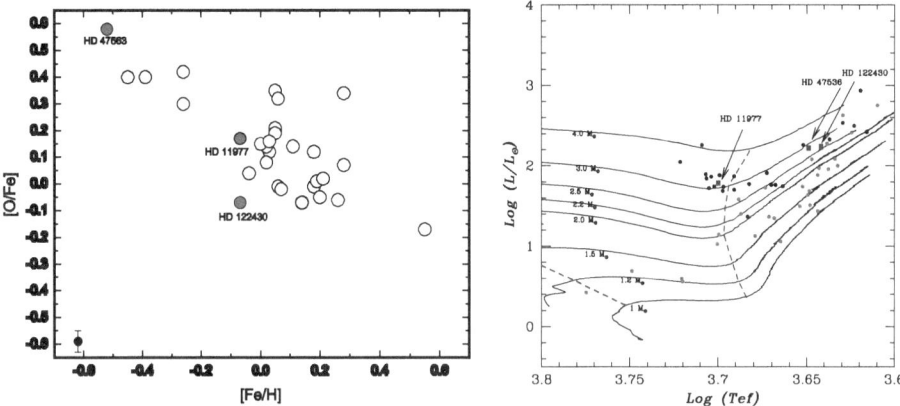

Fig. 1. (**a**) (*left*) Distribution of the [O/Fe] *vs.* [Fe/H] as derived from the forbidden oxygen line [OI] $\lambda 6300$ Å . Stars hosting giant planets are identified with *filled circles*. (**b**) (*right*) The HRD for the programme stars. Evolutionary tracks were computed from the Toulouse–Geneva code with masses between 1 and 4 M$_\odot$ and for [Fe/H] consistent with the sample stars (do Nascimento et al 2000 for a more detailed description). The dashed line indicates the evolutionary region where the subgiant branch starts, corresponding to hydrogen exhaustion in stellar central regions, and the beginning of the ascent of the red giant branch.

3   First Results

[O/Fe] abundances are presented for 30 G and K giants and subgiants, however, the total number of stars with precise radial velocities is around 70. *In the general, the HR position and the chemical properties of those stars showed a coherent behavior. The surface composition of the giant depends on the initial composition and mass.*

References

1. do Nascimento, J.D.Jr., Charbonnel, C., Lèbre, A., de Laverny, P., De Medeiros, J. R. 2000, A&A 357, 931
2. Kurucz, R. L., 1993, CD-ROMs, ATLAS9 Stellar Atmospheres Programs
3. Sneden C., 1973, Ph.D. thesis, University of Texas
4. Setiawan, J., Pasquini, L., da Silva, L., von der Luhe, O., Hatzes, A. 2003 A&A , 397, 1151

Oxygen in Metal-Rich Stars: Abundances from [O I] 6300, O I 7771–5 and Near-UV OH

A. Ecuvillon[1], G. Israelian[1], N.C. Santos[2,3], N. Shchukina[4], M. Mayor[3], and R. Rebolo[1]

[1] Instituto de Astrofisica de Canarias, 38200 La Laguna, Tenerife, Spain
[2] Observatorio Astronomico de Lisboa, Tapada da Ajuda, 1349-018 Lisboa, Portugal
[3] Observatoire de Geneve, 51 Ch. des Maillettes, CH-1290 Sauverny, Switzerland
[4] Main Astronomical Observatory, 03680 Kyiv-127, Ukraine

Abstract. Oxygen abundances of a large number of metal-rich stars, with and without known planets, were derived from the forbidden line [O I] 6300 Å, the O I 7771–5 Å triplet and from near-UV OH lines. Non-LTE corrections were calculated and applied to the LTE abundance results derived from the O I 7771–5 Å triplet. Spectral synthesis was performed for several OH lines. Results from different indicators are compared. We study abundance trends in planet host and comparison sample stars. We find for all the indicators that, on average, [O/Fe] clearly decreases with [Fe/H], with significantly negative slopes in all the linear fits.

1 Introduction

The discovery of the average metal-rich nature of planet-harbouring stars with regard to disc stars (i.e. [1],[2],[3]) has revealed the key role that metallicity plays in the formation and evolution of planetary systems. If the accretion processes were the main responsible for the iron excess found in planet host stars, volatile abundances should show clear differences in stars with and without planets, since volatiles (with low T_C) are expected to be deficient in accreted materials [4]. Previous studies of the abundance trends of the volatiles N, C, S and Zn [5, 6] have obtained no anomalies for a large sample of planet host stars.

We have determined oxygen abundances in two large samples, a set of planet-harbouring stars and a volume-limited comparison sample of stars with no known planets, using 3 different indicators: [O I] at 6300 Å, the O I 7771–5 Å triplet, and a set of 5 near-UV OH lines (see [7]). Non-LTE corrections were calculated and applied to the LTE abundance results for the triplet.

2 Results and Conclusions

The results from [O I] analysis are consistent with those from OH line synthesis, with discrepancies smaller than 0.2 dex in most cases. However, the abundances obtained from the triplet are systematically lower than the [O I] and OH results, with differences of ~ 0.3 dex. The reason of the triplet systematic underabundance could be that the NLTE corrections applied to triplet results correspond

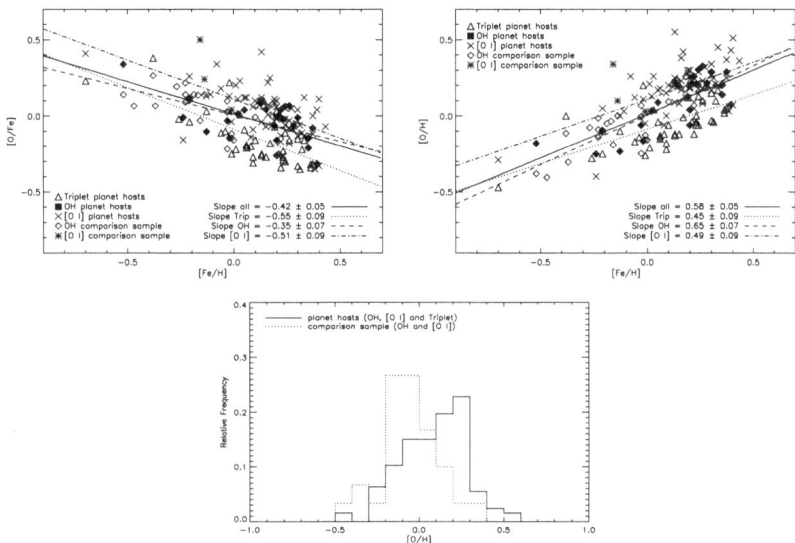

Fig. 1. Top: [O/Fe] and [O/H] vs. [Fe/H]. Linear least-squares fits to the two samples together, stars with and without planets, are represented for each indicator (*dotted, dashed and dotted-dashed lines* for triplet, OH and [O I], respectively). Bottom: [O/H] distributions for planet host (*solid line*) and comparison sample (*dashed line*) stars.

to the maximum effect, which does not take into account collisions with H atoms, and could produce an underestimation of the final abundances.

The [O/Fe] and [O/H] vs. [Fe/H] plots (fig. 1) show that the average trends resulting from different indicators present discrepancies, but similar kinds of behaviour. On average, [O/Fe] clearly decreases with [Fe/H] for all the indicators, with significantly negative slopes in all the linear fits. Planet host stars do not present anomalies in [O/Fe] ratios with respect to the comparison sample. The [O/H] distributions (fig. 1, bottom) for stars with and without planets presents an average value of 0.09 dex and −0.05 dex, respectively. If this difference had no link with the presence of planets, it should be reproduced and explained by Galactic chemical evolution models. Until then, we cannot completely exclude the possibility that those effects are related to the presence of planets.

References

1. Gonzalez, G.: MNRAS **285**, 403 (1997)
2. Santos, N.C., Israelian, G., & Mayor, M.: A&A **373**, 1019 (2001)
3. Santos, N.C., Israelian, G., & Mayor, M.: A&A **415**, 1153 (2004)
4. Smith, V.V., Cunha, K., & Lazzaro, D.: AJ **121**, 3207 (2001)
5. Ecuvillon, A., Israelian, G., Santos, N.C., Mayor, M., et al.: A&A **418**, 703 (2004)
6. Ecuvillon, A., Israelian, G., Santos, N.C., Mayor, M., et al.: A&A **426**, 619E (2004)
7. Ecuvillon, A., Israelian, G., Santos, N.C., Shchukina, N. et al.: A&A, in prep.

C I Non-LTE Spectral Line Formation in Late-Type Stars

D. Fabbian[1], M. Asplund[1], and M. Carlsson[2]

[1] Research School of Astronomy & Astrophysics, Australian National University, Mount Stromlo Observatory, Cotter Road, Weston ACT 2611, Australia
[2] ITA, University of Oslo, P.O. Box 1029, Blindern, N-0315 Oslo, Norway

Abstract. We present the results from our non-LTE investigation for neutral carbon, which was carried out to remove potential systematic errors in stellar abundance analyses. The calculations were performed for late-type stars and give substantial negative non-LTE abundance corrections. When applied to observations of extremely metal-poor stars, which within the LTE framework seem to suggest a possible [C/O] uprise at low metallicities (Akerman et al. 2004), these improvements will have important implications, enabling us to understand if the standard chemical evolution model is adequate, with no need to invoke signatures by Pop. III stars for the carbon nucleosynthesis.

Non-LTE C I Abundances

Our multi-level carbon model atom is adapted from D. Kiselman (private communication), with improved atomic data and better sampling of some absorption lines. The statistical equilibrium code MULTI (Carlsson 1986), together with 1D MARCS stellar model atmospheres for a grid of 168 late-type stars with varying $T_{\rm eff}$, log g, [Fe/H] and [C/Fe], were used in all C I non-LTE spectral line formation calculations, to solve radiative-transfer and rate equations and to find the non-LTE solution for the multi-level atom. We put particular attention in the study of the permitted C I lines around 9100 Å, used by Akerman et al. (2004).

Several tests were carried out to understand the driving non-LTE mechanisms for neutral carbon. The C I non-LTE line strength for the parameter grid explored is, with few exceptions at $T_{\rm eff}$= 7000 K, always larger than in LTE: the analysis thus gives significantly lower carbon abundances than in the 1D LTE case, by up to -0.8 dex and more for particular combinations of atmospheric parameters (notably at $T_{\rm eff}$= 6500 − 7000 K, log g= 2.00, [Fe/H]= −1.00, with corrections of up to ~ -1.0 dex). Different non-LTE effects are seen at solar and at low metallicities: the source function for the radiation field is not Planckian in the solar-metallicity case, while increased line opacity with the overpopulation of affected atomic levels is dominant in metal-poor stars. For each set of atmospheric parameters, the resulting non-LTE abundance corrections increase with line strength, in accordance to what found for the solar case (Asplund et al. 2004, A&A, in press). These calculations adopt a scale factor of $S_H = 0.001$ to the classical Drawin (1968) formula for inelastic collisions with hydrogen: the resulting abundance corrections are shown in Fig. 1. We also carried out all calculations with different values for S_H, finding smaller non-LTE abundance corrections with $S_H = 1$, typically ~ -0.2 dex for $T_{\rm eff} \simeq 6000$ K, log g= 4.50.

This improved analysis has a profound impact on the existing data of Akerman et al. (2004), where the observed relative abundance of carbon in respect to oxygen seems to rise at [O/H]< −1, a result obtained within a LTE analysis. If present, this trend could suggest high carbon yields from zero-metallicity stars. Our work aims at investigating whether the implicit assumption on which that analysis is based, that is that any non-LTE effect for C I and O I will balance out, is correct. We prove non-LTE effects for carbon to be larger than those typical for oxygen (∼ −0.2 dex), it is then likely that the [C/O] ratio is not increasing towards lower metallicities, implying that the standard chemical evolution model is adequate with no need to invoke Pop. III nucleosynthesis. A non-LTE reanalysis of new data for metal-poor stars should finally reveal if the [C/O] uprise at [O/H]< −1 is real or just an artifact due to neglection of non-LTE effects, with important implications for stellar evolution theories in either case.

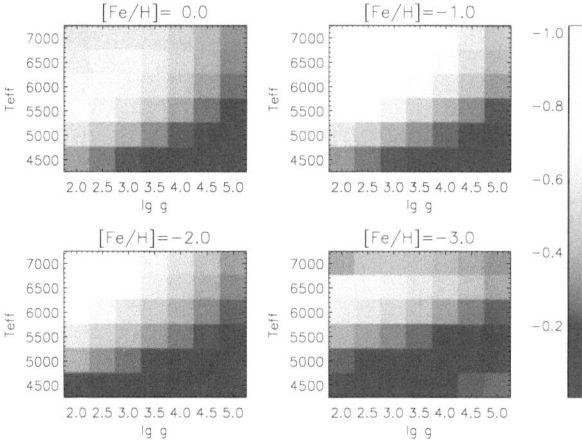

Fig. 1. Gray scale rendering of the non-LTE abundance correction for varying metallicity, for the grid of atmospheric parameters explored (with our standard choice for the collisions with hydrogen). Brighter parts of each plot correspond to larger non-LTE corrections, with intensity scaled as shown by the bar on the right hand side (dex units). The corrections refer to the absorption line of interest around 9100 Å

References

1. C.J. Akerman, L. Carigi, P.E. Nissen, M. Pettini, M. Asplund: A&A, 414, 931 (2004)
2. M. Asplund, N. Grevesse, A.J. Sauval, C. Allende Prieto, D. Kiselman: A&A, 417, 751 (2004)
3. M. Carlsson, 'A Computer Program for Solving Multi-Level Non-LTE Radiative Transfer Problems in Moving or Static Atmospheres'. In: *Uppsala Astronomical Report No. 33* (1986)
4. H.W. Drawin: Zeitschrift f. Physik, 211, 404 (1968)

Does the [α/Fe] Abundance Ratio Trend Reverse at Super-Solar Regime? A Test on the Role of Accurate Knowledge of Atmospheric Parameters

M. Franchini[1], C. Morossi[1], P. di Marcantonio[1], M. Lucia Malagnini[2], M. Chavez[3], and L. Rodriguez-Merino[3]

[1] INAF - Osservatorio Astron. di Trieste, Via G.B. Tiepolo, 11, I-34131 Trieste, Italy
[2] Dip. di Astron., Univ. Studi di Trieste, Via G.B. Tiepolo, 11, I-34131 Trieste, Italy
[3] Inst. Nacional de Astrofis., Opt. y Electronica, A.P. 51 y 216, 72000 Puebla, Mexico

1 Introduction

The trend of [α/Fe] abundance ratio with [Fe/H] is the subject of many papers (e.g. Bensby et al. 2003, Reddy et al. 2003). While there are clear indications of different behaviours in the thick disk and thin disk of the Galaxy, most of the studies show a decrease of [α/Fe] from +0.4 at [Fe/H]<-1.0 to [α/Fe]~0.0 at [Fe/H]=0.0 (e.g. Mishenina et al. 2004) in accord with predictions by nucleosynthesis theories and Galactic chemical evolution models (e.g. Chiappini et al. 1997). Recently, Allende Prieto et al. (2004, AP04) confirm that abundance ratios of α elements to iron become smaller as the iron abundance increases until approaching the solar values, but found that "the trends reverse for higher iron abundances" thus opening a new scenario. In discussing these results we should recall that abundance ratios are not directly observable quantities. They must be derived from observational quantities (i.e. equivalent widths) through stellar atmosphere models and the reliability of the derivation process depends on the accuracy of the main atmospheric parameters adopted in the analysis. In particular, the determination of $T_{\rm eff}$ in F, G and K stars is not straightforward since it is difficult to find a unique value which reproduces in a satisfactory way the observed photometry and, at the same time, provides the correct ionization balance of neutral and first ionized elements (see discussion in AP04).

The aim of this paper is to re-analyze the stars studied by AP04 with a technique which discriminates α-enhanced (Non Solar Scaled Abundance, NSSA) stars from those with Solar Scaled Abundances (SSA) without requiring any assumption about the stellar atmospheric parameter values.

2 Detection of α-Enhanced and SSA Stars

In Franchini et al. (2004) we introduced four Lick/IDS index-index diagrams, i.e. NaD vs Ca4227, NaD vs Mg2, NaD vs Mgb, and NaD vs CaMg, to identify SSA and α-enhanced stars irrespectively of their $T_{\rm eff}$, log g and [Fe/H]. By applying this method to the 84 normal (i.e. excluding binaries and variable) stars from the S^{4}N web site with [Fe/H] determined by AP04, it results that 8 stars are

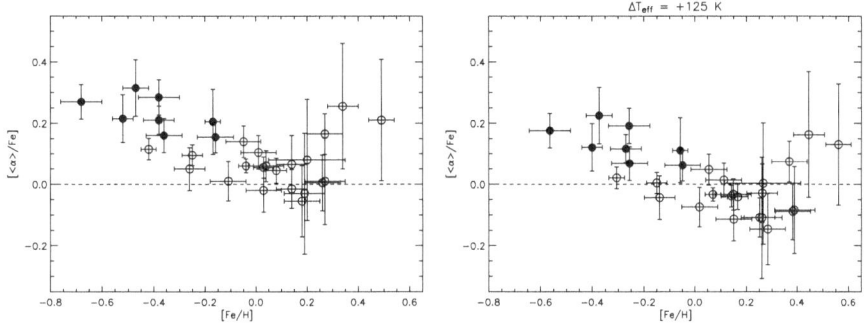

Fig. 1. Comparison between our classification and AP04 [α/Fe] ratios.

bona fide α-enhanced and 21 are SSA stars. Figure 1 (left panel) shows, for the detected NSSA (filled circles) and SSA (empty circles) stars, the average α-elements abundance ratio versus the [Fe/H] value derived by AP04. There is a clear agreement between our classification and the [α/Fe] ratios derived by AP04 as far as the metal poor and solar stars are concerned. On the other hand, we do not detect as α-enhanced any star with [Fe/H]>0.0. In particular, there are 12 stars which fall more than one sigma above the [α/Fe]=0 line, according to AP04, but are indicated as SSA by our method. To investigate if this discrepancy may be reduced by taking advantage of the dependency of the AP04 abundance ratios on the adopted $T_{\rm eff}$ values, we re-computed the [Fe/H] and the [α/Fe] values of the 29 stars by assuming a systematic off-set in their $T_{\rm eff}$'s on the basis of Table 5 in AP04. Figure 1 (right panel) shows the results obtained with an increase of 125 K. In this case the agreement with our classification, which is independent of the assumed $T_{\rm eff}$ values, is significantly improved with only one remaining discrepant case, namely HD 128621. On the basis of our analysis we can conclude that:

1. the [α/Fe] abundance ratio derived from the analysis of high resolution spectra is quite model dependent;
2. the AP04 statement of an increase of [α/Fe] at [Fe/H]>0.0 requires further checks since it is not confirmed by a $T_{\rm eff}$-independent method.

Acknowledgements
We acknowledge support from INAF-PRIN 2002 (PI Dr. Bertelli), from Università degli Studi di Trieste (ex 60% grants) and from MIUR COFIN-2003028039.

References

1. Allende Prieto, C., Barklem, P. S, Lambert, et al.: A&A **420**, 183 (AP04) (2004)
2. Bensby, T., Feltzing, S., Lundström I.: A&A **410**, 527 (2003)
3. Chiappini, C, Matteucci, F., Gratton, R.: ApJ **477**, 765 (1997)
4. Franchini, M., Morossi, C., Di Marcantonio, P., et al.: ApJ **601**, 485 (2004)
5. Mishenina, T.V., Soubiran, C., Kovtyukh, V.V., et al.: A&A **418**, 551 (2004)
6. Reddy, B.E., Tomkin, J., Lambert D.L, et al.: MNRAS **340**, 304 (2003)

Abundances and Ages of the Deconvolved Thin/Thick Disks of the Galaxy

P. Girard and C. Soubiran

Observatoire Aquitain des Sciences de l'Univers, L3AB, 2 rue de l'Observatoire, BP 89, 33270 Floirac, France

Abstract. We have investigated the abundance of several chemical elements in two large stellar samples kinematically representative of the thin and the thick disks of the Galaxy. Chemical, kinematical and age data have been collected from high quality sources in the literature. Velocities (U,V,W) have been computed and used to select stars with the highest probability to belong to the thin disk and the thick disk respectively. Our results show that the two disks are chemically well separated. Both exhibit a decline of [α/Fe] with increasing [Fe/H]. A transition between the thin/thick disks stars is observed at 10 Gyr

A sample of 823 stars with abundances of several elements (Fe,O,Mg,Ti,Si, Na,Ni,Al) was compiled from several papers (Ref 1 to 10) after checking the lack significant differences between their results. The velocities (U,V,W) and orbital parameters were computed for 640 Hipparcos stars having $\frac{\sigma_\pi}{\pi}$<0.25, to make a large database combining kinematics and detailed abundances. Ages of 442 stars were retrieved from Nordström et al(2004).

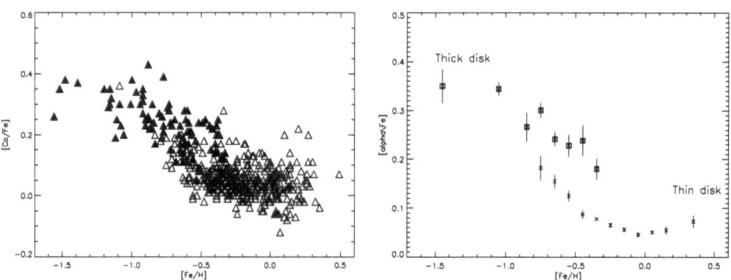

Fig. 1. Left : [Ca/Fe] vs [Fe/H] for the thin (empty triangles) and the thick (filled triangles) disk stars. Right : [α/Fe] vs [Fe/H].

In order to investigate the chemical and age properties of the thin and the thick disks separately we have performed the deconvolution of their velocity distributions. We show that about 25% of the sample has kinematics typical of the thick disk, adopting for its parameters $V_{lag} = -51\,\mathrm{km\,s^{-1}}$ and $(\sigma_U, \sigma_V, \sigma_W) = (63, 39, 39)\,\mathrm{km\,s^{-1}}$. Stars having a probability higher than 80% to belong to the thin and thick disks were selected. Plots on Fig.1 show nicely the separation between the thin and the thick disks. The thick disk is α-enhanced as compared to the thin disk but the decreasing trends are parallel. In the metallicity overlap, $[\alpha/Fe]$ of the thick disk exceeds by 0.08 dex that of the thin disk. No clear vertical gradient of abundance in the thick disk is seen on Fig.2. When only high precision ages (relative error $< 15\%$) are considered, a transition between ages of the thin and the thick disks stars at 10 Gyr is observed (Fig.2).

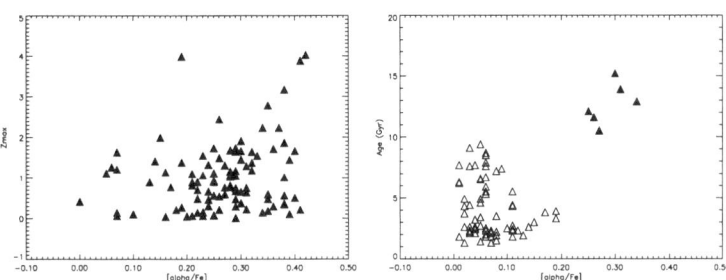

Fig. 2. Left : Zmax vs $[\alpha/Fe]$ for the thick disk stars. Right : Age distribution of the thin and thick disks stars.

Conclusion. Thanks to our large sample, the statistic is improved and the separation between the two disks is quantified. It is now clear that the thin and the thick disks are chemically well separated. We found a transition in the age distribution of the thin disk and the thick disk stars at 10 Gyr but no clear vertical gradient in the thick disk. These results constrain the formation scenarii of the Milky Way's disks.

References

1. Allende Prieto, C., Barklem, P.S., Lambert, D.L., Cunha, K., 2004, A&A, 420, 183
2. Bensby, T., Feltzing, S., & Lundström, I, 2003, A&A, 410, 527
3. Edvardsson, B., Andersen, J., Gustafsson, B., Lambert, D.L., Nissen, P.E., & Tomkin, J., 1993, A&A, 275, 101
4. Feltzing, S., Bensby, T., & Lundström, I., 2003, A&A, 397, L1
5. Fulbright, J.P., 2000, AJ, 120, 1841
6. Gratton, R.G., Carretta, Claudi, R., Lucatello,S., and Barbieri, M., 2003, A&A, 404, 187
7. Mishenina, T.V., Soubiran, C., Kovtyukh, V.V., Korotin, S.A., 2004, A&A, 418, 551
8. Nissen, P.E., Primas, F., Asplund, M., & Lambert, D.L., 2002, A&A, 390, 235
9. Prochaska, J.X., Naumov, S.O., Carney, B.W., McWilliam, A., & Wolfe, A.M., 2000, ApJ, 120, 2513
10. Reddy, B.E., Tomkin, J., Lambert, D.L., & Allende Prieto, C., 2003, MNRAS, 340, 304
11. Nordström, B., Mayor, M., Andersen, J., Holmberg, J., Pont, F., Jørgensen, B. R., Olsen, E. H., Udry, S., Mowlavi, N., 2004, A&A, 418, 989

Studying Old Open Clusters
with Detached Eclipsing Binaries

F. Grundahl[1], S. Meibom[2], H. Bruntt[3,4], H.R. Jensen[1], and J.V. Clausen[4], and S. Frandsen[1]

[1] Aarhus University, Institute of Physics and Astronomy, 8000 Aarhus C, Denmark
[2] University of Wisconsin, Madison, Wisconsin, USA
[3] US Air Force Academy, Department of Physics, Colorado, USA
[4] Copenhagen University, NBIfAFG, 2100 Copenhagen Ø, Denmark

1 Introduction

The single most important parameter which determines the evolution of a star is its mass. Detached eclipsing binaries offer the possibility to determine accurate masses and radii for their components. By studying such systems in open and globular clusters it is possible to determine cluster ages to higher precision than by "traditional" methods, such as main-sequence fitting. We have initiated a programme to study detached eclipsing binaries in old open and globular clusters and determine their masses and radii. This allows direct comparisons with models in the mass, radius plane thereby avoiding the troublesome conversion between model (bolometric luminosity, temperature) and observed (color, apparent luminosity) quantities.

2 Open Clusters

We are currently working on detached systems in the old open clusters NGC 2243, NGC 188 and NGC 6791. Presently we have the most complete and best data available for NGC 188 (see Fig. 1 and caption for more details) and NGC 2243. In both clusters the detached system is located close to the cluster turnoff and consists of two quite similar stars. In NGC 6791 we were able to determine the period for the system called V20 [1] and have subsequently obtained photometry for this system covering both eclipses. Due to the faintness ($V = 17.34$) we have not yet obtained radial velocities for the system which is comprised of a star very close to the cluster turnoff and a main–sequence star approximately 2.2 magnitudes (V) fainter.

References

1. H. Bruntt et al.: A&A, 410, 323 (2003)
2. I. Platais et al.: AJ, 126, 2922 (2003)
3. Zhang et al.: AJ, 123, 1548 (2002)

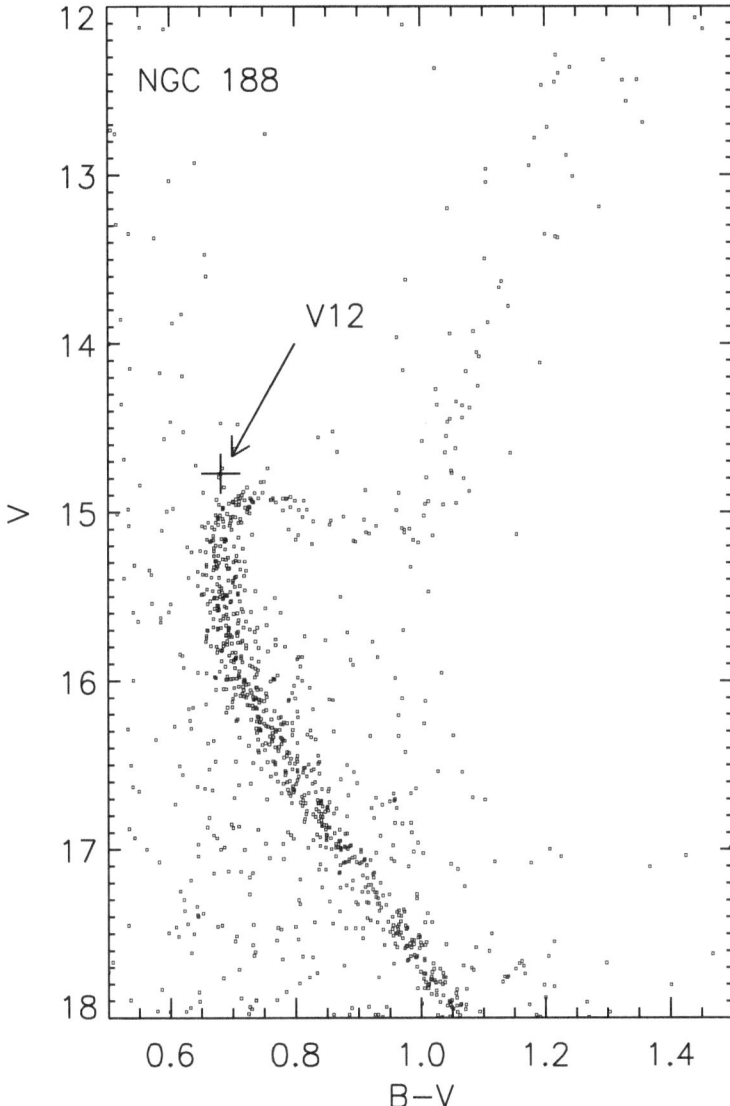

Fig. 1. The color–magnitude diagram of NGC 188 from [2] with the location of the detached eclipsing binary V12 [3] overplotted. From radial–velocity measurements we find (assuming an inclination of 90 degrees since we do not yet have photometry of the eclipses) that the masses of the two components are 1.06 and 1.08 solar masses. We estimate that we will be able to reach a precision of 1% in the mass estimate. We are in the process of acquiring eclipse photometry such that the radii and orbital inclination can be determined. Since both components are very close to the cluster turnoff their masses and radii can be directly used to give a very accurate age estimate for the cluster by comparing to isochrones in the (mass, radius) plane and requiring that they both lie on the same isochrone.

Isotopic Abundances in RGB- and AGB-Stars

H.U. Käufl[1], B. Aringer[2], S. Uttenthaler[1], and J. Hron[2]

[1] European Southern Observatory, D-85748 Garching bei München, Germany
[2] Institut für Astronomie der Universität Wien, Türkenschanzstrasse 17,
A-1180 Wien, Austria

Abstract. Asymptotic Giant Branch stars contribute significantly to the chemical evolution of their host galaxies. All stars with a main-sequence mass $\leq 8M_\odot$ will return eventually 30-80% of their mass to the interstellar medium. In parallel to the He-burning phase, which follows the original main sequence H-burning, AGB stars are known to produce heavy elements. He-burning is accompanied by a strong neutron flux which in turn breeds heavy nuclei in the so called s-process. Convective mixing of the entire star outside of the degenerate C/O core (dredge-up) ensures, that the s-process produced material becomes part of the outer shells which are being expelled in the mass-loss phase on the AGB. In spite of the importance of this process for understanding chemical evolution precious little observational constraints exist for the various models describing thermo-nuclear evolution. Fully resolved infrared spectroscopy of rotation-vibrational molecular transitions may contribute substantially to a better understanding of the thermo-nuclear evolution.

1 Introduction

In the last years there was substantial progress in modeling of infrared spectra from cool stars (e.g. [1]) even in the presence of pulsations (e.g. [2]). At ESO's VLT an adequate spectrograph, CRIRES[1], will soon become available. CRIRES is a cryogenic pre-dispersed long-slit (≈ 40") spectrograph with a nominal resolution $\frac{\lambda}{\Delta\lambda} \approx 10^5$ (i.e. $\approx 3\frac{km}{s}$). CRIRES covers the $1\text{-}5.5\mu m$ wavelength regime. A curvature sensing adaptive optics system feed is used to minimize slit losses. A mosaic of 4 Aladdin III InSb-arrays packaged on custom-fabricated ceramics boards provides for an effective 4096x512 pixel focal plane array, to maximize the free spectral range covered in each exposure. Stability, overall precision and calibration of CRIRES aim at a precision of $75\frac{m}{s}$. A detailed description of the instrument including sensitivities is given elsewhere ([3,4]).

2 Conclusion

Isotopic shifts for molecular transitions are of order of 1% and thus easy to observe. Suitable molecules with transitions accessible to CRIRES are CO, CN and SiO. The Si isotopes appear particularly interesting as they can provide for a neutron "dosimeter" during the AGB-phase. Models for the thermonuclear

---

[1] CRIRES stands for <u>CR</u>yogenic <u>I</u>nfrared <u>E</u>chelle <u>S</u>pectrograph.

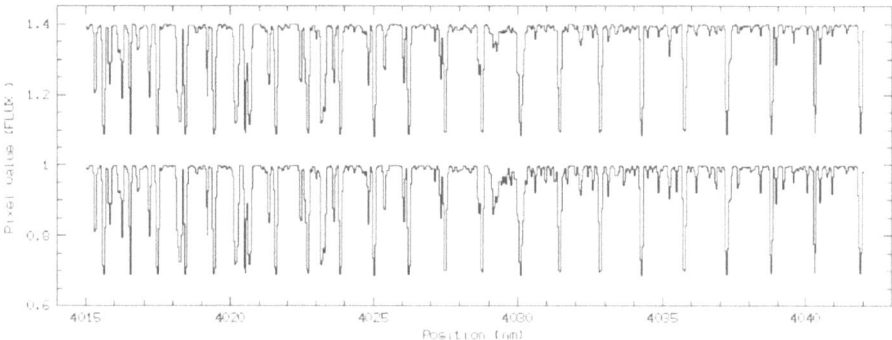

Fig. 1. Model Spectra re-binned to CRIRES Resolution: To demonstrate the potential for precise isotopic abundance determination two representative sample absorption spectra, normalized to unity, are shown. They result from a radiative transfer calculation using a hydrostatic MARCS model atmosphere for 3400 K. MARCS stands for $\underline{M}$odel $\underline{A}$tmosphere in a $\underline{R}$adiative $\underline{C}$onvective $\underline{S}$cheme; the methodology is described in detail e.g. in [1] and references therein. The models are calculated with a spectral bin size corresponding to a Doppler velocity of $1\frac{km}{s}$. They are re-binned to the nominal CRIRES resolution ($3\frac{km}{s}$), which even for the slowest rotators is sufficient to resolve absorption lines. The spectral range covers $\approx \frac{1}{3}$ of the CRIRES detector-array and has been centered at the band-head of a $^{29}Si^{16}O$ overtone transition at 4029 nm. In both spectra the band-head is clearly visible between the forest of well-separated low- and high-j transitions of the common isotope. The lower spectrum is based on the telluric ratio of the isotopes $^{28}Si/^{29}Si/^{30}Si$ (92.23:4.67:3.10) whereas the upper spectrum, offset by 0.4 in y-direction, has been calculated for a ratio of 96.00:2.00:2.00.

processes have been presented at these conference (Y. Fenner et al.). The SiO-overtone feature at $4\mu m$ is a prominent and easy to observe feature with little telluric interferences (sample spectra, at lower resolution, in [5]). At the VLT there will be sufficient sensitivity for measuring such lines in associations of stars, e.g. clusters, to statistically compare the abundances in RGB with AGB stars.

IR molecular spectra also allow for abundance determination of metals in a fundamentally different way as compared to optical spectroscopy. In that sense CRIRES can also provide an invaluable cross-calibration of 'optical' abundances.

References

1. Aringer, B., Kerschbaum, F. & Jørgensen, U.G. 2002 A&A 395, p. 915
2. Höfner, S., Gautschy-Loidl, R., Aringer, B. & Jørgensen, U.G. 2003 A&A 399, p. 589
3. Käufl, H.U., Ballester, B., Biereichel, P. et al 2004 "CRIRES: A High Resolution Infrared Spectrograph for ESO's VLT" in Proceedings SPIE **5492** part 3, p.'1218
4. Moorwood, A., 2003, *The Messenger* 114, p. 5
5. Aringer, B., Höfner, S., Wiedemann, G. et al. 1999 A&A 342, p. 799

Abundance Variations in the Galactic Disk: Planetary Nebulae, Open Clusters and Field Stars

W.J. Maciel and R.D.D. Costa

IAG/USP, Rua do Matão 1226, CEP 05508-900 São Paulo SP, Brazil

1 Introduction

Abundance gradients and their variations constitute one of the main constraints of chemical evolution models for the Galaxy. The time evolution of the gradients, in particular, is essential to distinguish between models involving different physical processes and time scales (see for example [6] and [8]). The gradients can be derived from different types of objects, but the study of their time variation requires the use of objects with a reasonably large age span, such as planetary nebulae (PN) and open clusters. In the present work, we compare the results obtained from PN with recent determinations from open clusters and cepheid variable stars. These objects offer some additional advantages compared with PN, such as more accurate distances and ages.

2 Planetary Nebulae

The time variation of the O/H radial gradient has been estimated by [9] from a sample of galactic PN. More recently, Lago and Maciel (in preparation) have extended these results for the elements S, Ne and Ar. From the observed abundances, the [Fe/H] ratio has been estimated, using a correlation valid for the galactic disk [7]. Adopting an age-metallicity relation which depends also on the galactocentric distance [3], the progenitor ages have been determined. Taking into account nebulae in different age groups, it was concluded that the gradients are flattening at an approximate rate of 0.005 dex kpc^{-1} Gyr^{-1}.

3 Open Clusters

Open clusters are favourite objects in the study of [Fe/H] radial gradients. Recently, new homogeneous results [4] and compilations [2] have become available. Such data allow us to estimate the time variation of the gradients, again resulting in some flattening for younger objects. In the present work, we have considered both sources and rederived the [Fe/H] gradients and their variations, taking into account samples composed of young, intermediate, and old clusters, as in [9] for planetary nebulae. We conclude that both samples [4] and [2] lead essentially to the same results. The main problem with this derivation refers to the youngest clusters, which are concentrated in the inner portions of the disk, increasing the uncertainties of their gradients.

4   Cepheids

Cepheid variables show several advantages for the determination of radial abundance gradients relative to the remaining objects. Their distances are accurately determined, as well as their ages. In a recent series of papers, Andrievsky et al. obtained accurate abundances for several elements from a spectroscopic analysis of a sample of cepheid variables (see [1] for the references). We have used this sample and rederived the O/H and [Fe/H] radial gradients. We have also determined the age distribution of the cepheids, which are concentrated in the young object group, so that the gradient at this age bracket is significantly improved.

5   Conclusions

A comparison of the gradients from PN, open cluster stars and cepheid variables shows that their temporal variations are similar, in the sense that the gradients seem to be flattening out with time for the last 8 Gyr approximately. The distribution of the youngest open clusters tends to be concentrated in the inner regions of the disk, which makes their gradient artificially flatter. For this reason, the results of the cepheids are especially important, since these objects have better determined distances and ages. The results confirm the time variation previously determined for PN and open clusters. The derived time variations of the gradient are in agreement with the predictions of theoretical models by Hou et al. [5]. These results do not allow an accurate determination of the flattening rate, but a rough estimate would be in the range $0.005 - 0.010$ dex kpc^{-1} Gyr^{-1} in the last few Gyr.

Acknowledgements. This work was partially supported by CNPq and FAPESP.

References

1. S.M. Andrievsky et al.: Astron. Astrophys. **413**, 159 (2004)
2. L. Chen, J.L. Hou, J.J. Wang: Astron. J. **125**, 1397 (2003)
3. B. Edvardsson et al.: Astron. Astrophys. **275**, 101 (1993)
4. E.D. Friel et al.: Astron. J. **124**, 2693 (2002)
5. J.L. Hou, N. Prantzos, S. Boissier: Astron. Astrophys. **362**, 921 (2002)
6. W.J. Maciel: The Evolution of the Milky Way, ed. F. Matteucci, F. Giovannelli, (Kluwer, Dordrecht, 2000), 81
7. W.J. Maciel: Rev. Mex. Astron. Astrophys. SC **12**, 207 (2002)
8. W.J. Maciel, R.D.D. Costa: IAU Symp. 209, ed. S. Kwok, M. Dopita, R. Sutherland, (ASP, San Francisco, 2003), 353
9. W.J. Maciel, R.D.D. Costa, M.M.M. Uchida: Astron. Astrophys. **397**, 667 (2003)

Some Concerns About the Reliability of LTE Abundance Analyses in Cool, Active Stars

T. Morel[1], G. Micela[1], and F. Favata[2]

[1] Istituto Nazionale di Astrofisica, Osservatorio Astronomico di Palermo
   G. S. Vaiana, Piazza del Parlamento 1, I-90134 Palermo, Italy
[2] Astrophysics Division - Research and Science Support Department of ESA,
   ESTEC, Postbus 299, NL-2200 AG Noordwijk, The Netherlands

Abstract. We discuss recent observational evidence illustrating the current limitations plaguing classical LTE abundance analyses of cool ($T_{\rm eff} < 5500$ K), chromospherically active stars. Although significant progress on this issue can be evidently expected from a more realistic atmospheric modelling and treatment of NLTE line formation, a homogeneous abundance study of a large sample of *inactive* K-type stars may also prove valuable in disentangling temperature and activity effects.

1 Background

A growing body of data on chemical abundances of cool, young open cluster stars is yielding intriguing results. A dramatically increasing oxygen overabundance with decreasing effective temperature (up to 1 dex) has been, for instance, reported in cool Pleiades members [5], while an order of magnitude difference between the iron abundances derived from Fe I and Fe II lines has been found in K-type dwarfs of the Hyades [6]. This schematically calls for two main culprits: (a) the largely unknown importance of NLTE effects in this type of objects and/or (b) difficulties in modelling the atmospheric structure of stars exhibiting a plethora of peculiarities, from a thermally inhomogeneous photosphere (large spot groups, etc.) to an overlying, prominent chromosphere. It is of relevance to note that similar conclusions, albeit perhaps to a lesser extent, also seem to hold for inactive K-type stars in the field [1].

2 Insights from an Analysis of a Sample of Tidally-Locked Active Binaries

We have recently derived the abundances of 13 chemical elements in a sample of 14 single-lined RS CVn systems from a differential, curve-of-growth LTE analysis of high-resolution FEROS spectra (the reader is referred to [2] and [3] for further details). A significant overabundance of several elements (e.g., Na, Al) compared to inactive, Galactic disk stars of similar metallicities is found. The case of oxygen is particularly illustrative [4]. As can be seen in Fig. 1, the O I triplet yields for the coolest, most active stars puzzlingly high overabundances relative to solar of up to 1.8 dex (note that this phenomenon is not observed for [O I] $\lambda 6300$

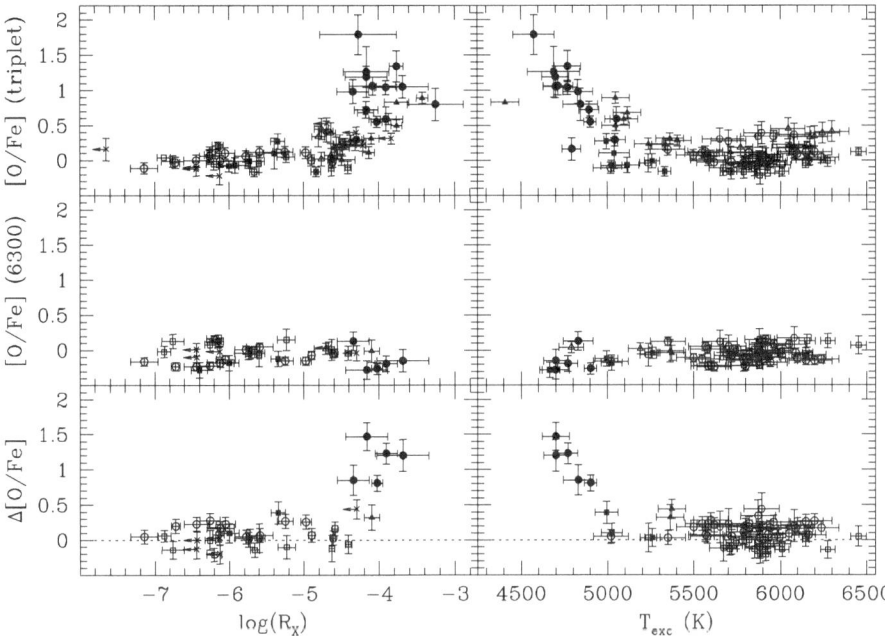

Fig. 1. Oxygen abundances as a function of the activity index, R_X, derived from X-ray data (*left-hand panels*) and the excitation temperature T_{exc} (*right-hand panels*). The bottom panels show the difference between [O/Fe] yielded by the O I triplet at about 7774 Å and the [O I] $\lambda 6300$ line. Filled circles: RS CVn binaries ([2] and [3]), filled squares: field subgiants [3], filled triangles: Pleiades stars, open triangles: Hyades stars, open circles, squares and hexagons: disk dwarfs. The source of the literature data for the open cluster and Galactic disk stars can be found in [4].

which is insensitive to departures from LTE). It is remarkable that the literature data exhibit similar temperature and activity trends, despite the heterogeneous nature of the analyses performed. Although it is not possible at this stage to clearly assess the relative importance of temperature and activity effects, it is hoped that our forthcoming abundance analysis of a large sample of inactive, K-type stars observed with FEROS will settle this issue.

References

1. C. Allende Prieto, P. S. Barklem, D. L. Lambert, K. Cunha: A&A **420**, 183 (2004)
2. T. Morel, G. Micela, F. Favata, D. Katz, I. Pillitteri: A&A **412**, 495 (2003)
3. T. Morel, G. Micela, F. Favata, D. Katz: A&A **426**, 1007 (2004)
4. T. Morel, G. Micela: A&A **423**, 677 (2004)
5. S. C. Schuler, J. R. King, L. M. Hobbs, M. H. Pinsonneault: ApJ **602**, L117 (2004)
6. D. Yong, D. L. Lambert, C. Allende Prieto, D. B. Paulson: ApJ **603**, 697 (2004)

Chemo-Dynamical Properties of F-G-K Stars in the −1.0<[Fe/H]< +0.50 Range

C. Morossi[1], M. Franchini[1], P. di Marcantonio[1], M.L. Malagnini[2], M. Chavez[3], and L. Rodriguez-Merino[3]

[1] INAF - Osservatorio Astron. di Trieste, Via G.B. Tiepolo, 11, I-34131 Trieste, Italy
[2] Dip. di Astron., Univ. Studi di Trieste, Via G.B. Tiepolo, 11, I-34131 Trieste, Italy
[3] Inst. Nacional de Astrofis., Opt. y Electronica, A.P. 51 y 216, 72000 Puebla, Mexico

1 Introduction and Observational Data-Sets

Differences in the chemical composition of galactic stars with similar [Fe/H], in particular in the range -1.0 to +0.50 dex, are attributed to the presence of different stellar components in the Galaxy, in general, and in the Galactic disk, in particular (e.g. Norris, 1999). Detailed high resolution (HR) analysis is the most appropriate approach to determine individual element abundances of individual stars. This kind of approach, when not feasible, may be substituted by the study of selected low resolution absorption features. In Franchini et al. (2004a) we introduced four combinations of Lick/IDS indices, namely NaD vs Ca4227, NaD vs Mg2, NaD vs Mgb, and NaD vs CaMg, to identify Solar Scaled Abundance (SSA) and α-enhanced (Non Solar Scaled Abundance, NSSA) stars without previous knowledge of atmospheric parameter values. Our method was applied in Franchini et al. (2004b) to 402 stars with accurate estimates of [Fe/H]. In this paper we extend the analysis to a larger number of stars to better investigate the metallicity and kinematical properties of NSSA and SSA stars.

We searched for NSSA and SSA stars the spectra observed by us at the INAOE "G. Haro" Observatory in Cananea (Mexico) and the following literature data-sets: the catalog by Worthey et al. (1994), the STELIB collection (Le Borgne et al. 2003) and the ELODIE collection (Baranne et al. 1996). By using our method we detect 175 NSSA and 356 SSA stars. The two sub-samples of NSSA and SSA stars with [Fe/H] in Taylor (1999, 2003) suggest an analogy of their kinematics with those of the Galactic Thick and Thin disk respectively (Franchini et al. 2004b). In this paper we improve the statistical validity of these results by adding stars with [Fe/H] from other sources.

2 Chemo-Dynamical Properties

Several estimates of [Fe/H] are available in the literature from analysis of HR spectra (e.g. Cayrel, Soubiran & Ralite, 2001) and from photometry calibrations. In this paper we use Taylor (1999, 2003) as primary sources of [Fe/H] but we complement them by using data from Montes et al. (1999), Valdes et al. (2004), and Nordström et al. (2004). Moreover, for stars not included in the above mentioned papers we derived [Fe/H] from Strömgren photometry by using

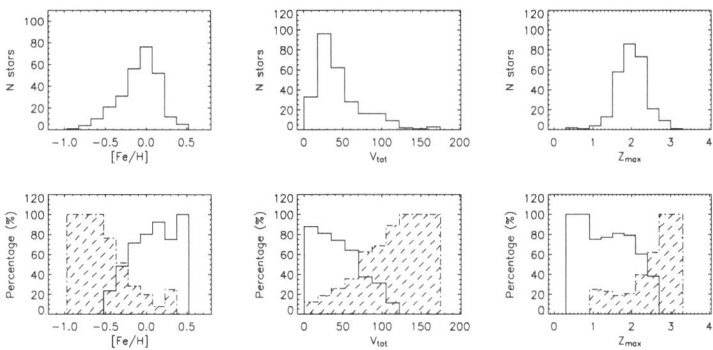

Fig. 1. Distributions in [Fe/H], V_{tot} and Z_{max} of the 268 stars: the 85 NSSA (dashed) and the 183 SSA (empty) stars are separated in the lower panels

the calibrations of Holmberg (2004), Schuster & Nissen (1999), and Kotoneva et al. (2002). In the case of more than one [Fe/H] determination available we select the adopted value following the order given above.

We were able to compute galactic velocities U_{LSR}, V_{LSR}, and W_{LSR} for 85 NSSA and 183 SSA stars. Figure 1 shows the [Fe/H], $V_{tot} = [U_{LSR}^2 + V_{LSR}^2 + W_{LSR}^2]^{1/2}$ and Z_{max} distributions of the 268 stars and the percentages of NSSA and SSA stars in each bin. Going from low to high [Fe/H] and from high to low V_{tot} and Z_{max} values the percentage of NSSA stars decreases while the percentage of SSA stars increases. These results confirm the similarity in chemo-dynamical properties of the NSSA and SSA samples with those of Galactic thick and thin disk already found by Franchini et al. (2004b). A work is in progress to check if these results can be extended to the whole populations of NSSA and SSA stars.

Acknowledgements

We acknowledge support from INAF-PRIN 2002 (PI Dr. Bertelli), from Università degli Studi di Trieste (ex 60% grants) and from MIUR COFIN-2003028039.

References

1. Baranne A., Queloz D., Mayor M., et al.: A&AS **119**, 373 (1996)
2. Cayrel de Strobel, G., Soubiran, C., Ralite, N.: A&A, **373**,159 (2001)
3. Franchini, M., Morossi, C. Di Marcantonio P., et al.: ApJ, **601**, 485 (2004a)
4. Franchini, M., Morossi, C., Di Marcantonio, P., et al.: ApJ,**613**, 312 (2004b)
5. Kotoneva, E., Flynn, C., Chiappini, C., et al.: MNRAS **336**, 879 (2002)
6. Le Borgne, J. F., Bruzual, G., Pelló, R., et al: A&A, **402**, 433 (2003)
7. Montes, D., Ramsey, L.W., Welty, A. D.: ApJS **123**, 283 (1999)
8. Nordström, B., Mayor, M., Andersen, J., et al: A&A **418**, 989 (2004)
9. Norris, J. E.: ApJS, **265**, 213 (1999)
10. Schuster, W.J., Nissen, P. E.: A&A **221**, 65(1989)
11. Taylor, B. J.: A&AS, **134**, 523 (1999)
12. Taylor, B. J.: A&A, **398**, 731 (2003)
13. Valdes, F., Gupta, R., Rose, J. A., et al.: ApJS **152**, 251 (2004)
14. Worthey, G., Faber, S. M., Gonzalez, et al.: ApJS, **94**, 687 (1994)

Does Rotation of B Stars Depend on Metallicity? Preliminary Results from GIRAFFE Spectra

F. Royer[1], P. North[2], C. Melo[3], J.-C. Mermilliod[2], E.K. Grebel[4], J.R. de Medeiros[5], and A. Maeder[6]

[1] ESO, Karl-Schwarzschild-Str. 2, D-85748 Garching, Germany
[2] Laboratoire d'Astrophysique de l'Ecole Polytechnique Fédérale de Lausanne, Observatoire, CH-1290 Chavannes-des-Bois, Switzerland
[3] ESO, Alonso de Cordova 3107, Casilla 19001, Santiago 19, Chile
[4] Astronomisches Institut, Universität Basel, Venusstr. 7, CH-4102 Binningen, Switzerland
[5] Universidade Federal do Rio Grande do Norte, 59072-970 Natal, R.N., Brazil
[6] Observatoire de Genève, CH-1290 Sauverny, Switzerland

Abstract. We show the $v \sin i$ distribution of main sequence B stars in sites of various metallicities, in the absolute magnitude range $-3.34 < M_V < -2.17$. These include Galactic stars in the field measured by [1], members of the h & χ Per open clusters measured by [6], and five fields in the SMC and LMC measured at ESO Paranal with the FLAMES-GIRAFFE spectrograph, within the Geneva-Lausanne guaranteed time. Following the suggestion by [5], we do find a higher rate of rapid rotators in the Magellanic Clouds than in the Galaxy, but the $v \sin i$ distribution is the same in the LMC and in the SMC in spite of their very different metallicities.

Introduction, Results and Conclusion

This work aims at testing the suggestion of [5] that stellar rotation is faster at lower metallicity by direct measurements, especially in the LMC and SMC, on stars with $-3.34 < M_V < -2.17$, i.e. spectral types B0-B6 or masses from ~ 6.7 to 14 $M_\odot$. This work is complementary to that of [4], which deals with slightly more massive stars. The results are shown on Fig. 1 and commented in the caption. There is an excess of slow rotators in the Galaxy relative to the MCs, but the $v \sin i$ distributions of the LMC and the SMC are surprisingly similar.

References

1. H.A. Abt, H. Levato, M. Grosso: ApJ **573**, 359 (2002)
2. N. Cramer: New Astronomy Reviews **43**, 343 (1999)
3. D. Erspamer, P. North: A&A **383**, 227 (2002)
4. S.C. Keller: Pub. Astron. Soc. Australia **21**, 310 (2004)
5. A. Maeder, E. Grebel, J.-C. Mermilliod: A&A **346**, 459 (1999)
6. P. North, F. Royer, C. Melo et al.: 'New homogeneous $v \sin i$ determinations for B stars in galactic open clusters'. In: *Stellar Rotation, IAU Symp. 215*, ed. by A. Maeder, P. Eenens (in press)
7. A. Slettebak, G.W. Collins, T.D. Parkinson et al.: ApJS **29**, 137 (1975)
8. J. Southworth, P.F.L. Maxted, B. Smalley: MNRAS **349**, 547 (2004)

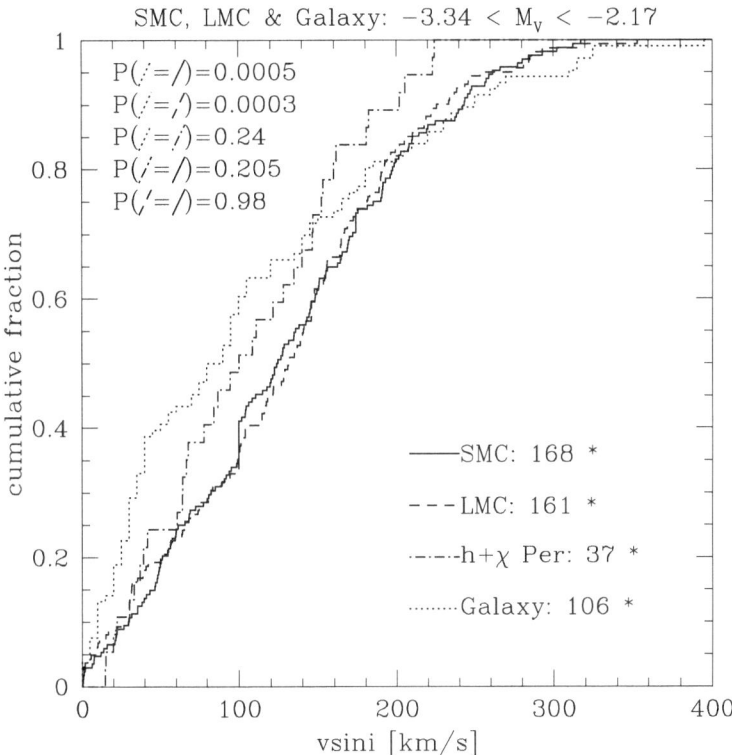

SMC, LMC & Galaxy: $-3.34 < M_v < -2.17$

$P(\dot{/}=/)=0.0005$
$P(\dot{/}=/)=0.0003$
$P(\dot{/}=\dot{/})=0.24$
$P(/=/)=0.205$
$P(/=/)=0.98$

cumulative fraction

vsini [km/s]

——SMC: 168 *

- - -LMC: 161 *

-·-·-h+χ Per: 37 *

········Galaxy: 106 *

Fig. 1. Cumulative $v \sin i$ distributions for Galactic stars in the field (dotted line), for members of the h & χ Per clusters (dash-dot) and for stars in the LMC (short dash) and in the SMC (solid line). The GIRAFFE spectrograph, attached to the UT2 telescope (VLT) and used in the L2 setup ($R = 6400$, $\lambda_c = 4272$ Å), was used on 3 fields in the LMC (centered on [$\alpha_{J2000} =$05:31:40,$\delta_{J2000} =$-66:59:48], on [05:30:40, -67:17:12] and on [05:03:48, -69:00:36]) and 2 fields in the SMC (centered on [00:56:12, -72:29:00] and on [00:49:26, -73:12:07]). We fitted synthetic spectra to observed ones in the range $4460 - 4490$ Å with the technique described by [3] using an average $T_{\rm eff} - M_V$ relation for the main sequence and assuming $\log g = 4.0$. The resulting $v \sin i$ values were then transformed to the scale of [7]. For the Galaxy, we defined the $v \sin i$ distribution using 1) the measurements made in the h & χ Per clusters by [6] and 2) the large sample of [1] of bright field B stars, Geneva photometry being used to determine M_V through the calibration of [2]. The SB2 systems were eliminated from this sample, which, although magnitude-limited, does not significantly differ from a volume limited one. The results are summarized in this Figure. Surprisingly, the overall $v \sin i$ distribution is almost exactly the same in the SMC (mean metallicity $Z \sim 0.008$) and in the LMC ($Z \sim 0.004$). There is only a marginal difference between h & χ Per (Z(h Per) ~ 0.01 according to [8]) and the MCs, but a very significant one ($P < 0.1$ %) between the Galactic field ($Z \sim Z_\odot = 0.018$) and the MC fields. Thus, either the metallicity effect saturates for $Z < Z$(LMC) ~ 0.008, or another cause affects rotational velocities, e.g. different rates and orbital parameters of SB1 binaries (not excluded from the samples), through tidal effects.

Metallicity in Open Clusters

G. Pace[1,2], L. Pasquini[3], P. François[4], and F. Matteucci[1]

[1] Dipartimento di Astronomia, Università di Trieste, Via G.B. Tiepolo 11,
   34131 Trieste, Italy
[2] Dipartimento di Astronomia, Università di Padova, Vicolo dell'Osservatorio 2,
   35122 Padova, Italy
[3] European Southern Observatory, Karl-Schwarzschild-Strasse 2,
   85748 Garching bei München, Germany
[4] Observatoire de Paris, 61 avenue de l'Observatoire, 75014 Paris, France

1 Introduction

We present here the results of abundance measurements of iron, calcium and nickel in four open clusters, from UVES spectra of solar–type stars. A code developed by one of the authors (François) performs line recognition, equivalent width measurements and finally obtains the abundances by means of OSMARCS LTE model atmosphere [4]. Temperature, gravity and microturbulence velocity have to be input to the program. This is made in an automatic way for a grid of values chosen on photometric basis. Those that best reproduce excitation and ionization equilibria are selected and used, namely when no significant trend of the computed abundances is seen, neither versus the excitation potential of the line nor versus its equivalent width, and for which the abundances obtained with lines of different ionization stages of the same specie give equal results within the errors. This check is made with iron lines, we have in fact at least thirty Fe I lines in each star, and six Fe II lines.

2 The Sample

We have analyzed high–resolution (R=100000) high signal to noise (S/N ~100) UVES spectra in 20 stars belonging to four open clusters. The advantage of our data, apart from the quality of the spectra, is that we have a homogeneous sample of solar–like stars, with B-V colours in the range between 0.51 and 0.72 mag (the solar value is evaluated to be around 0.62 mag). A UVES solar spectrum is analyzed first, and a line–by–line check is performed starting from a list of almost a hundred lines. Only those recognized in the sun are then used to measure the other stars' abundances, so that, when subtracting from the stellar–abundance measurements the solar values in order to obtain [M/H], many of the systematics and errors on the oscillator strengths cancel out. The high–quality data set here presented is meant to calibrate future works based on lower resolution spectra taken with multi–object facilities.

3   The Results

The results are summarized in the Table. Our metallicity value for **NGC 3680**, though still subsolar, is slightly higher than the values given in [1] and [13], [Fe/H]=-0.14 and -0.17 respectively. We need to be observe more stars in this cluster before claiming firm conclusions. The scantiness of the sample is due to the fact that most of the G dwarfs escaped the cluster during its lifetime.

Our results are in excellent agreement with some recent published metallicities concerning **IC 4651** ([14,10]), and with many about **M 67** ([15,8,11]), though [6] give a lower value for the latter cluster: [Fe/H]=-0.15.

[7,3] find for **Praesepe**'s metallicity values somewhat lower than ours, respectively 0.14 and 0.13, but the disagreement is only about 1σ. On the contrary [2,5] find almost solar values for [Fe/H]: 0.092 and 0.038 respectively.

Table 1. Table of the computed abundances: iron, and the two α elements calcium and nickel. All from the neutral–element lines. The standard deviation refers to the measurements in the different stars. In NGC 3680 we have only two targets, therefore instead of values σ we put the difference between the two stellar abundances obtained. The average number of lines identified for each specie is also indicated.

Cluster	N_{stars}	[Fe/H]	σ_{Fe}	N_{lines}	[Ca/H]	σ_{Ca}	N_{lines}	[Ni/H]	σ_{Ni}	N_{lines}
NGC 3680	2	-0.06	0.01	41	-0.01	0.13	4	-0.06	0.01	12
IC 4651	5	0.09	0.06	37	0.10	0.06	4	0.10	0.08	11
PRAESEPE	6	0.22	0.06	38	0.21	0.07	4	0.25	0.07	10
M 67	7	-0.04	0.07	41	-0.03	0.09	4	-0.02	0.06	11

References

1. Anthony Twarog, B., & Twarog, B.A. 1987, AJ 126, 1000
2. Boesgaard, A. M. 1989, ApJ 336, 798
3. Boesgaard, A. M., & Budge, K. G. 1988, ApJ, 332, 410
4. Edvardsson, B. et al. 1993, A&A, 275, 101
5. Friel, E.D., & Boesgaard, A. M. 1992, ApJ, 387, 170
6. Friel, E.D. et al. 2002, AJ 124, 2693
7. Hui–Bon–Hoa, A., & Alecian, G. 1998, A&A 332, 224
8. Hobbs, L. M., & Thorburn, J. A. 1991, AJ, 102, 1070
9. Janes, K.A., & Smith, G.H. 1984, AJ 89, 487
10. Meibom, S., Andersen, J., & Nordström, B. 2002, A&A 386, 187
11. Montgomery, K.A., Marschall, L.A., & Janes, K.A. 1993, AJ 106, 181
12. Nordström, B., Andersen, J., & Andersen, M.I. 1997, A&A 322, 460
13. Pasquini, L., Randich. S.,, & Pallavicini, R., 2001, A&A 374, 1017
14. Pasquini, L. et al. 2004, A&A 424, 951
15. Tautvaišiene, G. et al., I. 2000, A&A 360, 499

Light- and Heavy-Element Abundances from Mid-UV and Optical Spectral Syntheses

R.C. Peterson

UCO/Lick and Astrophysical Advances

Abstract. We are determining light- and heavy-element abundance ratios in a variety of standard stars of spectral types A through K by comparing their observed high-resolution echelle spectra to spectra calculated over broad wavelength regions. These calculations extend from 3740 Å to 9000 Å in all stars, and include the mid-ultraviolet for those stars with echelle spectra from the Hubble Space Telescope. We report on progress to date in comparing our calculations line-by-line to the observed spectra.

1 Synopsis

This is a brief update on the theoretical spectral synthesis of standard stellar spectra from the mid-ultraviolet (mid-UV) through the optical spectral regions. Peterson et al. (2004) [7] provide further details and color figures illustrating the comparison between our calculations and the observed high-resolution spectra of standard stars. A glance at those color plots reveals the multitude of absorption lines due to heavy elements, with lines of gallium and germanium in orange, s-process in red and green, and r-process in blue and purple. Although lines of many heavy elements appear in the optical, abundances of several elements such as germanium are shown to be best derived from the mid-UV.

In the optical, the primary standards are the Sun and Arcturus, which have digital observed spectra at very high signal-to-noise and resolution from Kurucz et al. [3] and Hinkle et al. [2]. We are redetermining the abundances of a wide variety of elements in each of these fundamental standards, then establishing astrophysical gf-values for as many optical and mid-UV lines as possible.

2 Procedure

We compare the observed high-resolution spectra with our spectral calculations, to develop better atomic and molecular gf-values for lines identified in the laboratory. In the mid-UV, and to a lesser extent in the optical, additional, unidentified lines appear in the spectra. We add these "missing" lines as either Fe I or Ti I features, based on their relative strength in two standards with similar iron abundance, one of which is a high-velocity star with enhancements of light elements such as magnesium and titanium. We estimate the lower excitation potentials of missing lines based on the rapidity with which their strength changes in cool versus hot standards. Peterson, Dorman, & Rood (2001) [6] present a complete description of these procedures.

The spectral calculations are run using the Kurucz program SYNTHE [5]. Although these calculations are static, one-dimensional, and in local thermodynamic equilibrium, the above references show that they provide excellent agreement with observed optical and mid-UV spectra of nearby mildly metal-poor standard stars. Moreover, they yield a consistent determination of the stellar temperature from all available diagnostics: profiles of Balmer line wings, the abundances determined from low- and high-excitation lines of the same species, the continuum slope of the mid-UV spectrum, and observed stellar colors versus those of the best-fitting models. Peterson et al. [6] found this agreement to hold only when photospheric models from Castelli are adopted in which convective overshoot is turned off, unlike those of Kurucz [4]. We now download the Castelli and Kurucz [1] ODFNEW models from http://kurucz.harvard.edu/grids. Along with an improved solar iron abundance, these ODFNEW models include new opacity distribution functions including millions of lines and ∼1200 opacity bins.

3   Setting the Continuum in the Mid-Ultraviolet

The Hubble mid-UV echelle spectra extend blueward to 2130 Å for hot stars, or to 2380 Å or 2885 Å for cooler ones. Redward they extend to 3120 Å for faint stars, to 2885 Å for bright hot stars, and to 3150 Å for bright cool stars. These spectra are flux-calibrated, which fixes the mid-UV continuum normalization.

The spectral observations and calculations are normalized over the entire ∼1000 Å mid-UV region with the choice of a single flux normalization constant for each star, which should be inversely proportional to the square of the stellar angular diameter. Peterson et al. [6] showed that the normalization constants deduced by fitting the spectra generally agreed to 10% with those determined from distances derived from Hipparcos parallaxes, when appropriate masses were assumed for each star. For the nearby F5IV-V standard Procyon, whose angular diameter has been measured, Peterson et al. [7] found an 8.5% discrepancy, marginally consistent with the observational flux uncertainties.

References

1. F. Castelli, R.L. Kurucz: In: *Modeling of Stellar Atmospheres*, IAU Symp. 210, ed. by N. Piskunov et al., CD-ROM poster A20 (2003); also astro-ph/0405087
2. K. Hinkle, L. Wallace, J. Valenti, D. Harmer: *Visible and Near Infrared Atlas of the Arcturus Spectrum 3727 – 9300 Å* (Astron. Soc. Pac., San Francisco 2002)
3. R.L. Kurucz, I. Furenlid, J. Brault, L. Testerman: *Solar Flux Atlas from 296 to 1300 nm*, Nat. Solar Obs. Atlas No. 1 (Smithsonian Astrop. Obs., Cambridge 1984)
4. R.L. Kurucz: CD-ROM 13, *ATLAS9 Stellar Atmospheres Program and 2 km/s Grid* (Smithsonian Astrop. Obs., Cambridge 1993)
5. R.L. Kurucz: CD-ROM 18, *SYNTHE Spectrum Synthesis Programs and Line Data* (Smithsonian Astrop. Obs., Cambridge 1993)
6. R.C. Peterson, B. Dorman, R.T. Rood: Ap. J. **559**, 372 (2001)
7. R.C. Peterson et al.: 'Mid-Ultraviolet Spectral Templates for Old Stellar Systems'. In: *Space Telescope Science Newsletter* **4**, 1 (2004)

FLAMES Observations
of the Star Forming Region NGC 6530

L. Prisinzano[1], F. Damiani[1], I. Pillitteri[2], and G. Micela[1]

[1] INAF-Osservatorio Astr. di Palermo, P.za del Parlamento, 1, 90134 Palermo, Italy
[2] Dip. Scienze Fisiche ed Astronomiche, Universitá di Palermo

Abstract. We use intermediate resolution ($R \sim 19\,300$) spectroscopic observations in the spectral region including the Li 6708 Å line to study 341 stars in the star forming region (SFR) NGC 6530. Based on the optical color-magnitude diagrams (CMD), they are G, K and early M type pre-main sequence (PMS) cluster candidates. 72% of them are probable cluster members since are X-ray sources detected in a Chandra-ACIS observation ([2]). We use our spectroscopic measurements to confirm cluster membership by means of radial velocities and to investigate the Li abundance of cluster members.

1 Observations and Data Reduction

Positions and photometry of our targets were selected from the cluster region of the CMD [4]. We used three spectroscopic observations taken with the multi-object spectrograph GIRAFFE-FLAMES of the ESO-VLT Telescope. These observations were obtained on May 27, 2003 and are part of the Guaranteed Time of the Ital-FLAMES Consortium. The data consist of 341 spectra of stars with V magnitude between 14.0 and 18.2. The exposure times were 2800 s, 5600 s and 5400 s. The data were reduced using the Geneva Giraffe data reduction system (Version girbldrs-1.08) released on July 01, 2003 and relative calibration files (version 2.0).

2 Radial Velocities

We computed relative radial velocities via Fourier cross-correlation of each target with a template spectrum of a relatively bright X-ray probable cluster star. The radial velocity distribution of the whole sample shows a significant peak which indicates the presence of the cluster. To distinguish cluster members from field stars, we fitted this distribution with a double gaussian using the maximum likelihood fitting. We find that the cluster gaussian is centered on 0.2 ± 0.3 km/s with a standard deviation $\sigma = 4.0 \pm 0.3$. The total number of possible cluster members within -3σ and $+3\sigma$ is 225 including 9 contaminating stars.

3 Lithium Abundances

To determine the Li abundance we considered only 147 slow rotator cluster members, with membership based on radial velocity and X-ray detection. We first

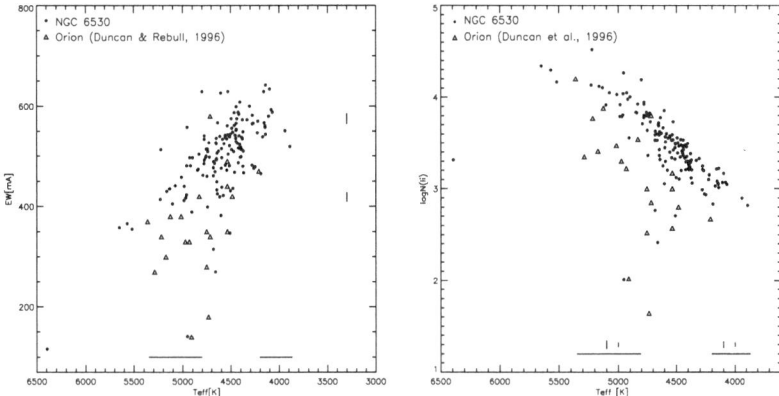

Fig. 1. Li EW (*left panel*) and abundances (*right panel*) of NGC 6530 stars compared to those of the coeval Orion Nebula cluster, using the same growth curves. Most of our targets show very strong Li lines increasing at lower temperatures. The results are not corrected for NLTE and veiling effects, as in Orion in [3]. Absolute values are overestimated but, relative values do not suffer from systematic errors.

obtained the continuum normalized spectra and therefore we used the standard IRAF's Gaussian fitting around the Li line to determine the equivalent width (EW). Errors on the EW, estimated by repeated measurements for an high and a low signal to noise stars, are indicated with the two vertical bars in Fig. 1.

The observed $(B - V)$ colors were dereddened by subtracting the mean reddening $E(B - V) = 0.35$ estimated in [6]. The dereddened $(B - V)_0$ colors were converted into effective temperatures using equation 4 of [1] with [Fe/H]=0 for giant stars, since our targets are low gravity stars. The errors on effective temperatures strongly depend on the unknown individual reddening, that is the largest source of error. Considering the reddening range $E(B - V)$=[0.25–0.50] given in [6] to estimate typical temperature errors at T=4000 K and T=5000 K, we obtained the errors indicated by the horizontal bars in Fig. 1. EW and temperatures were used to derive the Li abundances from linear interpolation of the growth curves of [5]. Errors on the Li abundances were derived with the same technique using typical errors in temperature and EW.

Li abundances of our sample are fully consistent with their PMS nature.

References

1. A. Alonso, S. Arribas, C. Martínez-Roger 1999, A&AS, 140, 261
2. F. Damiani, E. Flaccomio, G. Micela, et al. 2004, Ap, 608, 781
3. D. K. Duncan, L. M. Rebull 1996, PASP, 108, 738
4. L. Prisinzano, F. Damiani, G. Micela, et al., 2004, A&A, in press
5. D. R. Soderblom, B. F. Jones, S. Balachandran, et al.1993, AJ, 106, 1059
6. H. Sung, M. Chun, M. S. Bessell 2000, AJ, 106, 1059

The "Not-so-MAD" Coronal Abundances of Active Stars

J. Sanz-Forcada[1,2], F. Favata[1], and G. Micela[2]

[1] Astrophysics Division – Research and Space Science Department of ESA, ESTEC, Postbus 299, NL-2200 AG Noordwijk, The Netherlands
[2] INAF – Osservatorio Astronomico di Palermo, Piazza Parlamento 1, Palermo, Italy

Abstract. Coronal abundances have been a subject of debate in the last years due to the availability of high-quality X-ray spectra of many cool stars. Coronal abundance determinations have generally been compared to solar photospheric abundances; from this a number of general properties have been inferred, such as the presence of a coronal metal depletion with an inverse First Ionization Potential dependence, with a functional form dependent on the activity level. We report a detailed analysis of the coronal abundance of 4 stars with various levels of activity and with accurately known photospheric abundances. The coronal abundance is determined using a line flux analysis and a full determination of the differential emission measure. We show that, when coronal abundances are compared with real photospheric values for the individual stars, the resulting pattern can be very different; some active stars with apparent Metal Abundance Deficiency in the corona have coronal abundances that are actually consistent with their photospheric counterparts.

Introduction and Results

The comparison of coronal and photospheric abundances in cool stars is a very important tool in the interpretation of the physics of the corona. Active stars show a very different pattern to that followed by low activity stars such as the Sun, being the *First Ionization Potential* (FIP) the main variable used to classify the elements. The overall solar corona shows the so-called "FIP effect": the elements with low FIP (<10 eV, like Ca, N, Mg, Fe or Si), are enhanced by a factor of $\sim$4, while elements with higher FIP (S, C, O, N, Ar, Ne) remain at photospheric levels. The physics that yields to this pattern is still a subject of debate. In the case of the active stars (see [2] for a review), the initial results seemed to point towards an opposite trend, the so called "Inverse FIP effect", or the "MAD effect" (for *Metal Abundance Depletion*). In this case, the elements with low FIP have a substantial depletion when compared to the *solar* photosphere, while elements with high FIP have same levels (the ratio of Ne and Fe lines of similar temperature of formation in an X-ray spectrum shows very clearly this effect). However, most of the results reported to date lack from their respective photospheric counterparts, raising doubts on how real is the MAD effect.

Photospheric abundances of stars within the solar neighborhood may be quite different [1], therefore it is necessary to compare coronal and photospheric abundances of the same star in order to understand this phenomenon. We have conducted a research on a sample of stars of different activity levels, from which we

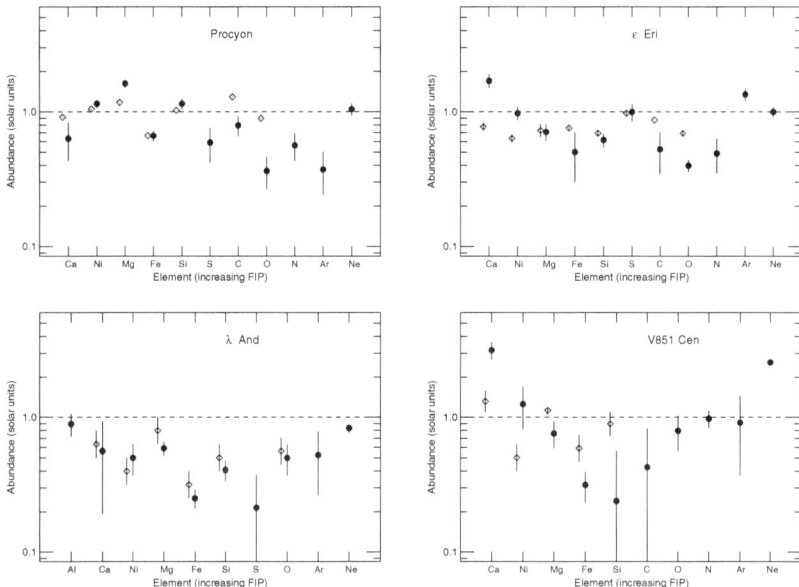

Fig. 1. Element abundances in the corona of Procyon, ϵ Eri, λ And and V851 Cen (filled circles) with respect to solar photospheric values. Open diamonds represent the stellar photospheric abundances. A dashed line indicates the adopted solar photospheric abundance

have both the coronal and photospheric abundances (see [3] for further details). These are the main conclusions from this work (see Fig. 1):

1. Low activity stars, such as The Sun or α Cen A show a FIP effect. However, Procyon does not follow any clear trend, although it could show some increased FIP effect within the temperature range.
2. Mid activity stars, such as ϵ Eri, show only a "mild" FIP effect, but no MAD seems to be present.
3. High activity stars have no FIP effect, but when they are compared to their own photospheric abundances, the MAD effect is not clearly present in the two cases (λ And and V851 Cen).
4. There are very few cases of active stars with known coronal and photospheric abundances. Therefore a detailed assessment of the presence of FIP or MAD biases can only be done in a small number of stars. More work is clearly needed.

References

1. G. Cayrel de Strobel, C. Soubiran, & N. Ralite, A&A, 373, 159 (2001)
2. F. Favata, & G. Micela, Space Science Reviews, 108, 577 (2003)
3. J. Sanz-Forcada, F. Favata, & G. Micela, A&A, 416, 281 (2004)

Fundamental Parameters of B Supergiants

S.C. Searle[1], R.K. Prinja[1], and P. Crowther[2]

[1] Department of Physics & Astronomy, University College London, Gower Street,
London WC1E 6BT England, UK
[2] Department of Physics & Astronomy, University of Sheffield, Hicks Building,
Hounsfield Road, Sheffield, S3 7RH

Abstract. To understand the chemical evolution of galaxies, we need a deep understanding of the evolution of massive stars and their role in processing and delivering chemical elements. B-type supergiants represent a substantial population in this context. We present here initial results from non-LTE, line-blanketed stellar atmosphere modelling [1] of a large sample of B0-B5 supergiants (see Fig.1). We focus in this report on revisions to the effective temperature scales, finding reasonable agreement with the temperatures of [2] for B1-B5 supergiants. For early type supergiants (B0-B1), the less luminous Ib's are hotter than Ia's as expected (excluding HD190603 (B1.5 Ia+) which is a hypergiant). The temperature discrepancy between B0 Ia's & B0 Ib's is up to 2500 K, decreasing to 1000 K for B0.5 and remains as 1000 K or less for later B type supergiants. Whilst some early B type supergiants are slightly cooler compared to the Humphreys et al. temperatures [2], a few B0.5 Ia & Ib stars appear to be slightly hotter. This discrepancy requires further investigation, since recent OB supergiant revised temperature scales have shown a trend for temperatures to be slightly cooler [3] than previously thought from [2].

Moreover, to improve our understanding of the evolution of massive stars and the galaxies that harbour them, we require more precise constraints on the effects of mass loss & rotation on their evolution. Mass loss rates have also been derived from Hα profiles for the sample of B supergiants presented here (using the same non-LTE, line blanketed stellar atmosphere model [1]) and will be presented in [4], along with further details of this work. Estimates of CNO abundances from the non-LTE models are however included here (see Fig.2) and will be compared to predicted CNO abundances from evolutionary tracks. This will allow us to determine how much CNO processing has occurred in these stars and if they have gone through a "blue loop" i.e. evolved from a blue to red supergiant phase (during which the star undergoes convective mixing of its core materials) then returned to blue supergiant status. Such an effect is predicted for some massive stars by stellar evolution models [5]. Early indications suggest that the B supergiants in this sample are partially processed and therefore have not gone through a red supergiant phase.

References

1. Hillier D. J. & Miller D. L. 1998, ApJ, **496**, 407
2. Humphreys R. et al., 1984, ApJ, **284**, 565
3. Evans C. J. et al., 2004, ApJ, **610**, 1021
4. Searle S. C., Prinja R. K. & P. A. Crowther, 2005 (in preparation)
5. Schaller G. et al. 1992, A&AS, **96**, 269
6. Lennon D. J. et al. 1992, A&A, **94**, 569

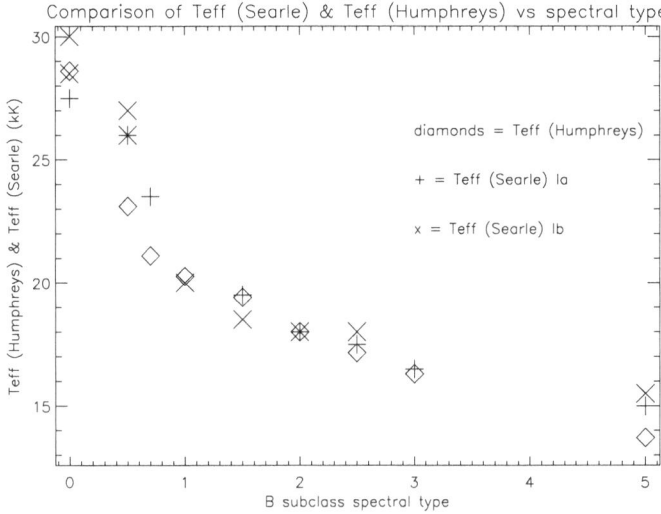

Fig. 1. compares the temperature scale derived here with that of [2]. In this work, temperatures were derived for individual stars, therefore Ia's & Ib's are treated separately, whereas in [2], the same temperature is assumed for both Ia's & Ib's. Spectral types taken from [6]

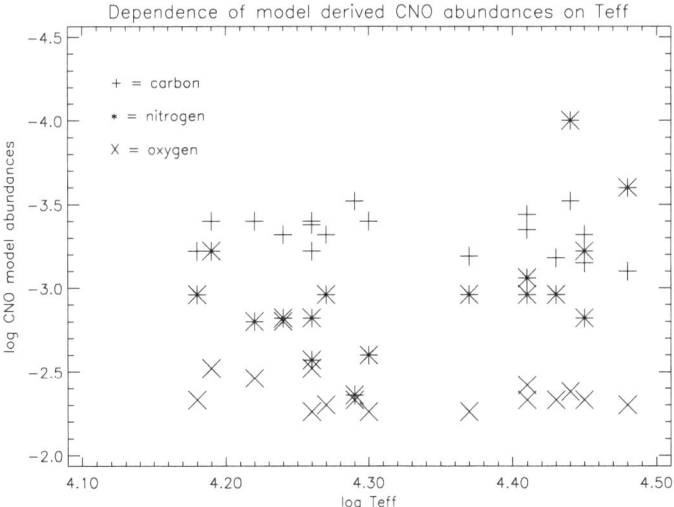

Fig. 2. shows CNO abundance variation with temperature. Abundances are expressed in mass fractions as before and were determined from non-LTE model fits to e.g. NII 3995 Å, CII 4267 Å, OII 4367 Å. Nitrogen shows an increase with increasing temperature, whereas oxygen implies a subtle decrease with increasing temperature and carbon does not display a definite trend

High Precision Effective Temperatures and New Abundances for a Large Sample of Disk Stars

T.V. Mishenina[1], C. Soubiran[2], O. Bienaymé[3], V.V. Kovtyukh[1], S.A. Korotin[1], and T.I. Gorbaneva[1]

[1] Astronomical Observatory of Odessa National University, Ukraine
[2] L3A Bordeaux, France
[3] Observatoire Astronomique de Strasbourg, France

Abstract. In an effort to determine accurate stellar parameters and abundances for a large sample of nearby stars, we have performed the detailed analysis of 350 high-resolution spectra of FGK dwarfs and giants. This sample will be used to investigate behavior of chemical elements and kinematics in the thick and thin disks, in order to better constrain models of chemical and dynamical evolution of the Galaxy.

1 Atmospheric Parameters

A set of 174 dwarfs and 171 giants spanning -1.0< [Fe/H]<+0.3 have been observed at S/N>100 with the ELODIE echelle spectrograph at Observatoire de Haute-Provence (R=42 000, $\lambda\lambda$ =385-680nm). High precision Teff (5–7 K for dwarfs, 10–15 K for giants) were obtained with the line-depth ratio technique, using $\sim$ 100 relations per spectrum (Kovtyukh et al. 2003). For dwarfs with [Fe/H] < –0.5, Teff was determined from the fitting of H_α line-wings. The gravity log g was derived from ionization equilibrium of neutral and ionized species for iron and also by fitting the wings of the line Ca I 6162 Å for giants. The microturbulent velocity Vt was determined by forcing the abundances determined from individual FeI lines to be independent of equivalent width.

2 Abundances

We have determined abundances of Fe, Si and Ni for dwarfs and Fe, Si, Ca, Ni for giants by WIDTH9, and O abundances by the method of synthetic spectrum (STARSP), under LTE approximation. Abundances of Mg have been determined through detailed NLTE calculations using equivalent widths of 4 lines (4730, 5711, 6318, 6319 Å) and profiles of 5 lines (4571, 4703, 5172, 5183, 5528 Å) on the basis of a model of the Mg atom consisting of 97 levels. Results concerning the set of dwarfs have been already published in Mishenina et al (2004), describing the details of the spectral analysis and showing some examples of profile fitting. Figure 1 shows how the synthetic and observed profiles of CaI and OI lines compare for two giants.

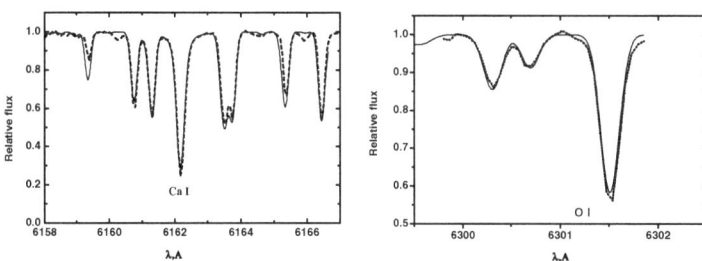

Fig. 1. Synthetic profile (continuous line) vs observed (dots) for HD180711 (Ca I) and HD127243 (O I).

3 Results

Figure 2 shows the obtained distributions of abundance ratios [Mg/Fe] and [O/Fe] versus [Fe/H] for giants, both exhibiting a decrease as metallicity increases. We have compared abundance trends of Mg, Si and Ni for giants and dwarfs. We observe similar trends for Si and Ni : an enhancement of Si as compared to the solar value, and an increase of [Ni/Fe] in the metal-rich regime. In the case of Mg, the trend seems to be different at solar and super-solar metallicities : the dwarfs exhibit a flat and positive distribution of [Mg/Fe] vs [Fe/H] whereas for giants the decrease continues and the mean value of [Mg/Fe] is around -0.10 at [Fe/H]=+0.10.

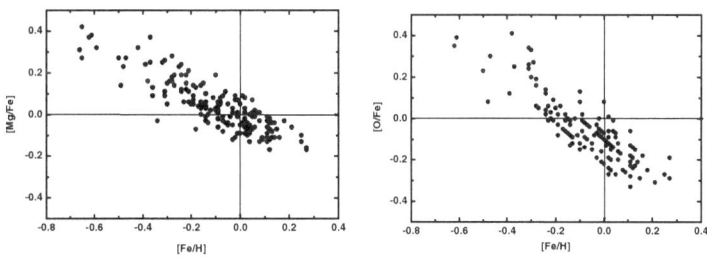

Fig. 2. Distribution of element ratios vs metallicity for the sample of giants

References

1. Kovtyukh V., Soubiran C., Belik S.I., Gorlova N.I., 2003, A&A 411, 559K
2. Mishenina T.V., Soubiran C., Kovtyukh V., Korotin S.A., 2004, A&A 418, 551M

Elemental Abundances in 10 Dwarfs of the Galactic Thick Disk

G. Tautvaišienė[1], A. Ivanauskas[1], M. Grenon[2], and I. Ilyin[3]

[1] Vilnius University Institute of Theoretical Physics and Astronomy, Gostauto 12, Vilnius 01108, Lithuania
[2] Observatoire de Genève, Chemin des Maillettes 51, CH-1290 Sauverny, Switzerland
[3] Astrophysikalisches Institut Potsdam, An der Sternwarte 16, D-14482 Potsdam, Germany

Abstract. The most recently discovered Galactic component – thick disk – still needs high-resolution spectral investigations since its origin and evolution is not understood enough. Elemental abundance ratios in the metallicity range $-0.68 \leq [\mathrm{Fe/H}] \leq -0.10$ were determined in a sample of 10 thick-disk dwarfs and compared with results of other stars investigated as well as with models of thin disk chemical evolution.

1 Introduction

More than twenty years have passed since the thick disk of the Galaxy was discovered (Gilmore & Reid 1983), however the origin and evolution of this population is still under discussion. In the high-resolution spectroscopic study by Barbuy & Erdelyi-Mendes (1989) it was suspected that a spread in [O/Fe] ratios at $-0.8 \leq [\mathrm{Fe/H}] \leq -0.5$ is connected to possible differences in the chemodynamical evolution of the thin and thick disks of the Galaxy. The first evidence of such a difference was offered by Fuhrmann (1998) in the analysis of [Mg/Fe] ratios of 9 stars belonging to the thick disk. A number of high-resolution spectral abundance analyses of thick-disk candidates are carried out (Bensby et al. 2004; Mashonkina et al. 2003; Tautvaišienė et al. 2001; Prochaska et al. 2000 and references therein), however the databasis of comprehensive abundance analyses still needs to be enlarged.

1.1 Selection of Stars

The representatives of the thick disk stars were selected among a large list of proper-motion stars from Luyten (NLTT, 1995). Photometric data were obtained in the Geneva photometric system and radial velocities measured with CORAVEL technique or taken from literature. For the identification of the thick-disk membership metallicity and kinematic criteria were used: (1) the photometric [Fe/H] in the range -0.5 and -0.7 dex, which is characteristic of the thick disk population; (2) the Grenon's kinematic age parameter in the range 0.40 to 1.0, which means a maximum height from the Galactic plane, derived in a Myamoto galaxy potential, $Z_{\max}$, higher than 0.5 Kpc.

Stars selected are: HD 133621, HD 142373, HD 148816, HD 156826, HD 157089, HD 192718, HD 200580, BD $+5°$ 3080, BD $+7°$ 4854, BD $+11°$ 4725.

2   Observations and Method of Analysis

The spectra for the stars were obtained at the Nordic Optical Telescope (La Palma) with the SOFIN échelle spectrograph in 2001. The 2nd optical camera ($R \approx 60,000$) was used to observe simultaneously 13 spectral orders, each of $40 - 60$ Å in length, located between 5600 Å and 8130 Å. The spectra were analyzed using a differential model atmosphere technique. The program packages, developed at the Uppsala Astronomical Observatory, were used to carry out the calculations of theoretical equivalent widths of lines, synthetic spectra and a set of plane parallel, line-blanketed, flux constant LTE model atmospheres. The effective temperatures for the stars were determined spectroscopically by requiring FeI lines with different lower excitation potentials to produce equal abundances. The preliminary effective temperatures in this procedure were evaluated from observations in the Geneva photometric system. The gravities were found by forcing FeI and FeII to yield the same iron abundances. The microturbulent velocities were determined by forcing FeI line abundances to be independent of the equivalent width. For more details on the method of analysis and atomic data see Tautvaišienė et al. (2001).

3   Results

In the sample of stars investigated the effective temperature range is from 5100 to 5870 K, $\log g$ is between 3.5 and 4.8 dex and the metallicity is from -0.10 to -0.68 dex. When compared to the models of the Galactic thin disk chemical evolution (Pagel & Tautvaišienė 1995, 1998), the elemental abundance results in the investigated stars supplement evidences that the thick disk population had a distinct chemical history from the thin disk – [El/Fe] ratios of α- s- and r-process elements at sub-solar metallicities are systematically higher in thick disk stars. Abundances of iron-group elements in thick and thin disk stars are similar. The onset of the bulk of SN Ia is suggested to appear at about [Fe/H]≈ -0.6 dex.

References

1. Barbuy B., Erdelyi-Mendes M. 1989, A&A **214**, 239
2. Bensby T, Feltzing S., Lundström 2004, A&A **415**, 155
3. Fuhrmann K. 1998, A&A **338**, 161
4. Gilmore G., Reid N. 1983, MNRAS **202**, 1025
5. Luyten W.J. 1995, VizieR On-Line Data Catalog I/98A.
6. Mashonkina I., Gehren T., Travaglio, C., Borkova, T. 2003, A&A **397**, 275
7. Pagel B.E.J., Tautvaišienė 1995, MNRAS **276**, 505
8. Pagel B.E.J., Tautvaišienė 1997, MNRAS **288**, 108
9. Prochaska J.X., Naumov S.O., Carney B.W., McWilliam A., Wolfe A.M. 2000, AJ **120**, 2513
10. Tautvaišienė G., Edvardsson B., Tuominen I., Ilyin I. 2001, A&A **380**, 578

Part II

Abundances in the Spheroidal Component

Abundances in the Galactic Bulge

B. Barbuy[1], M. Zoccali[2], V. Hill[3], A. Renzini[4], S. Ortolani[5], D. Minniti[2],
L. Pasquini[4], E. Bica[6], A. Gomez[3], Y. Momany[5], and J. Meléndez[1],
and A. Alves-Brito[1]

[1] Universidade de São Paulo, IAG, Rua do Matão 1226, 05508-090 São Paulo, Brazil
[2] Universidad Catolica de Chile, Casilla 306, Santiago 22, Chile
[3] Observatoire de Paris-Meudon, 92195 Meudon Cedex, France
[4] European Southern Observatory, 85748 Garching bei München, Germany
[5] Università di Padova, Vicolo dell'Osservatorio 2, I-35122 Padova, Italy
[6] Universidade Federal do Rio Grande do Sul, Porto Alegre 91501-970, Brazil

Abstract. A review is presented on abundance determinations in stars of the Galactic
bulge, both in the field and in globular clusters. Previous low-resolution spectroscopy
results are revised. Recent high resolution and high S/N spectroscopy results based
on Keck-Hires, Gemini-Phoenix and VLT-UVES data are presented. Finally, recent
analyses of FLAMES data are discussed.

The discussion on abundances will focus on metallicities, α- and r-process elements,
as probes of the nucleosynthesis history in the bulge, and timescale of bulge formation.

1 Introduction

The Galactic bulge is observable within $20° \times 20°$ around the Galactic center,
this angular size encompassing a region of radius around 2.8 kpc. In Barbuy et
al. (1998) 16 clusters contained within $5° \times 5°$ were listed, plus NGC 6553 at the
edge of this radius. Dutra et al. (2003a) have measured the extinction within
$10° \times 10°$ of the Galactic center using 2MASS data (Skrutskie et al. 1997). In
the high extinction region where no globular clusters are known, Dutra & Bica
(2000) have found a series of cluster candidates, some of which were confirmed
such as DB11 (Dutra et al. 2003b), these being young clusters. As a matter
of fact, in the inner Galaxy, a cluster will not survive for more than around
40 Myr. Out of 74 globular clusters projected within $20° \times 20°$, 60 of them are
within 4 kpc of the Galactic center. Therefore, the Galaxy contains 60 bulge
globular clusters, out of a total of 150 known clusters (Harris 1996, updated in
physun.physics.mcmaster.ca/Globular.html). Burkert & Smith (1997) classified
the bulge clusters in a few categories and Dinescu et al. (2003) measured proper
motions and classified known clusters in these categories, as for example, NGC
6304 and NGC 6553: disk, NGC 6528: bar, NGC 6522 and NGC 6723: halo,
NGC 6266: metal-poor thick disk, NGC 6316: bulge.

The metallicity distribution of globular clusters in the Galaxy has a metal-
rich peak at [Fe/H] $\approx$ -0.5 and a metal-poor peak at [Fe/H] $\approx$ -1.6 (e.g. Côté
1999), where most of the metal-rich ones are bulge clusters. Metallicities for samples of field stars were derived by McWilliam & Rich (1994, hereafter MR94),
Sadler et al. (1996), Ramirez et al. (2000). Zoccali et al. (2003) presented the

Colour Magnitude Diagram (CMD) of a region at -6° of the Galactic center, compared to template mean loci of clusters with known metallicity, and concluded that the mean metallicity is [Fe/H] ≈ -0.1. van Loon et al. (2003) analysed data from ISOGAL and DENIS and concluded that most bulge stars are old and metal-rich. Ortolani et al. (1995) and Zoccali et al. (2003) have shown that the metal-rich clusters NGC 6528 and NGC 6553 are nearly coeval with 47 Tucanae, and this was confirmed by Feltzing & Johnson (2002). Besides, Ortolani et al. (1995) have shown that the magnitude difference between TO and HB ΔV_{HB}^{TO} of Baade's Window and in these clusters is the same such that the clusters can be considered as templates of the field. Ortolani et al. (2001), using NICMOS on board HST, have confirmed the old age of the bulge clusters, where 47 Tuc having 15 Gyr, then Terzan 5 is 14 Gyr and NGC 6528 is 12 Gyr.

Among the 16 known globular clusters located within 5° of the Galactic center, six of them are metal-poor with [Fe/H] ≤ -1.0. In Barbuy et al. (1998) these clusters were reviewed based on CMD analyses: Terzan 4, HP-1, NGC 6522, NGC 6540, Terzan 9 and Terzan 10, and a CMD analysis of Terzan 9 was presented later in Ortolani et al. (1999). The very metal-poor cluster Terzan 4 was further analyzed in J vs. $J - H$ based on HST NICMOS images by Ortolani et al. (2001). In this work we present the first high-resolution analyses of two metal-poor bulge clusters. If the Milky Way formed in hierarchical clustering, then it can be predicted that the metal-poor stars and clusters within a few kiloparsecs of the Galactic center may be among the oldest objects in the Galaxy (van den Bergh 1993; Davidge 2001; Davidge et al. 2004).

2   Abundances in Bulge Globular Clusters

NGC 6553 appears to be the most well-studied metal-rich bulge globular cluster. Barbuy et al. (1999) showed a list of previous abundance determinations for this cluster, and derived a metallicity of [Fe/H]=-0.55 with pronounced α enhancements, based on relatively low resolution (R $\sim$ 20 000) and low S/N spectra. Further analyses have shown that the metallicity is higher, just below the solar metallicity: Cohen et al. (1999) analysed 5 Horizontal Branch (HB) stars observed with Keck-HIRES, at R $\sim$ 37 000, giving [Fe/H]=-0.16, [O/Fe]=+0.5 (based however on the rather unreliable OI triplet at 7771-4 Å, [Mg/Fe]=+0.4, [Si/Fe]=+0.14,[Ti/Fe]=+0.19, [Ca/Fe]=+0.26; Origlia et al. (2002) using Keck-Nirspec at R $\sim$ 25 000, obtained [Fe/H]=-0.3 and [α/Fe]=+0.3; Meléndez et al. (2003) used Gemini-PHOENIX with R$\sim$50 000 to analyse 5 giants, that resulted to show [Fe/H]=-0.2, [O/Fe]=+0.2, and made evident a mixing effect with [C/Fe]=-0.6, [N/Fe]=+1.3, and a [C+N/Fe]=+0.5 excess, probably of primordial origin. NGC 6528 is very similar to NGC 6553 in age and metallicity (Ortolani et al. 1995). Carretta et al. (2001, hereafter C01) analysed 6 HB stars, using Keck-HIRES at R$\sim$37 000 and obtained [Fe/H]=-0.06, [O/Fe]=0.07, [Mg/Fe]=0.14, [Si/Fe]=0.36, [Ca/Fe]=0.23 and [Na/Fe]=0.4. Zoccali et al. (2004, hereafter Z04) analysed 3 giants using VLT-UVES, at R$\sim$50 000, and obtained [Fe/H]=-0.11, [O/Fe]=0.15, [Mg/Fe]=0.07, [Si/Fe]=0.08, [Ca/Fe]=-0.4, [Ti/Fe]=-0.1, [Na/Fe]=

-0.43 and [Eu/Fe]=0.15. Origlia et al. (2004) analysed 4 stars in NGC 6528 and found [Fe/H]=-0.17 and [α/Fe]=+0.33. Other clusters studied at high resolution R∼ 25 000 using Keck-Nirspec are Liller 1, by Origlia & Rich (2002) with [Fe/H]-0.3, [α/Fe]=+0.3, and Terzan 4 (4 stars) with [Fe/H]=-1.6, [α/Fe]=+0.5 and Terzan 5 (6 stars) with [Fe/H]=-0.21, [α/Fe]=+0.3, where α's are O(from OH), Mg, Si, Ca and Ti, by Origlia & Rich (2004). Lee et al. (2004) studied stars in Palomar 6 with R∼ 42 000 obtained at the 3.5m NASA IRTF telescope, and found [Fe/H]=-1.0, [Ti/Fe]=0.5, [Si/Fe]=0.4; they suggest a Si/Ti ratio anticorrelation with Galactocentric distance. At low resolution (R∼1380 with IRS and 1650 with Osiris) at the 4m Blanco telescope, Stephens & Frogel (2004) derived metallicities for 7 bulge clusters: NGC 6256: -1.35, NGC 6539: -0.79, HP1: -1.3, Liller1: -0.36, Palomar6:-0.52, Terzan 2: -0.87, Terzan 4: -1.62. Davidge et al. (2004) used Gemini-Flamingos at R∼350 to observe 21 stars in NGC 6558 (not all members), and derived [Fe/H]=-1.5 for this cluster.

HP-1: A Metal-Poor Bulge Cluster Ortolani et al. (1997) presented the CMD of HP-1 that suggested a metallicity of [Fe/H]=-1.5. Using a 3 hour exposure VLT-UVES spectrum of a giant with V=17, Barbuy et al. (2004, in preparation) obtained [Fe/H]=-1.5, [O,Mg,Eu/Fe]=0.4, [Si/Fe]=0.7, [Ca,Ti/Fe]=0. This peculiar abundance pattern may be the signature of nucleosynthesis early in the Galactic bulge. It will be very important to further analyse metal-poor bulge stars in order to deepen our understanding on bulge formation.

3   A Bulge Survey with Flames

The FLAMES instrument has unique capabilities in terms of multi-object high resolution spectroscopy with large telescopes. The GIRAFFE instrument in the MEDUSA mode has 132 fibers, with 1.2" aperture each, and obtains spectra with R∼22 000 reaching V∼19. In parallel it is possible to observe 8 stars with UVES, at R∼45 000, that reaches V∼17 We have a program to observe 800 bulge stars (field and clusters) with 1 magnitude above the HB, a choice that avoids blending with TiO bands present in cooler giants. We observed four fields with latitudes from -3 to -12°. The spectral region observed ranges from 6120-6964 Å in 3 setups. The main aims of this project are those of defining subsets of stellar populations or *families* characterized by typical element ratios and kinematics, and showing the behaviour of [X/Fe] vs. [Fe/H] in order to have hints on the timescale of bulge formation. Preliminary results concern 6 giants of the bulge metal-poor globular cluster NGC 6558. Rich et al. (1998) obtained a CMD indicating -1.6<[Fe/H]<-1.2, and Davidge et al. (2004) estimated [Fe/H]= -1.5 from low resolution spectra. In Fig. 1 we show the spectrum of metal-rich bulge star observed with UVES and FLAMES, showing that the oxygen feature is clearly measurable in both spectra, and it is also compared with the spectrum of a metal-poor giant of NGC 6558. Our first results for NGC 6558 indicate [Fe/H]=-1.0, [Ca/Fe]=0.1, [Si/Fe]=0.4 and [O/Fe]=0.6. For NGC6558-42, we

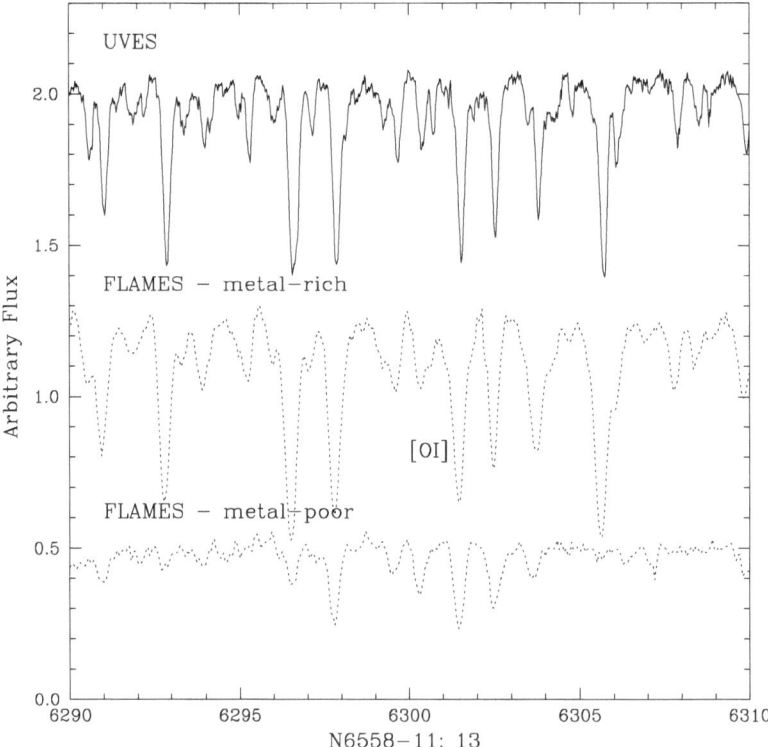

Fig. 1. Spectra of a metal-rich star observed with UVES and FLAMES, and a metal-poor star from NGC 6558 observed with FLAMES

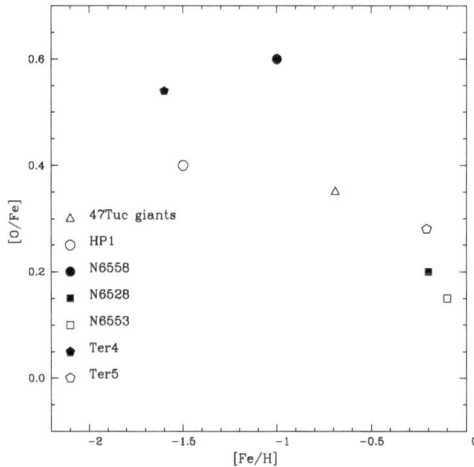

Fig. 2. [O/Fe] vs. [Fe/H] for bulge globular clusters in a range of metallicities, and including 47 Tuc for comparison

find [O/Fe]=+0.95. This peculiar abundance pattern will be further discussed in a paper in preparation (Zoccali et al. 2005, in preparation).

In Table 1 we present the elemental ratios for NGC 6528 from Zoccali et al. (2004) and Carretta et al. (2001), for field giants by McWilliam & Rich (1994), the present results on HP-1 and NGC 6558, and on 47 Tuc by Alves-Brito et al. (2004). The Ca and Ti abundances are low in our own calculations, relative to results by C01 and MR94. In fact, the use of damping constants based on the collisional broadening theory computed by Barklem et al. (1998 and references therein), together with gf-values from the NIST database, make results for these elements far more reliable presently. This will be further shown in a future paper on HP-1. McWilliam & Rich (2003, MR03) suggested that α-enhancements decrease with increasing metallicity. This seems indeed to be confirmed from our results for clusters, as shown in Figs. 2a,b with the behaviour of [O/Fe] vs. [Fe/H] and [Eu/Fe] vs. [Fe/H]. Finally, [O/Fe] vs. [Na/Fe] for the present clusters, compared to stars in 47 Tuc show clear anticorrelation is seen, showing effect of mixing or primordial variations (see review by A. Weiss, this volume).

For completeness, we would like to mention the recent work on bulge planetary nebulae by Górny et al. (2004) and Exter et al. (2004), that found an oxygen excess of 0.2dex in bulge planetary nebulae relative to disk ones.

4   Conclusions

Bulge studies in the near future will greatly benefit from derivations of abundance ratios in large samples of stars, together with kinematics, including radial velocities and proper motion determinations.

Table 1. Abundance ratios in NGC 6528, bulge field from MR94, HP-1 and NGC 6558 (present work), and 47 Tuc for comparison

[X/Fe]	NGC 6528 Z04	NGC 6528 C01	MR94	HP-1	NGC 6558	47 Tuc
[Fe/H]	-0.1	0.07	-0.2	-1.3	-1.0	-0.7
[O/Fe]	0.15	0.07	0.03	0.4	0.6	0.35
[Mg/Fe]	0.07	0.14	0.35	0.4	0.2	0.25
[Si/Fe]	0.08	0.36	0.14	0.7	0.4	0.25
[Ca/Fe]	-0.4	0.23	0.14	0.0	0.1	0.0
[Ti/Fe]	-0.1	0.03	0.37	0.0	0.1	0.0
[Eu/Fe]	0.15	–	–	0.4	0.4	0.33

References

1. Alves-Brito, A., Barbuy, B., Ortolani, S., Momany, Y., Hill, V., Zoccali, M., Renzini, A., Minniti, D., Pasquini, L., Bica, E., Rich, R.M. 2004, submitted
2. B. Barbuy, E. Bica, S. Ortolani, 1998, A&A, 333, 117
3. B. Barbuy, A. Renzini, S. Ortolani, E. Bica, M.D. Guarnieri, 1999, A&A, 341, 539
4. P.S. Barklem, S.D. Anstee, B.J. O'Mara, 1998, PASA, 15, 336
5. A. Burkert, G.H. Smith, 1997, ApJ, 474, L99
6. E. Carretta, J.G. Cohen, R.G. Gratton, B.B. Behr, 2001, AJ, 122, 1469 (C01)
7. J.G. Cohen, R.G. Gratton, B.B. Behr, E. Carretta, 1999, ApJ, 523, 739
8. P. Côté, 1999, AJ, 118, 406
9. T.J. Davidge, 2001, AJ, 122, 1386
10. T.J. Davidge, M. Ledlow, P. Puxley, 2004, AJ, 128, 300
11. D. Dinescu, T.M. Girard, W.F. van Altena, C.E. López, 2003, AJ, 125, 1373
12. C. Dutra, E. Bica, 2000, A&A, 359, L9
13. C. Dutra, B.X. Santiago, E. Bica, B. Barbuy 2003a, MNRAS, 338, 253
14. C. Dutra, S. Ortolani, E. Bica, B. Barbuy, M. Zoccali 2003b, A&A, 408, 127
15. K.M. Exter, M.J. Barlow, N.A. Walton, 2004, MNRAS, 349, 1291
16. S. Feltzing, R.A. Johnson, 2002, A&A, 385, 67
17. S.K. Górny, G. Stasińska, A.V. Escudero, R.D.D. Costa, 2004, A&A, 427, 231
18. W.E. Harris, 1996, AJ, 112, 1487
19. J.-W. Lee, B.W. Carney, S.C. Balachandran, 2004, AJ, 128, 2388
20. McWilliam, A., Rich, R.M. 1994, ApJS, 91, 749 (MR94)
21. A. McWilliam, R.M. Rich, 2003, in *Origin and Evolution of the Elements*, www.ociw.edu/ociw/symposia/series/symposium4/proceedings.html (MR03)
22. J. Meléndez, B. Barbuy, E. Bica, S. Ortolani, M. Zoccali, A. Renzini, V. Hill, 2003, A&A, 411, 417
23. L. Origlia, R.M. Rich, S. Castro, 2002, AJ, 123, 1559
24. L. Origlia, R.M. Rich, 2004, AJ, 127, 3422
25. L. Origlia, E. Valenti, R.M. Rich, 2004, astro-ph/0410519
26. S. Ortolani, A. Renzini, R. Gilmozzi, G. Marconi, B. Barbuy, E. Bica, R.M. Rich, 1995, Nature, 377, 701
27. S. Ortolani, E. Bica, B. Barbuy, 1997, MNRAS, 284, 692
28. S. Ortolani, E. Bica, B. Barbuy, 1999, A&AS, 138, 267
29. S. Ortolani, B. Barbuy, E. Bica, A. Renzini, M. Zoccali, R.M. Rich, S. Cassisi, 2001, A&A, 376, 878
30. S.V. Ramírez, K. Sellgren, J.S. Carr, S.C. Balachandran, R. Blum, D.M. Terndrup, A. Steed, 2000, ApJ, 537, 205
31. R.M. Rich, S. Ortolani, E. Bica, B. Barbuy, 1998, AJ, 116, 1295
32. M. Skrutskie, S.E. Schneider, R. Stiening, et al. 1997, in *The impact of Large Scale Near-IR Surveys*, ed. Garzon et al., Kluwer (Netherlands), 210, 187
33. E.M. Sadler, R.M. Rich, D.M. Terndrup, 1996, AJ, 112, 171
34. A.W. Stephens, J.A. Frogel, 2004, AJ, 127, 925
35. S. van den Bergh, 1993, ApJ, 411, 178
36. J. Th. van Loon, G. Gilmore, A. Omont, J.A. Blommaert, I.S. Glass, M. Messineo, F. Schuller, M. Schultheis, I. Yamamura, H.S. Zhao, 2003, MNRAS, 338, 857
37. M. Zoccali, A. Renzini, S. Ortolani, E. Bica, B. Barbuy, 2001, AJ, 121, 2638
38. M. Zoccali, A. Renzini, S. Ortolani, L. Greggio, I. Saviane, S. Cassisi, M. Rejkuba, B. Barbuy, R.M. Rich, E. Bica, 2003, A&A, 399, 931
39. M. Zoccali, B. Barbuy, V. Hill, S. Ortolani, A. Renzini, E. Bica, Y. Momany, L. Pasquini, D. Minniti, R.M. Rich, 2004, A&A, 423, 507 (Z04)

Abundance Ratios in the Galactic Bulge

J.P. Fulbright[1], R. Michael Rich[2], and A. McWilliam[1]

[1] Carnegie Observatories, Pasadena, California, USA
[2] University of California at Los Angeles, Los Angeles, California, USA

Abstract. We present abundance results from our Keck/HIRES observations of giants in the Galactic Bulge. We confirm that the metallicity distribution of giants in the low-reddening bulge field Baade's Window can be well-fit by a closed-box enrichment model. We also confirm previous observations that find enhanced [Mg/Fe], [Si/Fe] and [Ca/Fe] for all bulge giants, including those at super-solar metallicities. However, we find that the [O/Fe] ratios of metal-rich bulge dwarfs decrease with increasing metallicity.

1 Introduction

The Galactic Bulge contains about 20 percent of the Galaxy's stellar mass. Theories of its formation include a primordial free-fall collapse, remnants of accretion episodes, or secular evolution of bar instabilities ([1]). Accurate stellar abundance determinations can help distinguish between these models.

The limited quality and sample size of the McWilliam & Rich [2] data demand a renewed effort to study bulge abundances. We have been engaged in a long-term program to study the composition of stars in the Galactic bulge, with the aim of constraining the conditions of the bulge's formation and chemical evolution at a level of detail that is impossible to obtain with any other method.

2 Data and Analysis

The observational data consist of Keck/HIRES spectra of 27 bulge giants observed between 1998 and 2001. All of the target stars are located within Baade's Window, a region of low reddening located about 4 degrees from the Galactic center. The spectra have a resolving power of 45,000 to 60,000, covering the red region of the spectrum. The S/N levels range from about 50 to 100. The abundance analysis employs the LTE 1-D spectrum synthesis program MOOG [3] and the grid of LTE model atmospheres from the Kurucz web page (http://kurucz.harvard.edu). Additional details on the analysis method, including the line list, continuum determination, stellar parameters, and the details of the abundance analysis of the non-iron lines will be presented in a series of future papers.

2.1 Metallicity Distribution Function

The metallicity distribution function (MDF) is an indicator of the enrichment history of a system. Large samples can be obtained [4,5] using low resolution spectra in order to measure the MDF of Baade's Window. Our high-resolution sample has been used to recalibrate the earlier low-resolution data of Sadler et al. We find an exponential distribution of $Z/Z_\odot$, similar to that seen in earlier studies. However, this recalibration assumes Z goes as [Fe/H], which is incorrect since Baade's Window giants do not show solar ratios of many of the alpha elements. Correcting for this would increase the values of Z for bulge stars.

2.2   Oxygen and Magnesium Abundances

The bulge sample shows enhancements in [α/Fe] at all [Fe/H] values, with the exception of a decreasing trend in [O/Fe] at high [Fe/H]. The difference in [O/Fe] and [Mg/Fe] trends creates a quandary: Type II models (e.g. [6]) predict that O and Mg are produced in similar mass progenitors. There are no major producers of Mg other than Type II supernovae. Therefore, why do O and Mg not show similar distributions?

A possible solution may be due to the fact that while O and Mg should come from similar stars, they are not formed in similar layers within those stars. Oxygen is produced during hydrostatic helium burning while magnesium comes from hydrostatic carbon and neon burning. In high-metallicity high-mass stars, post main-sequence mass loss can remove a large fraction of the star's original mass. During this Wolf-Rayet phase it may be possible that the final hydrostatic He-burning layer could be greatly reduced or even lost completely while the C-burning layer could be preserved. The Wolf-Rayet progenitor models of ref. [6] do not predict this amount of mass loss, so further work is necessary to confirm this hypothesis.

3   Future Directions

The analysis of the bulge data continues. The spectra exhibit absorption lines of many elements not discussed here, including C, N, Na, Al, Fe-group elements and many elements heavier than the Fe-group, including several of the traditional indicators of the s- and r-processes. In addition, we have obtained data in three other fields at different Galactic latitudes. This will allow for the investigation of the homogeneity of the chemical properties of the bulge to be studied, which may reveal evidence of galactic accretion or multiple star formation events.

References

1. I. Ferreras, R. F. G. Wyse, and J. Silk, MNRAS 345 (2003) 1381.
2. A. McWilliam and R. M. Rich, ApJS 91 (1994) 749.
3. C. Sneden, ApJ 184 (1973) 839.
4. R. M. Rich, AJ 95 (1988) 828.
5. E. M. Sadler, R. M. Rich, and D. M. Terndrup, AJ 112 (1996) 171.
6. S. E. Woosley and T. A. Weaver, ApJS 101 (1995) 181.

Abundances in Globular Cluster Dwarfs/Subgiants

E. Carretta

INAF - Osservatorio Astronomico di Bologna, via Ranzani 1, I-40127 Bologna, Italy

Abstract. Unevolved or scarcely evolved stars of globular clusters do not have either high enough temperatures in their cores or large enough convective envelopes to dredge up to the surface the products of high temperature proton-capture reactions. Hence, these stars represent an ideal diagnostic to study the pattern of the chemical anomalies observed in the C, N, O, Na, Mg, Al elements. I will review the status of researches on this presently debated subject, presenting also the latest results from high resolution UVES spectra on dwarf and subgiant stars in NGC 6397, NGC 6752 and 47 Tuc.

1 Introduction

I will review the status of abundances in scarcely evolved stars in globular clusters (GCs), proceeding along a pattern of increasing Coulomb barrier. Stars of GCs offer an ideal diagnostic in order to understand stellar evolution for low and intermediate stellar masses. However, since the pioneering study of [1] it is known that a spread in the light elements (C,N,O, but also Na, Al, Mg) is present among cluster stars of similar evolutionary phase. The observational and theoretical progresses are summarized in a number of reviews, most recently in [2].

Theory predicts two main scenarios: episodes of very deep mixing, possible only along the red giant branch (RGB), and a scenario of primordial pollution from a previous generation of stars, of course not dependent on the evolutionary status. However, things are getting complicate, since it appears that *both* contribution are required, with mixing effects acting as a "noise" over a floor of primordial inhomogeneities.

In the interpretation of the observations the guideline is given by the stellar structure itself: stars on the main sequence (MS) and in the post-MS phases belong to two different realms. In fact, MS stars build up an increasing molecular weight barrier in the inner regions, which is smoothed only after the so called RGB-bump. Only then evolving giants may onset some kind of induced mixing, able to connect the outer envelope with regions near the energy-generating shell. Moreover, the extension of the convective envelope dramatically differs for the two kind of stars.

How to disentangle the two above scenarios? Two approaches are useful: first, to compare cluster stars with undisturbed analogs evolving in much less environments and second, to look where one of the two mechanism is not allowed to exist.

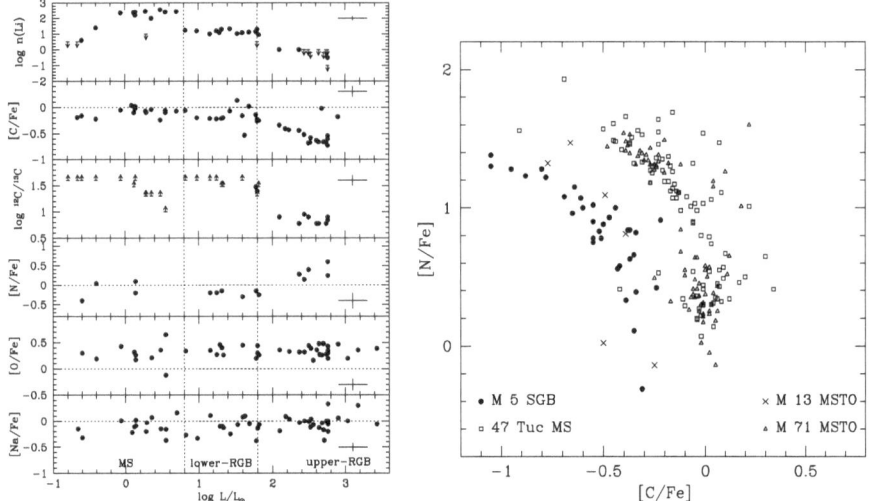

Fig. 1. Left: run of abundances of Li, [C/Fe], [N/Fe], [O/Fe], [Na/Fe] and carbon isotopic ratios with luminosity from [4]. Right: run of the [C/Fe] ratio as a function of [N/Fe] from the compilation by [6].

2   The Lesson from Field Stars

The first step forward was made by [3] who successfully tested on a small sample the idea of rotation-induced mixing as a way to explain the much too low values of Li, carbon isotopic ratios and carbon-to-nitrogen ratios in field stars. Using a much larger sample of stars of well known evolutionary status, [4] were able to show that in field stars two mixing episodes (the first dredge-up and a second one after the RGB-bump) are well evident for the lightest elements (Li, C, N; see left panel of Fig. 1). They also showed that Na and O remains untouched in field stars, but not in clusters stars: something was clearly missing. Moreover, even looking only at Carbon, there were hints of a primordial range in GCs, as suggested by giants in M 92 ([5]). The key role of unevolved cluster stars is then to quantify the composition of stars before any mixing might onset, accessing to a phase where likely only one mechanism is in action.

3   The Key Role of Unevolved Cluster Stars

As a summary of several studies for light elements in unevolved cluster stars (based on photometric or low dispersion spectroscopic indexes), the right panel of Fig. 1 nicely shows the main features. There are star-to star abundance variations, with large spreads in N anticorrelated with much smaller spreads in C (as found all the way from MS to RGB). The implications are that we are not looking simply at a conversion of C into N, and that primordial inhomogeneities are likely to be responsible for the bulk of alterations. Moreover, we have to take into account not simply a surface contamination: if this was the case then the effects would be washed out by the deepening of the convective envelope during the RGB evolution.

The drawback of this kind of studies is that no reliable indicator of the O abundance is available; moreover, informations on the Na abundances are difficult to obtain. However, thanks to the new generation of efficient spectrographs mounted at 8-10m class telescopes, high resolution spectra for unevolved stars in GCs are feasible.

In the following, I will present results from a ESO Large Program (PI Gratton) for C, N, O, Na and isotopic ratios $^{12}C/^{13}C$ in unevolved stars of NGC 6397, NGC 6752 and 47 Tuc ([7]). Abundances of C and isotopic carbon ratios were derived by spectrum synthesis of UVES spectra in the CH G-band regions, whereas N abundances are derived from the CN violet band at 3880 Å. In all three clusters we found a C-N anticorrelation, and the comparison with literature results (Fig. 2) allows us to draw a few conclusions about light elements: (i) again, we can follow two step-like depletions in C, as in field stars, but (ii) in cluster stars we see a more extreme spread of C depletions and N enhancements, anticorrelated with each other; (iii) C is not as low as expected for CN-cycled material; (iv) the isotopic ratios $^{12}C/^{13}C$ are somewhat lower in GCs, compared to field stars, but *not* at the CN equilibrium value, not even for very N rich stars; (v) finally a large fraction of dwarfs and subgiants have N abundances as large as those observed in bright giants.

The conclusion is that no unevolved star shows the composition of evolved RGB stars, but rather one similar to that predicted for yields of intermediate-mass AGB stars (IM-AGB; see [8]). However, since dilution with material not contaminated by CNO burning could explain the observed trends for C and N, we have to dig deeper.

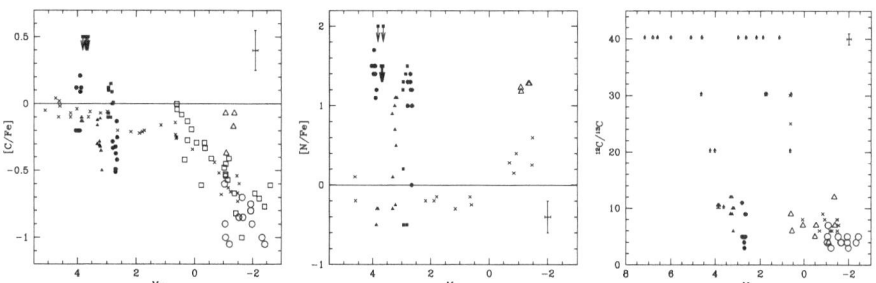

Fig. 2. Left, middle and right panel: run of the [C/Fe], [N/Fe] and carbon isotopic ratios with absolute magnitude. Filled triangles, circles and squares are dwarfs/subgiants in NGC 6397, NGC 6752 and 47 Tuc, respectively ([7]). Open symbols are RGB stars in the same clusters, from a number of literature studies (see [7] for references). Crosses are the field stars from [4].

4   The Na-O Anticorrelation and CNO Arithmetic

The Na-O anticorrelation studied among giants by the Texas-Lick group (e.g. the recent summary in [2]) tells us that we are in presence of proton-capture reactions at high temperature, where ON and NeNa cycles operates, explaining the observed O depletions and Na enhancements ([9],[10]). The left panel of Fig. 3 shows the Na-O anticorrelation among unevolved stars in the three clusters studied by [7]. As stated in [11], our findings rule out the possibility of internal mixing, since these stars do not reach the high temperature required to forge Na and destroy O; moreover, they do not have an efficient convective envelope to dredge to the surface the ashes of these reactions. The implication is that these reactions must take place in *other* stars.

Having at hand the complete set of C,N,O abundances, it is possible to test whether the observed pattern is due to CNO-cycle alone. From the large sample in literature (see right panel in Fig. 3) we know that since the sum C+N increases as C decreases, either the ON-cycle is involved (raising the N abundance) or there is a variable amount of N. On the other hand, over about 2 dex of spread in N, the sum C+N+O is almost constant, signature of the action of complete CNO-cycle (left panel of Fig. 4). Again, the source has to be found outside the observed stars.

Two classes of stars are presently suggested, IM-AGB stars and RGB stars slightly more massive than the present turn-off mass ([12]). Both are able to provide the ON and NeNa cycles in H burning (in shell in RGB stars and at the base of the convective envelope in AGB) and a mechanism to bring to the surface and then eject the matter, polluting other stars.

To test if dilution of the products of CNO burning may explain the difference in abundance pattern with evolved giants and a possible excess in ^{12}C visible in N-rich stars (see left panel of Fig. 4), we use simple models in the plane [C/N] vs [O/N] (right panel of Fig. 4). Starting from the approximate composition of N-poor stars, the trend for different fractions of gas processed in the complete CNO-cycle (solid line) reproduces fairly well the data, albeit it predict too low C abundances for N-rich dwarfs. Pollution from RGB stars with composition N-rich from very deep mixing (complete CNO and Na enrichment involved, dotted line) reproduces also rather well the data, apart for N-rich dwarfs. On the other hand, the N-poor case, typical of the chemical composition of field RGB stars, is a very poor match (dashed line). Moreover, in this case, the model would predict C-poor, Na-poor stars, whereas no one is observed among over 40 dwarfs/subgiants in the 3 clusters.

Thus, no unevolved star seems to be polluted by cluster giants with composition typical of field stars. On the other hand, we are not aware of any physical mechanism able to select only N-rich giants in order to form binary systems and pollute their companions, as in the scenario devised by [12].

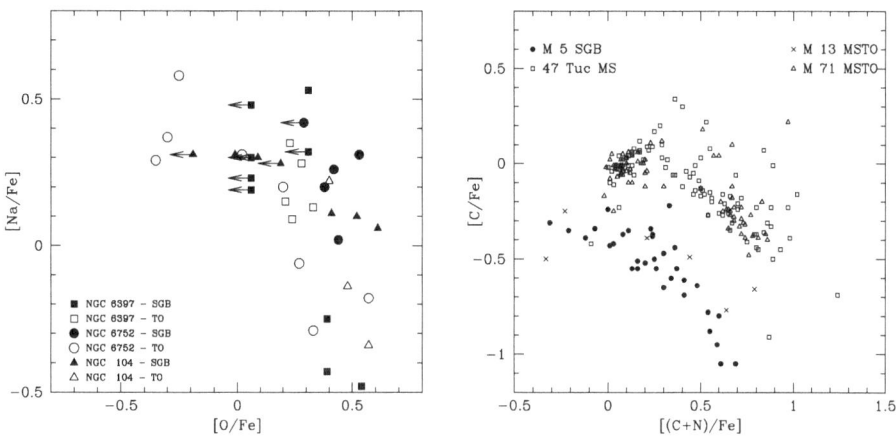

Fig. 3. Left: run of the [Na/Fe] ratio as a function of [O/Fe] for dwarfs and subgiants in NGGC 6397, NGC 6752 and 47 Tuc ([7]). Right: the [C/Fe] ratio as a function of the sum [(C+N)/Fe] for stars of the compilation by [6].

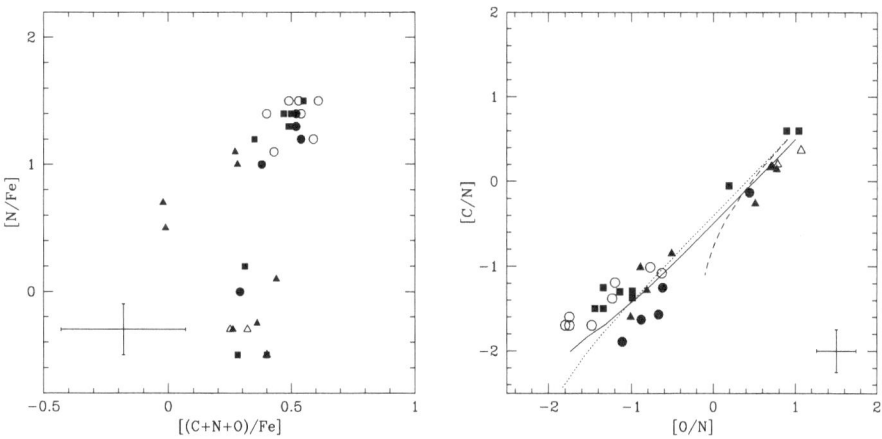

Fig. 4. Run of [N/Fe] as a function of the sum [(C+N+O)/Fe] (left panel) and run of the [C/N] ratio vs [O/N] ratio (right panel) for unevolved stars in NGC 6397, NGC 6752 and 47 Tuc. Symbols are as in left panel of previous figure. Right panel: superimposed to the data are models for dilution with matter processed by complete CNO cycle (solid line), contamination from N-poor RGB stars (dashed line) and contamination from N-rich upper-RGB stars experiencing very deep-mixing (dotted line).

5 Conclusions

There are additional concerns in a mechanism involving pollution by RGB stars. Since the contamination cannot be a simple surface alteration and most un-evolved stars in a cluster are N-rich, this require that an huge amount of mass is lost. This in turn seems unlikely, because RGB stars have a large fraction of their mass locked up in the degenerate core.

Moreover, the observed excess of C compared to the dilution models in the N-rich dwarfs suggests that maybe an additional source of ^{12}C does exist; a similar suggestion is provided by the ^{12}C/^{13}C ratios. The observed ratios of 7-8 in NGC 6752 are compatible with about 60% of the observed C being produced by triple-α burning.

Summarizing, we can consider various mass ranges of candidate polluters. Stars with $1.2 < M < 3$-5 $M_\odot$ are the classical donors of CH stars, but this is unlikely since they are very scarce in GCs and moreover they would produce also $s-$process elements, not observed to vary in unevolved GC stars ([13]).

Slightly less massive RGB stars, on the other hand, have to face the above shortcomings.

Hence, we are apparently left with stars more massive than 3 $M_\odot$, the IM-AGB stars, considered good candidate also due to their short lifetimes: they provide a good gas reservoir, able to pollute a second generation of stars before the stripping action due to disk crossing. However, this scenario has still to face severe problems in the yields and in some poorly known input physics ([14]). Further progress in stellar modeling is strongly encouraged.

References

1. W. Osborn: Observatory **91**, 223 (1971)
2. R. Gratton, C. Sneden, E. Carretta: ARA&A **42**, 385 (2004)
3. C. Charbonnel: A&A **282**, 811 (1994)
4. R. Gratton, C. Sneden, E. Carretta, A. Bragaglia: A&A **354**, 169 (2000)
5. S. Bellman, M.M. Briley, G.H. Smith, C.F. Claver: PASP **113**, 326 (2001)
6. M.M. Briley, J.G. Cohen, P.B. Stetson: AJ **127**, 1579 (2004)
7. E. Carretta, R.G. Gratton, S. Lucatello, A. Bragaglia, P. Bonifacio: A&A, in press
8. P. Ventura, F. D'Antona, I. Mazzitelli: A&A **393**, 215 (2002)
9. P. Denisenkov, S.N. Denisenkova: A.Tsir., **1538**, 11 (1990)
10. G.E. Langer, R. Hoffman, C. Sneden: PASP **105**, 301 (1993)
11. R.G. Gratton et al.: A&A **369**, 87 (2001)
12. P.A. Denissenkov, A. Weiss: ApJ **603**, 119 (2004)
13. G. James, P. François, P. Bonifacio, E. Carretta, R. Gratton, F. Spite: A&A in press (2004)
14. P. Ventura, F. D'Antona, I. Mazzitelli: Mem.S.A.It. **75**, 335 (2004)

Unraveling the Origins and History of the Outer Halo Cluster System

I.I. Ivans

California Institute of Technology, Pasadena, CA 91125, USA

Context

The nucleosynthetic histories of the outer halo globular clusters, as inferred by their chemical abundance patterns, are crucial tracers of the early formation of the Galactic halo and subsequent evolution of the Galaxy. Some outer halo clusters have been purported to be associated with galactic merger events such as the present-day merger with the Sgr dwarf galaxy. Reported here are preliminary abundance results for a sub-sample of the data acquired for this investigation.

Preliminary Red Giant Star Abundance Results

Among the sample of outer halo stars being investigated are Keck I HIRES observations of four globular clusters with $[Fe/H] \simeq -2$. Information regarding these clusters and comparison objects is presented in Table 1.

Table 1. Abundances of Observed Clusters and Comparison Sample

Object	R_{GC}[1] (kpc)	[Fe/H]	$<\alpha>$	Fe-peak	Parent Galaxy	Abundance References	$<N_{Ia}/N_{II}>$
NGC 6293	1.4	−2.0	+0.31	+0.02	MWG	LC02 [2]	0.08 ± 0.18
NGC 6287	1.7	−2.0	+0.36	+0.00	MWG	LC02 [2]	0.00 ± 0.03
NGC 6541	2.2	−1.8	+0.35	−0.04	MWG	LC02 [2]	0.00 ± 0.03
NGC 6535	3.9	−1.8	+0.24	−0.06	MWG	This study	0.03 ± 0.02
NGC 5053	16.9	−2.3	+0.22	−0.08	Sgr	This study	0.03 ± 0.01
NGC 5634	21.2	−1.9	+0.33	−0.09	Sgr	This study	0.04 ± 0.08
NGC 4147	21.3	−1.8	+0.31	−0.05	Sgr	This study	0.00 ± 0.02
Rup 106	18.5	−1.3	−0.05	−0.08	Sgr	BWZ97 [3]	0.37 ± 0.35
Pal 12	29.4	−0.9	−0.01	−0.35	Sgr	BWZ97 [3]	0.34 ± 0.27

Comparisons to Other Stellar Populations: α Elements

While the $<\alpha>$-abundances relative to iron (i.e., $<[(Si,Ca,Ti)/Fe]>$) I derive for the most metal-poor cluster (NGC 5053) in Table 1 are 0.1 dex lower than those of other metal-poor clusters purported to be associated with the Sgr dSph (NGC 4147 and NGC 5634), the abundances are in good agreement with those of NGC 6535, a cluster born within the MWG.

In addition to results from this study, Table 1 includes two of the relatively more metal-rich globular clusters associated with the Sgr dSph. There appears to be little in common between the two metallicity groups in their $<\alpha>$-abundances relative to iron. Abundances reported so far for *in situ* Sgr dSph field stars of comparable metallicities [4] are in accord with those of its metal-rich clusters.

Iron-Peak Elements & Supernovae Histories

The derived supernovae ratios of the metal-rich systems in Table 1 indicate that both the low-α and low iron-peak abundances are consistent with significant contributions from Type Ia supernovae. In general, when compared with the abundances of stars born within the MWG, the metal-rich stars associated with the Sgr dSph possess low iron-peak abundances relative to iron (e.g., [3], [4]).

By comparing the observed chemical abundance ratios to supernova model yields, one can calculate $<N_{Ia}/N_{II}>$, the ratio of the number of SNe Ia to SNe II events that fit the observations and the synthesized mass of the elements from the model yields. In a study adopting the same analysis techniques as those performed here, [5] found large values of $<N_{Ia}/N_{II}>$ for a trio of low-α stars of [Fe/H] $\simeq$ –2. Employing the abundances derived in this study of stars with comparable metallicities, I find that the metal-poor systems presented here possess α- and iron-peak abundances (and $<N_{Ia}/N_{II}>$ based on Na, Mg, Si and Fe) consistent with those observed in metal-poor stars of the MWG (e.g., [6]).

Future Prospects

Abundance results for additional clusters are currently underway and include analyses of the neutron-capture elements (in order to trace the onset of contributions from low-mass Type II SNe as well as AGB stars). Combined with their ages, the nucleosynthetic histories of the outer halo clusters will better constrain the timescales of formation and construction of the Galaxy.

This research has been supported by NASA through Hubble Fellowship grant HST-HF-01151.01-A from the Space Telescope Science Inst., operated by AURA-USA, under NASA contract NAS5-26555.

References

1. W. E. Harris: AJ, **112**, 1487 (1996); Feb. 2003 on-line version
2. J.-W. Lee, B. W. Carney: AJ, **124**, 1151, (2002); LC02
3. J. A. Brown, G. Wallerstein, D. Zucker: AJ, **114**, 180, (1997); BWZ97
4. T. A. Smecker-Hane, A. McWilliam: astro-ph/0205411 (2002); Feb. 2003 version
5. I. I. Ivans, C. Sneden, C. R. James, G. W. Preston, *et al*: ApJ, **592**, 906 (2003)
6. C. Sneden, I. I. Ivans, J. P. Fulbright: 'Globular Clusters and Halo Field Stars'. In: *Origin and Evolution of the Elements: Volume 4, Carnegie Observatories Astrophysics Series*, ed. by A. McWilliam, M. Rauch (Cambridge, 2004)

Chemical Abundance Inhomogeneities in Globular Cluster Stars

J.G. Cohen

California Institute of Technology, Pasadena Ca 91125, USA

Abstract. It is now clear that abundance variations from star-to-star among the light elements, particularly C, N, O, Na and Al, are ubiquitous within galactic globular clusters; they appear seen whenever data of high quality is obtained for a sufficiently large sample of stars within such a cluster. The correlations and anti-correlations among these elements and the range of variation of each element appear to be independent of stellar evolutionary state, with the exception that enhanced depletion of C and of O is sometimes seen just at the RGB tip. While the latter behavior is almost certainly due to internal production and mixing, the internal mixing hypothesis can now be ruled out for producing the bulk of the variations seen. We focus on the implications of our new data for any explanation invoking primordial variations in the proto-cluster or accretion of polluted material from a neighboring AGB star.

Over the past two decades the upper giant branches of the nearer GCs have been well studied with 4-m class telescopes by, among others, the Lick-Texas group (see Sneden et al 2004 and references therein) or the Padua group (Gratton and his collaborators, see, e.g. Carretta & Gratton 1997) (see also the early review of some of the issues to be discussed here by Kraft 1994 and the recent review of Gratton, Sneden & Carretta 2004). Recently 10-m class telescopes coupled with efficient spectrographs have enabled us to explore detailed abundance ratios and chemical history ever deeper in the stellar luminosity function in galactic globular clusters (GCs). We can now reach with considerable precision the RGB in *all* galactic GCs (see, for example, the study of NGC 7492, at a distance of 26 kpc, by Cohen & Melendez 2005b). For the nearer GCs, abundance analyses for the brightest main sequence stars in the nearest GCs, and for the subgiant branch for those slightly more distant, are now feasible.

The chemical analyses within the past 5 years in which the author has been involved include the GCs NGC 6528 (Carretta et al 2001), NGC 6533, M71, M5 (Ramírez & Cohen 2003), M3, M13 (Cohen & Melendez 2005a and reference therein) and shortly M15 and M92 as well as NGC 7492 (Cohen & Melendez 2005b) and Pal 12 (Cohen 2004), all observed with HIRES (Vogt et al. 1994) at Keck. Our approach is to study stars over the full range of luminosity from the RGB tip to the faintest possible that can be reached in 2 to 4 hours of integration. The large program carried out with UVES at the VLT described in Gratton et al (2001) concentrate on comparing subgiants with a small number of main sequence stars in the nearest southern clusters.

Viewing in totality the collective effort of these and other groups, we find that star-to-star abundance variations from star-to-star among the light elements, particularly C, N, O, Na and Al, are ubiquitous. They are seen whenever data of sufficient quality is obtained for a sufficiently large sample of stars within a galactic globular cluster. Our most recent Na-O anti-correlation, for M13, with a sample of 25 stars reaching almost to the main sequence turnoff can be seen in Fig. 16 of Cohen & Melendez (2005a), not shown here due to lack of space.

The correlations and anti-correlations among these elements for stars in GCs and the range of variation of each element resemble those of proton-burning. They appear to be independent of stellar evolutionary state, with the exception that enhanced depletion of C and of O is sometimes seen just at the RGB tip. This extra depletion of O just near the RGB tip is seen in our M13 data shown (see also Sneden et al 2004). Metal poor halo field stars, however, show no evidence for O burning (Gratton et al 2000) or Na enhancement. The variations seen in the field stars are much closer to those predicted by classical stellar evolution that those seen in the GC stars.

At the same time, the abundance ratios among the elements heavier than Al, at least through the Fe peak, do not show any detectable variation in any known GC (except, of course, ω Cen). The rock steady abundances for these elements requires explanation as well, and places important constraints on the formation mechanisms of GCs.

Interesting as this is, we are still plagued, when observing at high dispersion, with small samples, at least until FLAMES came into use. Small samples trying to discern small variations is not the ideal combination.

My approach to this issue has been to use the molecular bands of CH, CN and now NH to study the star-to-star abundance variations of C and of N. Since these bands are strong enough to be observed at moderate resolution, I can use the multiplexing capability of the Low Resolution Imaging Spectrograph at Keck (Oke et al 1995) to build up large samples. This effort is being undertaken jointly with Michael Briley of the University of Wisconsin at Oshkosh and with Peter Stetson of the National Research Council, Victoria, Canada.

We have now analyzed large samples of low luminosity stars (subgiants or main sequence turnoff region stars) in each of four GCs spanning a wide range in metallicity. In each cluster we have a sample of ~70 stars. Our most recent work, a study of M15, is being written up for publication. Our analyses of M71, M5 and M13 are already published (see Cohen, Briley & Stetson 2002, Briley, Cohen & Stetson 2002 and 2004b, and references therein). We derive C/Fe from the CH band, and N/Fe from the CN band. For M15 we must use the NH band at 3360 Å; the CN bands are too weak. This has the advantage of achieving a N/H ratio which is to first order independent of the C abundance, which would not hold were CN to be used for this purpose.

Fig. 10 from our study of a large sample of stars in M5 (Cohen, Briley & Stetson 2002) (not reproduced here due to limits of length) is typical of the GCs studied thus far in such detail. It shows a strong anti-correlation between C and N abundances, i.e. conversion of C into N, with strong-to-star variations in derived C and N abundances seen at all luminosities probed. (This sample contains mostly stars at the base of the RGB and just below the main sequence turnoff,

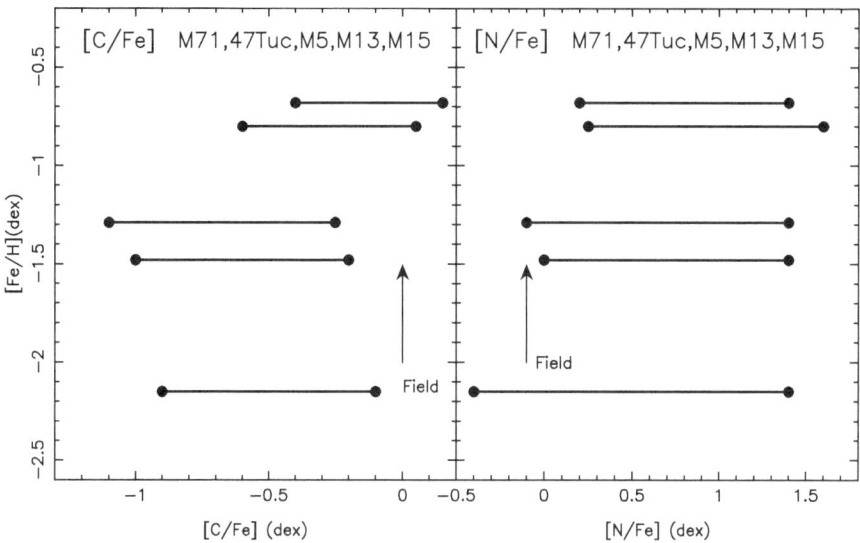

Fig. 1. The range of [C/Fe] (left panel) and [N/Fe] (right panel) is shown as a function of metallicity ([Fe/H]) for the globular clusters from our work on M71, M5, M13, and M15 as well as for 47 Tuc (from Briley et al 2004a). Large samples of stars, mostly subgiants, were used in each case. Each GC is represented by a horizontal line. The characteristic field star ratio, from Carretta, Gratton & Sneden (2000) for C and from Henry, Edmunds & Koppen (2000) for N, are indicated by vertical arrows in each panel.

with V ∼ 16.5 to 19 mag, where the turnoff of M5 is at about 18.2 mag.) We also find ON burning is required to reproduce the most extreme N enhancements, which are very large. In our paper, we commented that external pollution from a nearby AGB star, presumably a binary companion, perhaps can match the star-to-star abundance patterns seen in GCs, but does not seem capable of producing the highly organized abundance variations as such an "external" mechanism is stochastic in nature. Furthermore, the amount of "polluted" mass that needs to be accreted becomes a significant fraction of the total mass of the low luminosity star. The popular AGB companion hypothesis is beginning to crumble at this point.

Combining all our data with the study of 47 Tuc by Briley et al (2004a), we determine roughly the range in [X/Fe], where X is either C or N, in each GC. The results are shown in Fig. 1; the vertical arrows indicate the mean location of metal poor field stars. Note that the field star mean coincides roughly with the GC high end of the range for C and with the low end of the GC range for N. These GCs span a range in metallicity of a factor of ∼40. Yet we find that this range is approximately constant for both C and for N. Thus the additional material is not from some primary process which dumps a fixed amount of N into the GC gas. Instead it behaves like a secondary process, increasing as [Fe/H] increases. Since the production of C and N in AGB stars is to first order a primary process, this strongly suggests that ejecta from AGB stars do not cause the star-to-star variation in the abundance of these elements in GCs. This leaves some kind of variation imprinted in the proto-cluster before the present generation of stars we now observe were formed as the only viable scenario. Furthermore the source of this cannot have been some previous generation of AGB stars, unless mass loss rates vary proportionately to metallicity. While it is believed that they do increase with metallicity, the factor generally discussed is far smaller than the factor of 40 range in our GC sample.

So after more than 20 years of searching for an answer, we have much better data on the nature of star-to-star variations in the abundances of the light elements in GCs, but still no definitive understanding of the physical mechanism(s) responsible, nor of why metal poor field halo stars do not show these phenomena.

References

1. Briley, M. M., Cohen, J. G. & Stetson, P. B., Ap.J.L., **579**, L17 (2002)
2. Briley, M. M., Harbeck, D., Smith G. H. & Grebel, E. K., AJ, **127**, 1588 (2004a)
3. Briley, M. M., Cohen, J. G. & Stetson, P. B., AJ, **127**, 1579 (2004b)
4. Carretta, E. & Gratton, R. G., A&AS, **121**, 95 (1997)
5. Carretta, E., Cohen, J. G., Gratton, R. G. & Behr, B. B., AJ, **122**, 1469 (2001)
6. Carretta, E., Gratton, R. & Sneden, C., A& A, **356**, 238 (2000)
7. Cohen, J. G., AJ, **127**, 1545 (2004)
8. Cohen, J. G., Gratton, R. G., Behr, B. B. & Carretta, E., ApJ, **523**, 739 (1999)
9. Cohen, J. G., Behr, B. B. & Briley, M. M., AJ, **122**, 1420 (2001)
10. Cohen, J. G., Briley, M. M. & Stetson, P. B., AJ, **123**, 2525 (2002)
11. Cohen, J. G. & Melendez, J., AJ (in press) (2005a)
12. Cohen, J. G. & Melendez, J., AJ (in press) (2005b)
13. Gratton, R. G., Sneden, C., Carretta, E. & Bragaglia, A., A&A, **354**, 169 (2000)
14. Gratton, R. G., et al, A&A, **369**, 87 (2001)
15. Gratton, R. G., Sneden, C. & Carretta, E., Ann. Review Astron. & Astrophys, **42**, 385 (2004)
16. Henry, R.C.B., Edmunds, M.G. & Koppen, J., ApJ, **541**, 660 (2000)
17. Kraft, R. P., 1994, PASP, **106**, 553 (1994)
18. Oke, J. B., et al., PASP, **107**, 375 (1995)
19. Ramírez, S. V. & Cohen, J. G., AJ, **123**, 3277 (2002)
20. Ramírez, S. V. & Cohen, J. G., AJ, **125**, 224 (2003)
21. Sneden, C., Kraft, R. P., Guhathakurta, P., Peterson, R. C. & Fulbright, J. P., AJ, **127**, 2162 (2004)
22. Vogt, S. E. et al, SPIE, **2198**, 362 (1994)

Red Giants Survey in ω Cen:
Preliminary FLAMES GTO Results

E. Pancino

INAF - Osservatorio Astronomico di Bologna,
via Ranzani 1, I-40127 Bologna, Italy

Abstract. I present preliminary results for a sample of $\sim$700 red giants in ω Cen, observed during the Ital-FLAMES Consortium GTO time in May 2003, for the Bologna Project on ω Cen. Preliminary Fe and Ca abundances confirm previous results: while the metal-poor and intermediate populations show a normal halo α-enhancement of $[\alpha/\mathrm{Fe}]\simeq+0.3$, the most metal-rich stars show a significantly lower $[\alpha/\mathrm{Fe}]\simeq+0.1$. If the metal-rich stars have evolved within the cluster in a process of self-enrichment, the only way to lower their α-enhancement would be SNe type Ia intervention.

1   Introduction

The treatment of the large data volumes that can be obtained with large telescopes and multi-objects spectrographs, such as FLAMES at the ESO VLT, requires a major upgrade of the methods commonly used to reduce and analyze high and medium resolution spectra.

We are therefore developing a set of routines for an automatic or semi-automatic abundance analysis of stellar spectra based on equivalent widths (EW). The first product is DAOSPEC, a code developed by P. B. Stetson for automatic EW measurement (http://cadcwww.hia.nrc. ca/stetson/daospec/). The preliminary abundance analysis presented here is the first step of an iterative and automatic procedure under development at the Bologna Observatory.

2   Analysis and Results

Part of the Ital-FLAMES Guaranteed Time has been devoted to the study of ω Cen, within the framework of the Bologna Project on ω Cen [1]. 700 giants have been observed with the ESO-VLT GIRAFFE spectrograph in May 2003, with a resolution of R$\sim$20,000 and a S/N$\simeq$50–150 per pixel, spanning more than 4 magnitudes and covering all known sub-populations in the cluster. The spectral ranges observed (Setups HR09, HR13, HR11 and HR15) allow the determination of iron peak, α, s-process elements and of Eu.

The spectra have been reduced with the GIRAFFE BLDRS pipeline developed at the Geneva Observatory. EW have been measured with DAOSPEC [2], based on a linelist produced with the Vienna Atomic Line Database (VALD) [3]. Preliminary estimates of the stellar parameters T_{eff}, $\log g$, v_t and [M/H] have been obtained from the WFI photometry published by [4] and the color-temperature calibration by [5]. MARCS model stellar atmospheres [6] have been

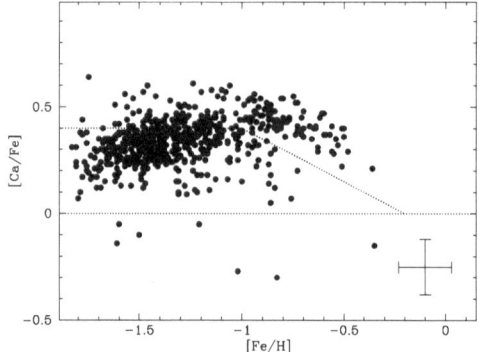

Fig. 1. [Ca/Fe] versus [Fe/H] for the ~700 stars observed with GIRAFFE. Typical errorbars are plotted in the lower-right corner. The typical disk-halo behaviour is represented by dotted lines.

employed together with the abundance analysis code by [7] to produce a first guess abundance of Fe and Ca, strictly based on the photometric parameter estimates.

The resulting [Ca/Fe] versus [Fe/H] plot is shown in Fig. 1, where except for a few outliers that will have to be manually inspected, a clear trend appears: [Ca/Fe] slowly rises with [Fe/H] until it reaches a maximum and then declines again for the most metal-rich stars (RGB-a according to [4]). This nicely confirms a previous finding by [8] and [9]. If the metal-rich stars have evolved within the cluster in a process of self-enrichment, the only way to lower their α-enhancement would be SNe type Ia intervention. No simple explanation is provided for the rise of [Ca/Fe] at low [Fe/H], although a series of star formation bursts should be the likely cause.

References

1. Ferraro, F. R., Bellazzini, M., Origlia, L., Pancino, E., & Sollima, A. 2003, ASP Conf. Ser. 296: New Horizons in Globular Cluster Astronomy, 215
2. Stetson, P. B. & Pancino, E., *in preparation*
3. Kupka, F., Piskunov, N., Ryabchikova, T. A., Stempels, H. C., & Weiss, W. W. 1999, A&AS, 138, 119
4. Pancino, E., Ferraro, F. R., Bellazzini, M., Piotto, G., & Zoccali, M. 2000, ApJ, 534, L83
5. Alonso, A., Arribas, S., & Martínez-Roger, C. 1999, A&AS, 140, 261
6. Edvardsson, B., Andersen, J., Gustafsson, B., Lambert, D. L., Nissen, P. E., & Tomkin, J. 1993, A&A, 275, 101
7. Spite, M. 1967, Annales d'Astrophysique, 30, 685
8. Pancino, E., Pasquini, L., Hill, V., Ferraro, F. R., & Bellazzini, M. 2002, ApJ, 568, L101
9. Origlia, L., Ferraro, F. R., Bellazzini, M., & Pancino, E. 2003, ApJ, 591, 916

Galactic Evolution of Carbon and Nitrogen

G. Israelian

Instituto de Astrofisica de Canarias, Via Lactea s/n, La Laguna 38200, Tenerife, Spain; gil@ll.iac.es

Abstract. Abundance analysis of Carbon and Nitrogen has been performed in a sample of 32 late F and early G type dwarf metal poor stars in the metallicity range $-3.3 < \text{[Fe/H]} < 0$ using molecular lines of CH and NH in the near-UV. We find that [C/Fe] decreases slowly with increasing [Fe/H] while [N/Fe] remains flat. Furthermore we derived uniform and accurate C/O and N/O ratios using oxygen abundances from near-UV OH lines employed in our previous studies. We confirm the metallicity dependence of C/O ratio known from previous studies and caused by the metallicity dependence of the C yields from massive stars with mass loss. [C/O] does not remain constant below [O/H]$=-0.5$ but increases again with a large scatter. We find that a primary component is required in order explain the observations of N/O and that the N production history is similar in our Galaxy and DLAs.

1 Introduction

Observations of CNO in metal poor stars helps to understand the chemical evolution of the Galaxy because and their key role in the chain of nucleosynthesis. Carbon comes from the Triple Alpha reaction. Already in 1957 Burbidge et al. (1957) proposed that carbon is provided by mass loss from evolved stars. However, Arnett and Schramm (1973) suggested that massive stars may be an important contributor. Different views with respect the relative role of massive, intermediate and low mass stars have reflected the different uncertainties concerning the dredge-up, production of carbon and mass loss from different types of stars. The situation is far from being clear and there are many uncertainties. The stable isotope of Nitrogen is synthesized from C and O through the CNO cycle in a hydrogen-burning shell of the stellar interior. Various studies have accumulated evidence that the production of nitrogen at low [Fe/H] proceeds principally as a primary rather than a secondary process (Pagel & Edmunds (1981); Bessel & Norris (1982); Carbon et al. (1987); Henry et al. (2000)) while the secondary process takes over at high metallicities. Two primary sources of nitrogen have been proposed. The first is intermediate mass (4–8 $M_\odot$) stars during their thermally pulsing asymptotic giant branch phases (van den Hoek & Groenewegen 1997). The second source is rotating massive stars (Meynet & Maeder 2002). Several authors support the origin of primary nitrogen in massive stars because of the low scatter of (N/O) ratios in galaxies observed at different stages of their evolution, which would imply no time delay between the injection of nitrogen and oxygen.

High resolution and high S/N spectra of 32 metal poor stars observed with various 4-m class telescopes (Israelian et al. 1998, 2001, Bihain et al. 2004) have been used to analyse N/O and C/O ratios from near-UV molecular lines of CH, NH and OH. We have carried out an independent study of the (N/O) ratio using the NH band at 3360 Å and the OH lines employed by Israelian et al. (1998, 2001). Details of the analysis and stellar parameters are provided in Israelian et al. (2004).

2   Carbon

A weak but useful carbon line [CI] 8727.13 Å disappears in halo dwarfs with metallicities below −1. To measure carbon abundance in halo stars one can use four CI high excitation lines near 9100 Å and the CH band at 4300 Å. The CI lines at 9100 Å together with the OI triplet at 7771 Å have been used by Tomkin et al. (1992) and Akerman et al. (2004) to study the behaviour of C/O versus metallicity. However, CI and OI lines employed in these papers are sensitive to a non-LTE effects and one has to bare in mind that this sensitivity is different for C and O. The CH band at 3145 Å used by Israelian et al. (1999) is almost saturated in disk stars and several blends makes the abundance analysis less accurate. To ensure a homogeneous analysis of the C/O and N/O ratio from NH,CH and OH lines in the near-UV, we used the same model atmospheres and tools as in our previous studies. The oxygen abundances were compiled from Israelian et al. (1998, 2001) and Boesgaard et al. (1999).

We stress that any possible abundance errors produced by uncertainties in the near-UV continuum, 3D and/or non-LTE effects are not critical for the present analysis. These errors will cancel out when forming abundance ratios from lines formed in the same atmospheric layer. This philosophy led Tomkin and Lambert (1984) to derive C/O ratio from the bands NH 3360 and CH 4300 Å.

We have selected several unblended CH lines located in the near-UV between 3145–3190 Å modified their oscillator strengths (gf-values) by fitting the solar high-resolution spectrum and assuming solar abundance of carbon 8.56 from Anders and Grevesse (1989). These lines are measurable in dwarfs down to the metallicities −3. Our results are shown in the Fig. 1. We confirm the metallicity dependence of the C/O ratio (Tomkin et al. 1992, Akerman et al. 2004). On the other hand, our plot for C/O shows a steep rise at [O/H]< −1. It is not clear if this effect is real. The work is in progress to address this issue using other abundance indicators such as CH 4300 Å and better quality spectra.

3   Nitrogen

Various spectral features can be used to derive the nitrogen abundance in dwarfs. Unfortunately weak high excitation (χ=10.34 eV) near-infrared NI lines at 7468.31, 8216.34, 8683.4, 8703.25 and 8718.83 Å disappear at metallicities [Fe/H]< −1 and for the analysis of N in metal-poor stars we are left with the CN and NH molecular bands at 3883 and 3360 Å, respectively. It must be mentioned

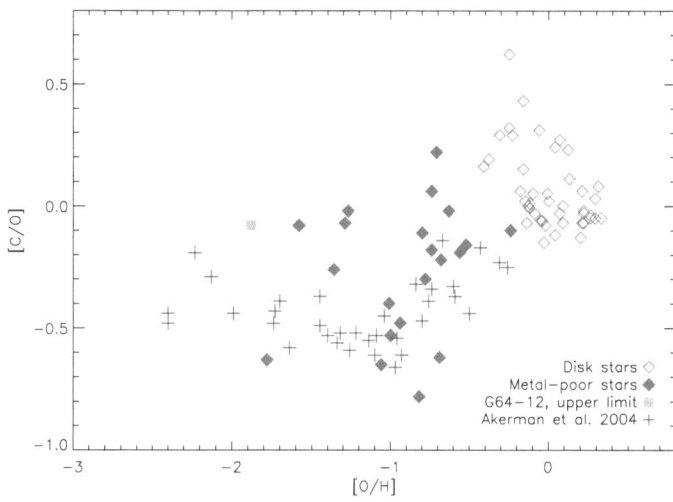

Fig. 1. Comparison of C/O ratios for programme stars (filled diamonds) with observations of Akerman et al. (2004) (crosses). Empty boxes are metal-rich stars from Israelian et al. (2004).

that the abundances derived from the NH 3360 Å and NI line at 7468.31 Å agree very well in metal rich stars (Ecuvillon et al. 2004). Observations of the strong NH band at 3360 Å allowed to delineate the Galactic evolution of N down to [Fe/H]~ -3 (Laird 1985; Carbon et al. 1987; Tomkin & Lambert 1984; Israelian, García-López & Rebolo 2000). These authors found a flat trend for [N/Fe] while they obtained different constant values of [N/Fe].

Our own analysis (Israelian et al. 2004) based on the NH band at 3360 Å shows that the N/O ratio has an increasing trend together with both oxygen and nitrogen abundance while [N/Fe] is flat down to [Fe/H]~ -3. Our observations show that a primary component is required in order explain the observations of N/O (Fig. 2). We have also discovered few Nitrogen rich halo dwarfs that have a N/O ratio higher than expected from the trend (Fig. 2). Comparing N/O ratios in galactic stars and DLAs, we found that the N production history is similar in our Galaxy and DLAs, but different in BCGs (see Israelian et al. 2004).

Acknowledgements

The author is pleased to acknowledge fruitful discussions and collaboration with A. Ecuvillon, R. Rebolo, R. García López, P. Bonifacio, P. Molaro, A. Maeder and G. Meynet.

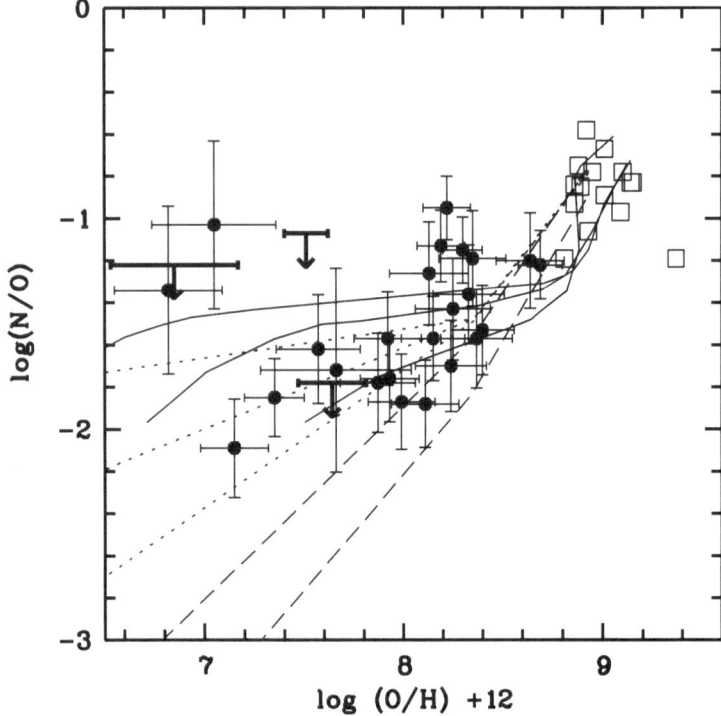

Fig. 2. Comparison of N/O ratios for programme stars (filled circles) with the simple models of Meynet & Maeder (2002), both including stellar rotation (dotted lines) and with no stellar rotation (dashed lines) and with the analytical models of Henry et al. (2000, solid lines) which were computed to fit the N/O ratios observed in extragalactic H II regions (blue compact galaxies). Empty boxes are metal-rich stars. See Israelian et al. (2004) for details.

References

1. Akerman, C. J., Carigi, L., Nissen, P.E., Pettini, M. & Asplund, M. 2004, A&A, 414, 931
2. Anders, E., & Grevesse, N. 1989, Geochim et. Cosmochim. Acta 53, 197
3. Arnett, W.D. & Schramm, D.N. 1973, ApJ, 185, L47
4. Bessell, M. S. & Norris, J. 1982, ApJ, 263, L29
5. Bihain, G., Israelian, G., Rebolo, R., Bonifacio, P. & Molaro, P. 2004, A&A, 423, 777
6. Boesgaard, A.M., King, J.R., Deliyannis, C. P., & Vogt, S.S. 1999, AJ, 117, 492
7. Burbidge, E.M., Burbidge, G.R., Fowler, W.A. and Hoyle, F. 1957, Rev. Mod. Phys., 29, 547
8. Carbon, D. F., Barbuy, B., Kraft, R. P., Friel, E. D., & Suntzeff, N. B. 1987, PASP, 99, 335
9. Ecuvillon, A., Israelian, G., Santos, N., Mayor, M., García López, R. J. & Randich, S. 2004, A&A, 426, 619

10. Edmunds, M. G. & Pagel, B. E. 1978, MNRAS, 185, 78P
11. Henry, R. B. C., Edmunds, M. G., & Köppen, J. 2000, ApJ, 541, 660
12. Israelian, G., García López, R. J. & Rebolo, R. 1998, ApJ, 507, 805
13. Israelian, G., García López, R. J. & Rebolo, R. 2000, in "The Evolution of the Milky Way: stars versus clusters", Eds. F. Matteucci and F. Giovannelli, (Dordrecht: Kluwer) ,p. 35
14. Israelian, G., Rebolo, R., García López, R., Bonifacio, P., Molaro, P., Basri, G., & Shchukina, N. 2001, ApJ, 551, 833
15. Israelian, G., Ecuvillon, A., Rebolo, R., García López, R., Bonifacio, P., Molaro, P. 2004, 421, 649
16. Laird, J. B. 1985, ApJ, 289, 556
17. Meynet, G. & Maeder, A. 2002, A&A, 390, 561
18. Pagel, B. E. J., & Edmunds, M. G. 1981, Ann. Rev. Astr. Ap., 19, 77
19. Tomkin, J., & Lambert, L. 1984, ApJ, 279, 220
20. Tomkin, J., Lemke, M., Lambert, D. L., & Sneden, C. 1992, AJ, 104, 1568
21. van den Hoek, L. B. & Groenewegen, M. A. T. 1997, A&AS, 123, 305

First Stars: Abundance Patterns from O to Zn and Derived SNe Yields in the Early Galaxy

R. Cayrel[1] and M. Spite[2], presenter at the workshop

[1] Observatoire de Paris, 61 av. de l'Observatoire, F-75014 Paris, France
[2] Observatoire de Paris, Section de Meudon, place Jules Janssen,
   Meudon Cedex F92125, France

1   The Team

Most of the work reported here has been conducted within the ESO Large Programme 165.N-0276 "Galaxy Formation, Early Nucleosynthesis, and the First Stars", which has covered 4 periods 65-68, from April 2000 to November 2001, with a total of 38 nights in visitor mode. The team had R. Cayrel as PI, and 13 CoIs:

F. and M. Spite, P. François, V.Hill and E. Depagne (Obs. de Paris, F)

B. Plez (Univ.Montpellier 2, F)

P. Molaro and P. Bonifacio (Trieste Obs., I)

J. Andersen, Copenhagen Obs, DK. and B. Nordström (Copenhagen Obs. DK and Lund Univ. SE)

F. Primas (ESO)

Timothy C Beers (Michigan State Univ. USA)

Beatriz Barbuy (Univ. São Paulo, BR)

2   Aim of the Project

The aim of our large programme was to obtain the abundances of the nuclides present in the early Galaxy, by studying these abundances in matter still very little affected by astration. By astration (a word coined by Hubert Reeves) we mean presence of nuclides built by nuclear reactions inside stars, and expelled into the interstellar medium by explosions (SNe), or winds (massive stars, asymptotic giant branch stars in terminal evolution). The least astrated matter is that left after the Big Bang, by the primordial nucleosynthesis. A star having this chemical composition is called population III star, but, so far, none has been seen. The present interstellar matter has between 1 and 2 per cent (by mass) of elements built by astration. But a few stars are known to have only 0.001 to 0.0001 per cent of those elements. It is these stars which were our targets in the LP, the nearest to pop. III, still observable today. In the following sections, we successively speak of the construction of the sample, of the observations, of the technique of analysis, of the results obtained, and of their interpretation.

3   Construction of the Sample

Most of the stars of our sample have been selected from the H&K BPS survey (Beers, Preston & Shectman [1]. First, stars were selected from the weakness of their H&H lines for the Balmer lines intensity on prism-objective Schmidt telescope plates. Then, the candidate stars were observed with a slit spectrograph in order to have a quantitative estimate of their metallicity. The survey has operated on about 7000 square degrees of the sky, mostly on the polar caps. It has supply a vast amount of metal-poor stars, with hundreds of them more metal-poor than the most metal-poor globular clusters. We selected from this sample stars with metallicities estimated to have [Fe/H] < -2.7. The actual metallicity histogram is given for the sample on fig. 1.

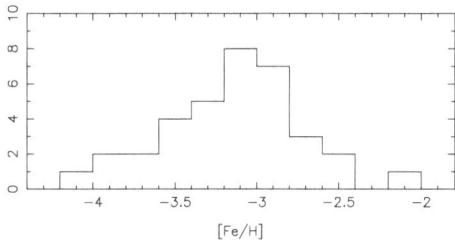

Fig. 1. Metallicity distribution of our sample of 35 giants. Bins are 0.2 dex wide

   It is important to note that we have tried to avoid carbon-rich stars, because they have a rich molecular line spectrum, mostly CN, CH and C_2, obliterating many interesting atomic lines of rare elements. This is why we had in our sample a star, CS 31082-001, in which we were able to measure the 385.97 nm line of U II, whereas in the similar r-process element enriched star CS 22892-052, but carbon rich, a CN line obliterates the U II line.

4   Observations

The observations were made at Paranal, with the VLT Kueyen and the UVES spectrograph, used in its dichroic mode. So spectra were simultaneously taken in the red and in the blue, allowing an almost complete coverage from 330 to 980 nm. The standard setting was a slit width of 1.0" on the sky, with a spectral resolution of 43 000. The signal/noise per pixel was typically of 200 in the visible and the red, 140 in the blue, and falling to 30-50 at 340 nm. Reduction was made using the ESO pipeline dedicated software for UVES. A sample spectrum of the star CS 22186-025 (V=14.2) is given in fig. 2 and 3.
   Comparison of our equivalent widths with those of the literature, when available, shows a satisfactory agreement.

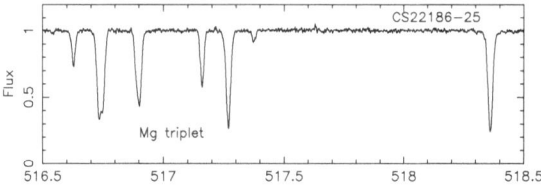

Fig. 2. sample spectrum in the region of the Mg I b lines for one of our target stars

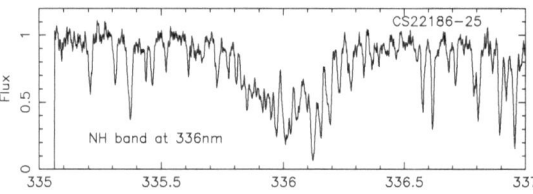

Fig. 3. sample spectrum in the region of the UV NH band, for one of our target stars

5   Analysis

Our analyses are based on 1-D LTE opacity sampling models (OSMARCS) from the Uppsala group (Gustafsson et al. [5]; Plez et al. [7]; Asplund et al. [2]) computed by one of our CoI Bertrand Plez. The computation of the synthetic spectrum to be compared to the observations was done with the code TUR-BOSPECTRUM (Alvarez & Plez [3]). For a few elements (Na,Al,K) NLTE corrections have been applied. The atmospheric parameters have been determined as follows. Effective temperatures were mostly derived from photometry, with reddening corrections from maps of Burstein & Heiles [4]. Colours B,V,R,I,J, K were generally available, and we estimate our adopted effective temperatures accurate to 80 K, except for three stars. Gravities were determined from the ionisation equilibria of Fe and Ti.

The microturbulent velocity was set by the need of having an iron abundance independent of the equivalent width of the lines.

6   Results and Interpretation

We present here the results for 14 elements from C to Zn. Results for heavier elements are presented by Patrick François at this meeting, and a more detailed discussion about C and N abundances versus mixing, in the stars of the sample, is presented by Monique Spite. Here we are interested in the initial composition of the stars, so we have eliminated the stars which have undergone evolutionary alteration of their initial composition, for the elements affected by this process, of C,N and to some extend Na. This cleaning up is evidenced in fig. 4 by the comparison of of the left and right panels.

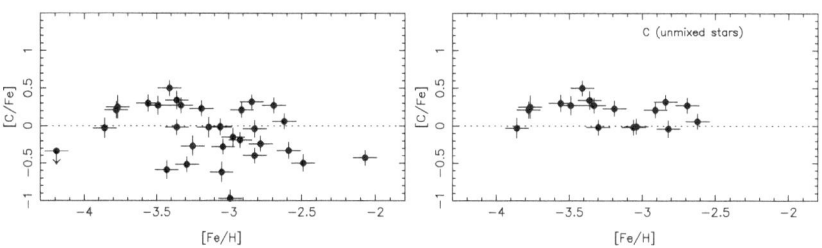

Fig. 4. left:[C/Fe] versus [Fe/H] for the full sample. The scatter on [C/Fe] is very large. right: same for the non-mixed stars. All stars with [C/Fe] < 0 have gone

C, N and Na are the only elements presenting evidence for an intrinsic scatter. Other elements show a scatter which is not significantly larger than the observational scatter. The best example is the plot of [Cr/Fe] versus [Fe/H] (figure 5). A full set of diagrams with either [Fe/H] or [Mg/H] in abscissa can be found for the 16 elements in the reference Cayrel et al. 2004[9]. We give here two examples of those diagrams (figure 6). In the next section we further discuss the meaning of these diagrams. The choice of [Mg/H] as astration indicator is safer than [Fe/H] as an element less affected by the "cut" position or fallback.

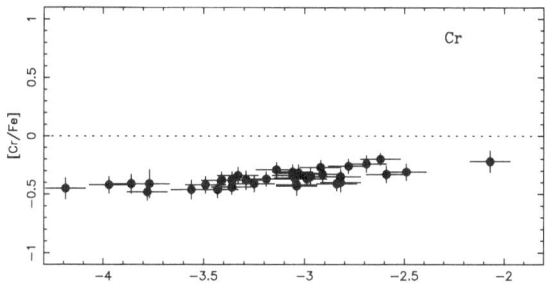

Fig. 5. [Cr/Fe] versus [Fe/H]. Note the small scatter, and the flat behaviour below [Fe/H]= -3.5

7   Yields in the Early Galaxy

The abundance ratios found in the photospheres of our target stars are imprints of the explosions of the first SNe II or even more massive stars. At very low metallicities there is a reasonable hope that the SNe which have polluted the environment were themselves primordial objects. In former papers on the chemical composition of very metal poor stars, some accent was put on trends of abundance ratios with metallicity (McWilliam et al.[6], Norris et al. [8]). Such was the case for [Mn/Fe], or [Cr/Fe] decreasing with decreasing metallicity, or

[Co/Fe] with a reverse trend. At first view our diagram of [Cr/Fe] versus [Fe/H] presents the same trend, but a more careful examination of the diagram shows that, if the trend is clear between -2.5 and -3.5 it may well have disappeared below -3.5. Another startling feature of our diagram is the smallness of the scatter, not larger than the error bars of observations, typically 0.05 dex, much smaller than what was obtainable before instruments as the VLT-UVES were available. In another paper (Cayrel et al.[9] we discuss in more detail the origin of the slope, which may have nothing to do with the yields, but more with the assumption of local thermodynamical equilibrium made in the analysis of the spectrum.

Actually, we are interested in reaching a behaviour in which no trend exist anymore with metallicity, situation expected if the polluting SNe are primordial, i.e. without elements heavier than lithium. From our diagrams such a regime seems to hold for [Mg/H] < -2.9. We then consider the mean of the [X/Mg] ratios for [Mg/H] < -2.9 as representative of the yields in the primeval Galaxy. They are given in table 1.

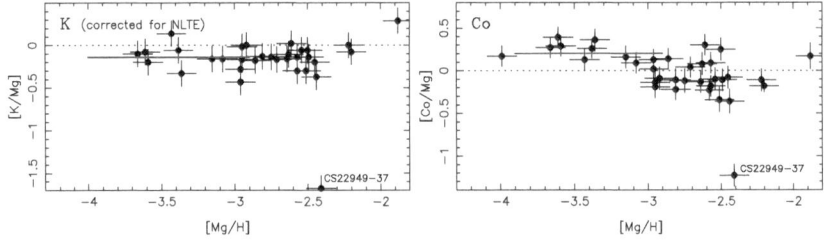

Fig. 6. Left:[K/Mg] versus [Mg/H]. Right: [Co/Mg] versus [Mg/H]

Table 1. Mean values of [X/Mg] for stars with [Mg/H] < -2.9, representing the yields of SNe in the early Galaxy. The column N is the number of objects in the mean

Elem	dex	rms	N	Elem	dex	rms	N
[C/Mg]	-0.01	0.19	10	[Sc/Mg]	-0.17	0.14	14
[N/Mg]	-0.05	0.24	8	[Ti/Mg]	-0.01	0.09	14
[O/Mg]	0.32	0.21	13	[Cr/Mg]	-0.63	0.09	14
[Na/Mg]	-0.91	0.16	9	[Mn/Mg]	-0.65	0.20	14
[Al/Mg]	-0.59	0.05	10	[Fe/Mg]	-0.21	0.10	14
[Si/Mg]	0.21	0.14	14	[Co/Mg]	0.13	0.17	14
[K/Mg]	-0.14	0.14	13	[Ni/Mg]	-0.23	0.13	14
[Ca/Mg]	0.06	0.09	14	[Zn/Mg]	0.15	0.19	14

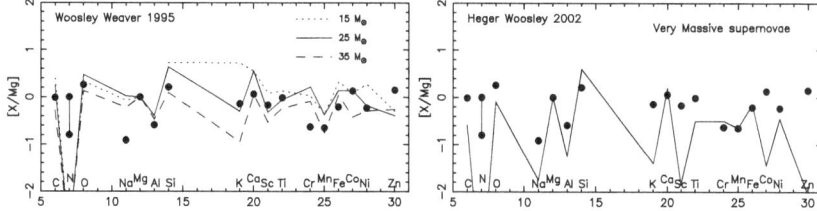

Fig. 7. Left: comparison of our yields with theoretical yields computed for a zero metallicity SNe by Woosley and Weaver 1995 [10]. Right: same but with theoretical yields computed for zero metallicity pair-instability very massive SNe by Heger & Woosley [11]. [N/Mg] has the largest scatter, within the two points

It is interesting to compare our yields with theoretical yields, computed for primordial SNe. Fig. 7 shows this. Clearly, SNe in the mass range 15 to 35 $M_\odot$ perform better than Pair-Instability (very massive) SNe.

8 Conclusion

Extremely metal-poor stars have a well defined nuclide composition for elements from C to Zn, except for C, N and Na, displaying some intrinsic scatter. This composition is in fairly well agreement with theoretical yields of primordial SNe of masses in the range 15 to 35 $M_\odot$, but not with those of Pair-Instability SNe.

References

1. Beers, T.C., Preston, G.W., & Shectman, S.A. 1992, AJ **103**, 1987
2. Asplund, M., Gustafsson, B., Kiselman, D., & Eriksson, K., 1997, A&A **318**, 521
3. Alvarez, R. & Plez, B. 1998, A&A **330**, 1109
4. Burstein, D. & Heiles 1982, AJ **87**, 1165
5. Gustafsson B., Bell R. A., Eriksson K., Nordlund Å., 1975, A&A **42**, 407
6. McWilliam, A., Preston, G.W., Sneden, C., Searle, L., 1995, AJ **109**, 2757
7. Plez B. Brett J.M., Nordlund Å., 1992, A&A **256**, 551
8. Ryan, S.G., Noris, J.E., Beers, T.C., 1996, ApJ **471**, 254
9. Cayrel, R., Depagne, E., Spite M. et al. 2004, A&A **416**, 1117
10. Woosley, S.E. & Weaver, T.A. 1995, ApJS **101**, 181
11. Heger, A., Woosley, S.E. 2002, ApJ **567**, 532

Nitrogen in the Early Universe

J.A. Johnson[1], F. Herwig[2], T.C. Beers[3], and N. Christlieb[4]

[1] DAO/HIA/NRC, 5071 West Saanich Road, Victoria, BC, V9E 2E7, Canada
[2] LANL, Theoretical Astrophysics Group, T-6, MS B224, Los Alamos, NM 87545, USA
[3] Dept. of Physics and Astronomy and Joint Institute for Nuclear Astrophysics, Michigan State University, East Lansing, MI 48824, USA
[4] Hamburger Sternwarte, University of Hamburg, Gojenbergsweg 112, D-21029 Hamburg, Germany

1 Introduction

Theoretical models for nucleosynthesis in asymptotic giant branch stars predict a large contribution to the cosmic nitrogen abundance from intermediate-mass stars [1]. In particular, hot-bottom-burning in stars above a certain mass produces $[C/N] \lesssim -1$ [2]. However, observations of C and N abundances in C-rich, metal-poor stars, usually using the CH and CN bands, show $[C/N]$ values that vary between -0.5 and 1.5. (Fig. 1). If any of these stars have been polluted by intermediate mass AGB stars, then they should have lower $[C/N]$ ratios. However, most of the CH stars with detailed abundances have $[C/Fe] > 1.0$, and it is more likely than stars mildly enhanced in C have been polluted by N-rich stars.

2 Current Study

We observed the NH band at 3360Å in 18 metal-poor stars with $0 < [C/Fe] < 1$. These stars had $[Fe/H]$ and $[C/Fe]$ values from the follow-up to the HK [3] and HES [4] surveys. We exposed for 1-2 hours per star on the RCSpec at KPNO and CTIO at 2-2.5 Å resolution. The signal-to-noise was ~ 50 in the NH region. We used MOOG [5] to synthesize the NH and CH bands to get $[C/N]$ ratios. The CH lines are from Kurucz, while the NH gf values were calculated from lifetime measurements and rotational constants in the literature (see refs in [5]). Spite et al [6] found that N from CN was 0.5 dex lower than N from NH, and we found the same effect while testing our linelist on high resolution spectra. Our values of nitrogen from NH have been adjusted down by ~ 0.5 dex to put them on the CN scale for this preliminary analysis.

We do not find any stars with low $[C/N]$ ratios and the range of $[C/N]$ we find is much more restricted than predicted. If a standard initial mass function is used, we would expect about one in five AGB stars to undergo hot-bottom-burning, a ratio not seen in Fig. 1. It is possible that some of the C-rich stars in our sample were polluted by nucleosynthesis sources other than AGB stars, and we are currently observing a larger sample of stars. There are also many assumptions which need to re-evaluated, such as mass loss being unaffected by

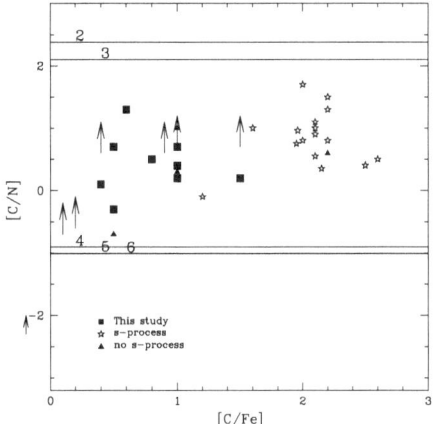

Fig. 1. C/N] from the literature [7–11] and our observations for stars with [Fe/H] <-2.0 and [C/Fe] > 0. The theoretical predictions [2] are labeled by the mass of the AGB star. Our sample fills in the mildly carbon enrichment gap. The disagreement with the predictions is clear, both at high [C/N] and low [C/N]. The literature values are from CN measurements, while our sample is from NH measurements. Adopting the NH scale would increase the N values. Adopting a higher dissociation energy or the Kurucz gf values decrease the N abundance.

binary evolution and the mass ratio of binaries being independent of mass. We are also planning follow-up studies of Li and Na abundances and radial velocity variations to help explain the very narrow range in observed [C/N] in C-rich stars.

References

1. A. Renzini and M. Voli. A&A. 94 (1981) 175
2. Herwig, F. 2004, ApJS, 155, 651.
3. T. C. Beers, G. W. Preston, and S. A. Shectman. AJ 103 (1992) 1987.
4. N. Christlieb, P. J. Green, L. Wisotzki D. Reimers. A&A 375 (2001) 366.
5. C. Sneden. PhD dissertation (1973)
6. M. Spite, et al. A&A (2004), submitted
7. B. Barbuy, R. Cayrel, M. Spite, T. C. Beers, F. Spite, B. Nordstrom, and P. E. Nissen. A &A 317 (1997) 63.
8. W. Aoki, J. E. Norris, S. G. Ryan, T. C. Beers, H. Ando, N. Iwamoto, T. Kajino, G. J. Mat hews, and M. Y. Fujimoto. ApJ 561 (2001) 346.
9. W. Aoki, J. E. Norris, S. G. Ryan, T. C. Beers, and H. Ando. ApJ 567 (2002) 1166.
10. G. W. Preston & C. Sneden. ApJ 122 (2001) 1545
11. S. Lucatello, R. Gratton, J. G. Cohen, T. C. Beers, N. Christlieb, E. Carretta, and S. Ra mirez. AJ 125 (2003) 875.

Abundance of Heavy Elements in Extremely Metal-Poor Stars

P. François[1], R. Cayrel[1], E. Depagne[1], M. Spite[1], V. Hill[1], F. Spite[1], B. Plez[2], T. Beers[3], F. Primas[4], J. Andersen[5,9], B. Barbuy[6], P. Bonifacio[7], P. Molaro[7], and B. Nordström[5,8]

[1] GEPI, Observatoire de Paris-Meudon, F-92125 Meudon Cedex, France,
[2] GRAAL, Université de Montpellier II, F-34095 Montpellier Cedex 05, France,
[3] Department of Physics & Astronomy, Michigan State University, East Lansing, MI 48824 USA,
[4] European Southern Observatory (ESO), Karl Schwarzschild-Str. 2, D-85748 Garching bei München, Germany,
[5] Astronomical Observatory, NBIfAFG, Juliane Maries Vej 30, DK-2100 Copenhagen, Denmark,
[6] IAG, Universidade de Sao Paulo, Departamento de Astronomia, CP 3386, 01060-970 Sao Paulo, Brazil,
[7] Osservatorio Astronomico di Trieste, INAF, Via G.B. Tiepolo 11, I-34131 Trieste, Italy,
[8] Lund Observatory, Box 43, SE-221 00 Lund, Sweden,
[9] Nordic Optical Telescope Scientific Association, Apartado 474, ES-38 700 Santa Cruz de La Palma, Spain.

The [Sr/Ba] Ratios in Extremely Metal Poor Stars

In the framework of the VLT Large program "First Stars" (165.N-0276(A)), we have measured the abundance of 13 heavy elements using high quality UVES spectra. In this paper, we report on the abundance of Sr and Ba in this sample of stars. In 1995, McWilliam et al. [4] showed that the [Sr/Fe] and [Ba/Fe] ratios exhibited a large dispersion in metal-poor stars. If these 2 elements are produced by the same nucleosynthetic process, then the variation of [Sr/Ba] as a function of metallicity should be constant . However, it is known (see Arlandini et al. 1999 [1] for example) that a significant part of Sr is built by s-process in massive stars. As this process is a secondary process, it is unlikely that this process is fully in operation at the early stages of the chemical evolution. On figure 1a , the [Sr/Ba] vs [Fe/H] are plotted together with some data found in the literature.

We can clearly see on this figure a spread of the [Sr/Ba] ratio. This spread is larger than the expected errors, confirming the results found by previous authors (see Honda et al. 2004 [2] and reference therein). Such a large scatter found in the [Sr/Ba] ratio can be explained by inhomogeneous models of chemical evolution which predict the existence of such a large variation (see for example Ishimaru et al. 2004).

It is therefore more interesting to plot the ratios [Sr/Ba] as a function of an heavy element (formed by neutron capture) instead of [Fe/H] which is built under completely different conditions. If the r-only parts of Sr and Ba are built

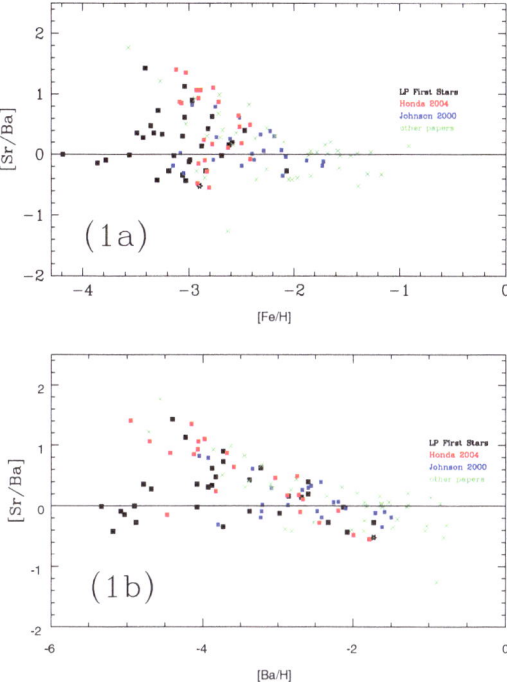

Fig. 1. [Sr/Ba] as a function of [Fe/H] (1a) and [Ba/H](1b)

by the same process, therefore, we should observe a constant [Sr/Ba] ratio in these extremely metal poor stars. On figure 1b, we have plotted the [Sr/Ba] as a function of Ba.

Figure 1b shows clearly that:

- There is a clearly visible trend of the ratio [Sr/Ba] as a function of [Ba/H], showing a decreasing [Sr/Ba] when [Ba/H] increases.
- The spread of [Sr/Ba] at a given [Ba/H] decreases as a function of the increasing barium abundance.

This large scatter in this diagram can be interpreted by invoking the existence of 2 r-processes, one favoring the synthesis of the lighter n-capture elements (weak r-process, see for example Wanajo et al. 2003 [5])

References

1. Arlandini et al. 1999, ApJ 525, 886
2. Honda S. et al. 2004, ApJ 607, 474
3. Ishimaru Y. et al. 2004, ApJ 600L, 47
4. McWilliam et al. 1995, AJ 109, 2757
5. Wanajo, S. et al. 2003, ApJ 593, 968

The Chemical Compositions
of Carbon-Enhanced Metal-Poor Stars

W. Aoki[1], S.G. Ryan[2], J.E. Norris[3], T.C. Beers[4], N. Christlieb[5],
S. Tsangarides[2], and H. Ando[1]

[1] National Astronomical Observatory, Mitaka, Tokyo 181-8588, Japan
[2] The Open University, Walton Hall, Milton Keynes, MK7 6AA, UK
[3] The Australian National University, Cotter Road, Weston, ACT 2611, Australia
[4] Michigan State University, East Lansing, MI 48824-1116
[5] Hamburg University, Gojenbergsweg 112, 21029 Hamburg, Germany

Recent surveys of metal-deficient stars have discovered a large number of carbon-rich objects, with a marked increase in their frequency at [Fe/H] < -2.5. In order to constrain the origin(s) of their carbon excesses, we have performed elemental abundance analyses for 40 objects selected from candidate metal-poor stars with strong CH G bands identified in the HK and Hamburg/ESO surveys. High-resolution spectroscopy has been obtained with AAT/UCLES and Subaru/HDS; a portion of these studies have already been published [1–3].

The abundance analyses for metal lines and molecular bands of CH and C_2 demonstrates that (1) our sample covers the metallicity range from [Fe/H]$= -3.5$ to -1.7, and (2) 39 objects have [C/Fe]$\geq +0.5$. Here we regard these 39 stars as carbon-enhanced, metal-poor (CEMP) stars.

Fig. 1a shows the abundance ratio [Ba/Fe] for this sample as a function of [C/Fe]. Thirty stars (77% of the sample) have [Ba/Fe]$> +0.7$, while the others have [Ba/Fe]< 0.0. There is a clear gap in the Ba abundances between the two groups, suggesting at least two different origins of the carbon excesses.

Ba-enhanced stars: The Ba-enhanced stars exhibit a correlation between the Ba and C abundance ratios (Fig. 1a). This fact suggests that carbon was enriched in the same site as Ba. The Ba excesses in these objects presumably originated from the s-process, rather than the r-process, because (1) nine stars in this group for which detailed abundance analysis is available clearly show abundance patterns associated with the s-process [2], and (2) there is no evidence of an r-process excess in the other 21 objects. Hence, the carbon enrichment in these objects most likely arises from Asymptotic Giant Branch (AGB) stars, which are also the source of the s-process elements.

Since most (if not all) low-metallicity objects that are currently observed in the halo are not in the AGB phase, material enriched in carbon and the s-process elements is assumed to have accreted from the companion AGB stars, which have already evolved to faint white dwarfs, to the surface of the surviving companion. This scenario is the same as that applied to classical CH stars [4]. Unfortunately, long-term radial velocity monitoring has been obtained for only a limited number of objects; a clear binarity signature has been established for six objects in our sample to date. However, there exists additional support for the mass-accretion scenario for the Ba-rich CEMP stars. Fig. 1b shows [C/H] as a function of luminosity, roughly estimated from the effective temperature

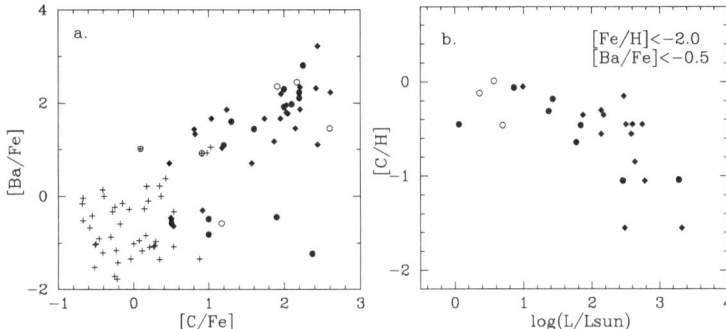

Fig. 1. a: [Ba/Fe] as a function of [C/Fe] for our sample (filled symbols) and others from literature. b: [C/H] as a function of luminosity for Ba-rich stars(see text)

and gravity determined in the abundance analyses. The [C/H] values are high in low-luminosity stars, i.e. dwarfs and subgiants, while the value decreases with increasing luminosity. A similar trend is also found in [Ba/H]. These facts can be naturally interpreted by asserting that the outer atmospheres of the low-luminosity, unevolved stars preserve the material accreted from their companion AGB stars, while, in the more-luminous evolved stars, the accreted material has been diluted by internal mixing during evolution along the red-giant branch.

It should be noted that at least three of the Ba-rich stars in our sample exhibit no clear variation in their radial velocities over the last 8-10 years. Either their periods are quite long, or the mass-accretion scenario may not apply to these objects. Further investigation of the binarity for these objects is clearly required. **Ba-normal stars:** The other nine objects in our sample have relatively low Ba abundances ($-1.0 \leq$ [Ba/Fe] ≤ -0.5). These values are typical in metal-deficient stars that show no carbon excess ([C/Fe]$< +0.5$), hence the scenario of carbon enrichment by AGB stars cannot be simply applied to these stars.

One object in this group, CS 29498-043, also shows large excesses of oxygen and the α-elements relative to Fe [3]. Another star with a similar abundance pattern is CS 22949-037 [5]. The excess of α-elements suggests a large contribution from core-collapse supernovae; the excess of carbon is presumably produced by the same process. One candidate for the source of such objects is a so-called "faint supernova" which ejects little material from the vicinity of its iron core at the time of its explosion [6,7]. No enhancement of α-elements is observed in the other eight CEMP stars in our sample with normal Ba abundances (note that oxygen abundances are not yet determined for these objects). Hence, for these stars, the origin of the carbon excess remains unclear.

References

1. W. Aoki, J. E. Norris, S. G. Ryan, T. C. Beers, H. Ando, ApJ, 567, 1166 (2002a)
2. W. Aoki, S. G. Ryan, J. E. Norris, et al., ApJ, 580, 1149 (2002b)
3. W. Aoki, J. E. Norris, S. G. Ryan et al., ApJ, 608, 971 (2004)
4. R. D. McClure, ApJ, 280, L31 (1984)
5. A. McWilliam, G. W. Preston, C. Sneden, L. Searle, AJ, 109, 2757 (1995)
6. T. Tsujimoto, T. Shigeyama, ApJ, 584, L87 (2003)
7. H. Umeda, K. Nomoto, Nature, 422, 871 (2003)

Lead Stars at Low Metallicity:
Observation versus Theory

D. Delaude[1], R. Gallino[1], S. Cristallo[2], and O. Straniero[2]

[1] Dipartimento di Fisica Generale dell'Universitá di Torino, Via P. Giuria 1,
   10125 Torino, Italy
[2] Osservatorio Astronomico di Collurania, Via M. Maggini, 64100 Teramo, Italy

The s process in AGB stars is mainly driven by the $^{13}C(\alpha,n)^{16}O$ reaction. During a third dredge up episode, penetration of a small amount of protons from the envelope into the top layers of the ^{12}C-rich and He-rich zone gives rise to the formation of a so-called ^{13}C pocket [3]. At any given metallicity, a large range of ^{13}C-pocket efficiencies is required for the interpretation of the s-process distributions observed in s-enhanced stars in the Galactic disk [4]. Stellar models predict wide ranges of [hs/Fe], [ls/Fe], [hs/ls], where ls=ls(Y, Zr) represents the first s-peak at neutron magic N = 50 and hs=hs(Ba, La, Nd, Sm) the second s-peak at neutron magic N = 82. With decreasing the metallicity, the s-process fluence tends to accumulate at the termination point of the s process, feeding the double magic nucleus ^{208}Pb. Consequently, a large range of [Pb/Fe] ratios is predicted at low metallicity, depending on the ^{13}C-pocket efficiency. In Fig. 1 AGB predictions for the [Pb/hs] versus [Fe/H] for different ^{13}C-pocket efficiencies are compared with the recent spectroscopy data reported in Table 1. The

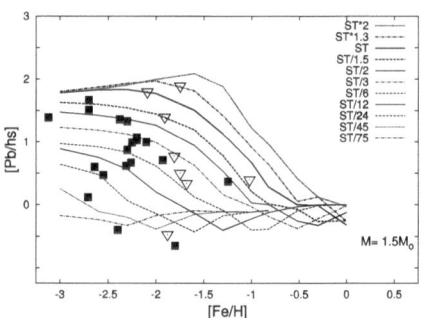

Fig. 1. AGB predictions for [Pb/hs] versus [Fe/H] for different ^{13}C-pocket efficiencies (see the text) as compared with high-resolution spectroscopy data. Data are taken from the references of Table 1 (*full squares*) and from Lucatello (where hs=hs(Ba), these Proceedings) (*empty triangles*).

header symbols of Table 1 are: $M_{\odot}$: initial mass of the adopted model; *pocket*: adopted efficiency of the ^{13}C pocket; $\star$: spectroscopic [Eu/Fe] observed; *dil*: dilution factor. The ST case is from [7], the other cases are obtained by multiplying or dividing the ^{13}C amount of the ST case by the indicated factors. In AGB models, $[Eu/hs]_s \sim -0.7$ is expected. For a few stars, a different ratio is observed within the stated uncertainty. A possibility to overcome this discrepancy is to account for different values of $[Eu/Fe]^{ini}$ in the parental cloud [12]. Grant Italian FIRB Project Astrophysical Origin of the Heavy Elements beyond Fe.

Table 1

Object	[Fe/H]	[Pb/Fe]	[Pb/hs]	ref.	M⊙	pocket	dil	[Eu/Fe]^{ini}
CS 22183-015	-3.12	3.17	1.4	[9]	1.5	ST/1.5	0.65	1.0 ⋆
CS 22880-074	-1.93	1.90	0.7	[2]	1.2	ST/6	0.80	0.5 ⋆
CS 22898-027	-2.26	2.84	0.7	[2]	1.5	ST/3	0.30	1.5 ⋆
CS 29497-030	-2.70	3.55	1.5	[11]	1.3	ST*2	0.40	0.5 ⋆
CS 29526-110	-2.38	3.3	1.4	[2]	1.5	ST/1.5	0.20	1.5 ⋆
CS 30301-015	-2.64	1.7	0.6	[2]	1.3	ST/12	1.25	0.0 ⋆
CS 31062-012	-2.55	2.4	0.5	[2]	1.2	ST/18	0.00	1.5 ⋆
CS 31062-050	-2.31	2.9	0.6	[2]	1.3	ST/6	0.00	1.5 ⋆
LP 625-44	-2.71	2.6	0.1	[1]	1.3	ST/12	0.00	1.5 ⋆
HD 26	-1.25	1.9	0.4	[13]	1.5	ST/2	0.80	0.5
HD 187861	-2.3	3.3	1.3	[13]	1.5	ST/2	0.25	0.5
HD 189711	-1.80	0.7	-0.6	[13]	1.3	ST/24	0.40	0.5
HD 196944	-2.25	1.7	1.0	[2]	1.5	ST/6	1.40	0.0 ⋆
HD 198269	-2.20	2.2	1.1	[13]	1.2	ST/9	0.40	0.5
HD 201626	-2.10	2.4	1.0	[13]	1.5	ST/3	0.70	0.5
HD 224959	-2.2	3.1	1.0	[13]	1.5	ST/2	0.35	0.5
HE 0024-2523	-2.72	3.3	1.7	[10]	1.5	ST	0.50	0.5 ⋆
HE 2148-1247	-2.30	3.12	0.9	[5]	1.5	ST/3	0.15	2.0 ⋆
V Ari	-2.40	1.0	-0.4	[13]	1.2	ST/30	0.05	0.5

References

1. W. Aoki, S.G. Ryan et al.: ApJ **561**, 346 (2001)
2. W. Aoki, S.G. Ryan et al.: ApJ **580**, 1149 (2002)
3. M. Busso, R. Gallino, G.J. Wasserburg: ARA&A **37**, 239 (1999)
4. M. Busso, D. L. Lambert, et al.: ApJ **557**, 802 (2001)
5. J. G. Cohen, N. Christlieb, Y. Z. Qian, G. J. Wasserburg: ApJ **588**, 1082 (2003)
6. D. Delaude, R. Gallino et al.: Mem. Soc. Astron. It. (in press)
7. R. Gallino, C. Arlandini et al.: ApJ **497**, 388 (1998)
8. R. Gallino, D. Delaude et al.: Nucl. Phys. A (in press)
9. J. A. Johnson, M. Bolte: ApJ **579**, L87 (2002)
10. S. Lucatello, R. Gratton et al.: AJ **125**, 875 (2003)
11. T. Sivarani, P. Bonifacio et al.: A&A **413**, 1073 (2004)
12. C. Travaglio, D. Galli et al.: ApJ **521**, 691 (1999)
13. S. Van Eck, S. Goriely, A. Jorissen, B. Plez: A&A **404**, 291 (2003)

Sulphur in the Early Ages of the Galaxy

E. Depagne[1], V. Hill[2], M. Spite[2], P. François[2], F. Spite[2], B. Plez[3],
T.C. Beers[4], B. Barbuy[5], R. Cayrel[2], J. Andersen[6], P. Bonifacio[7],
B. Nordström[6,8], and F. Primas[9]

[1] European Southern Observatory, Chile
[2] Observatoire de Paris-Meudon, France
[3] Université de Montpellier II, France
[4] Michigan State University, USA
[5] Universidade de São Paulo, Brazil
[6] Astronomical Observatory, Denmark
[7] Istituto Nazionale di Astrofisica, Italy
[8] Lund Observatory, Sweden
[9] European Southern Observatory, Germany

Abstract. We present here the results of the measurement of the sulphur abundance in very metal-poor stars. Our sample covers the [-4;-2] range of metallicity, and thus allows us to constraint the chemical evolution models and also to put some key constraints on the supernovae models.

We measured the IR triplet of the sulphur, that is to say the following lines: 9212.863, 9228.093 and 9237.538. The abundance determination leads to the confirmation that sulphur behaves like an α element, and we can confirm that the [S/Fe] ratio reaches a plateau when the metallicity decreases and becomes lower than -2. The value of this plateau is close to $0,4$ dex.

1 Sulphur: α or not α?

Sulphur is generally considered as an *alpha*-element. Thus we expect the sulphur over iron ratio to behave as do other [a/Fe] ratios, that is to say that this ratio reaches a plateau when metallicity goes beyond -1.0. But unfortunately, measuring sulphur abundance can be quite tricky. We can measure S abundance using two different line systems: one system lies around 800nm and the other is around 920 nm. The first system has very weak lines, and the second is blended by atmospheric lines. Thus, the S abundance is difficult to derive from EW measurement.

2 Measures

We have used the infra-red triplet (921.286 nm, 922.809 nm and 923.754 nm) to determine the sulphur abundance in the 32 giants from ESO's Large Program "Galaxy Formation, Early Nucleosynthesis, and First Stars".

The equivalent widths were determined using gaussian fit and the atmospheric models were computed using OSMARCS code improved by [6,3]. When it was not possible to measure equivalent width, the abundance was directly determined by using spectrum synthesis.

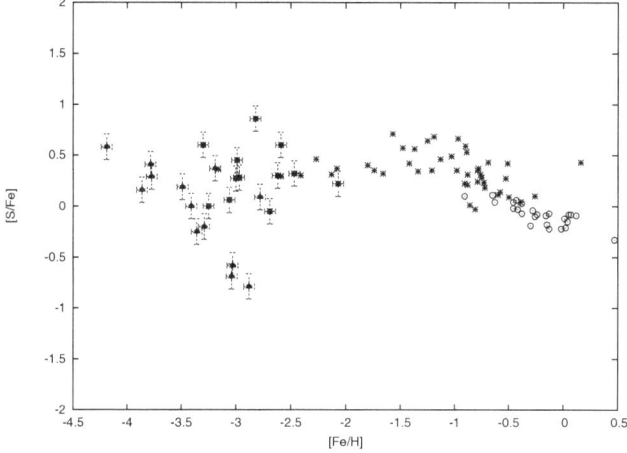

Fig. 1. [S/Fe] ratio versus metallicity. See text for details

3   What's Already Known?

The first study on sulphur in halo stars is from [4]. And since then, the metallicity of the stars in which we could measure sulphur lines has decreased.

It seems that the [S/Fe] ratio follows the behaviour of the other α-elements. But we lack measurements at very low metallicities.

4   Results

Figure 1 shows the global view on sulphur. The filled symbols are our measurements or the upper limits we can determine. The crosses are the measures from Nissen et al. [5] and the circles are the values from Chen et al. [2]

5   Conclusions

Figure 1 shows that the plateau appears when the metallicity is beyond -1.0. But the spread of the [S/Fe] ratio for metallicities lower than -2.5 is quite large. This large spread is quite surprising, if we compare the behaviour of this element to the 17 others presented in [1] and this needs further investigations.

References

1. Cayrel, R., Depagne, E., Spite, M. et al 2004, A&A 416, 1117
2. Chen, Y. Q., Nissen, P. E., Zhao, G., Asplund, M. 2002, A&A 390, 225
3. Edvardsson B., Andersen J., Gustafsson B. et al. 1993 A&A 275, 101
4. François, P. 1986, A&A, 195, 226
5. Nissen, P.E., Chen, Y.Q., Asplund, M., Pettini, M. 2004, A&A 415, 993
6. Plez B., Brett J.M., Nordlund Å. 1992, A&A 256, 551

Star Clusters in the Galactic Anticenter Stellar Structure: New Radial Velocities & Metallicities

P.M. Frinchaboy[1], R.R. Muñoz[1], S.R. Majewski[1], E.D. Friel[2], R.L. Phelps[2,3], and W.B. Kunkel[4]

[1] University of Virginia, P.O. Box 3818, Charlottesville, VA 22903, USA
[2] National Science Foundation, 4201 Wilson Boulevard, Arlington, Virginia 22230
[3] C. S. U., Sacramento, 6000 J Street, Sacramento, CA 95819
[4] Las Campanas Observatory, Casilla 601, La Serena, Chile

Abstract. The Galactic Anticenter Stellar Structure (GASS) has been identified with excess surface densities of field stars in several large area sky surveys, and with an unusual, string-like grouping of star clusters. Some members of the cluster grouping have radial velocities (RVs) consistent with the observed GASS velocity-longitude trend. We provide new RV measurements of stars in six clusters that have been suggested to be associated with the GASS. We show that the RVs of at least four clusters are consistent with the previously measured RV trend for GASS. We also derive spectroscopic metallicities for four clusters, and provide an improved age-metallicity relation for the clusters apparently associated with GASS.

We have found that the outer most open clusters found to date seem to be strung along the Galactic anticenter stellar structure (GASS) [5]. This structure [8], [10], [6], [7], [9], [4] was discovered as an excess of stars beyond the apparent limit of the Galactic disk. Previous work [4] has resulted in a number of parameters of the stream including: (1) a velocity-longitude trend indicating a slightly non-circular orbit, (2) a velocity dispersion smaller than even that of disk stars, (3) a wide metallicity spread from $-1.6 <$ [Fe/H]< -0.4.

Spectra for stars in the clusters Berkeley 29, BH 176, and Saurer 1 were collected with the Hydra Spectrograph and clusters Berkeley 22, BH 144(ESO 096-SC04), and ESO 093-SC08 were collected using the RC Spectrograph on the Blanco 4-meter telescope. The data were reduced using standard IRAF reduction methodology with RVs determined using `fxcor`. The CaII infrared triplet was measured using the index definitions of [1], transformed to the common system derived in [3]. The resulting RVs and metallicities are presented in Table 1. The typical error in the metallicities ($\sigma_{[Fe/H]}$) are ~ 0.3 dex. All clusters measured are shown with the GASS longitude-velocity (l vs. V_{gsr}) trend (Fig. 1a) and members are used to produce an age-metallicity relation (Fig. 1b). We find that the clusters Berkeley 29, Saurer 1, ESO 0903-SC08, and possibly BH 144 & 176, have RVs consistent with being part of the GASS cluster system; however, further work is needed to confirm actual membership in GASS.

We acknowledge funding by NSF grant AST-0307851, NASA/JPL contract 1228235, the David and Lucile Packard Foundation, Robert J. Huskey Travel fellowship, AAS International Travel Grant, and the F.H. Levinson Fund of the Peninsula Community Foundation.

Table 1. GASS Cluster Candidate Radial Velocities and Metallicities

Cluster	l	b	# stars Metallicity(RV)	V_r (km/s)	V_{gsr} (km/s)	[Fe/H]	GASS?
Berkeley 22	199.8	−8.1	4(5)	106 ± 9	+16	−0.97	N
Berkeley 29	198.0	+8.0	8(8)	26 ± 6	−52	−0.62	Y
Berkeley 20	203.5	−17.3	3(3)	78 ± 6	−19	−0.68	(Cal)
Saurer 1	214.3	−6.8	2(4)	98 ± 9	−39	−0.49	Y[1]
Berkeley 39	223.5	+10.1	19(19)	59 ± 4	−103	−0.27	(Cal)
ESO093−SC08	293.5	−4.0	0(1)	86 ± 10	−125	......	Y
BH 144 (ESO096−SC04)	305.3	−3.2	2(3)	40 ± 10	−146	−0.51	?[2]
47 Tuc	305.9	−44.9	5(.)	....	.....	−0.78	(Cal)
BH 176	328.4	−4.3	0(3)	13 ± 5	−100	......	Y[2]

[1] We find Saurer 1 to be a member due to its V_{gsr}; [2] did not correct this RV to V_{gsr} before incorrectly excluding Saurer 1 from the GASS member clusters. [2] It is noted from the spatial distribution of the clusters that GASS should have an elliptical orbit. If this change is made, BH 144 and BH 176, along with NGC5286 and NGC 2808, should fit the corrected $l - v_{GSR}$ trend.

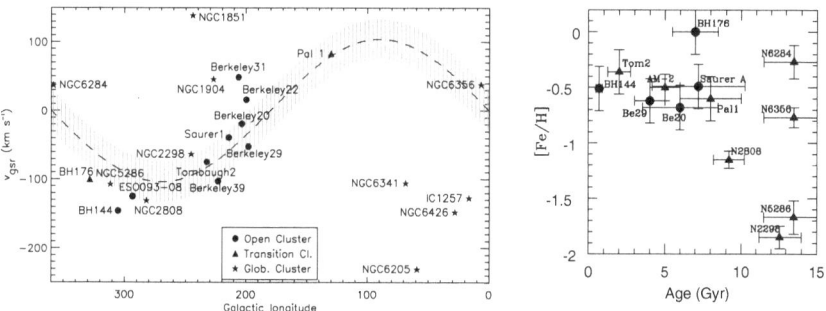

Fig. 1. (a,left) The $l - v_{GSR}$ distribution of objects lying within 2.35 kpc of the GASS cluster plane. The hash marks represent a velocity dispersion of 30 km s^{-1} about the v_{gsr} of a circularly orbiting object at $r_{GC} = 18$ kpc with $v_{circ} = 220$ km s^{-1}, which reasonably matches GASS M giant velocities[4]. (b,right) AMR of all GASS candidate clusters. Circles denote values measured and listed in table, triangles are data from [5].

References

1. Armandroff, Zinn, AJ, 96, 92 (1988)
2. Carraro, Bresolin, Villanova, Matteucci, Patat, Romaniello, (astro-ph/0406679)
3. Cole, Smecker-Hane, Tolstoy, Bosler, Gallagher, MNRAS, 347, 367 (2004)
4. Crane, Majewski, Rocha-Pinto, Frinchaboy, Skrutskie, Law, ApJL, 594, L119 (2003)
5. Frinchaboy, P., et al., ApJL, 602, L21 (2004)
6. Ibata, Irwin, Lewis, Ferguson, Tanvir, MNRAS, 340, L21 (2003)
7. Majewski, Skrutskie, Weinberg, Ostheimer, AJ, 128, 245 (2003)
8. Newberg, H., et al., ApJ, 569, 245 (2002)
9. Rocha-Pinto, Majewski, Skrutskie, Crane, ApJL, 594, L115 (2003)
10. Yanny, B., et al., ApJ, 588, 824 (2003)

Abundance Variations in NGC 288, NGC 362 and NGC 1851

F. Grundahl[1] and H. Bruntt[2,3]

[1] Aarhus University, Institute of Physics and Astronomy, 8000 Aarhus C, Denmark
[2] US Air Force Academy, Department of Physics, Colorado, USA
[3] Astronomisk Observatorium, Copenhagen University, Copenhagen, Denmark

1   Introduction

Strömgren $uvby$ photometry is a powerful tool to study abundance variations in globular clusters through the c_1 and m_1 indices. This is because the NH (3360Å) and CN–bands (4215Å) are present in the u and v filters, respectively. In [1] it was shown that in NGC 6752 the observed star–to–star variations in the c_1 index correlates very strongly with the abundance of nitrogen as measured through the strength of the NH band. As part of a project to study globular clusters with Strömgren photometry we have observed 20 clusters from the Nordic Optical Telescope on La Palma and the Danish 1.54m telescope on La Silla. All clusters in our sample show star–to–star variations in their c_1 index indicating that they all have large variations in their nitrogen abundance. The variations are seen for stars with luminosities ranging from the cluster turnoff to the brightest RGB stars.

We will here discuss results for the three 2'nd parameter clusters NGC 288, NGC 362 and NGC 1851, which have very similar metallicities but very different HB morphologies.

2   Nitrogen Variations – c_1 Scatter

From our Strömgren photometry we find that also NGC 288, NGC 362 and NGC 1851 exhibit the variations in the c_1 index. By drawing histograms of the distribution of c_1 (NH strength) we find that there is a significant difference between the three clusters. In particular, we find that in NGC 288 the distribution has a clear hint of bimodality and is skewed towards low NH strengths, whereas in NGC 362 there is also a bimodality but skewed towards high NH strengths. For NGC 1851 it is not clear what the distribution shows, except that the range in star–to-star scatter in c_1 is slightly larger than for NGC 288 and NGC 362 where it appeared to be of equal magnitude. Curiously NGC 1851 is the only cluster in out sample which shows m_1 scatter, see Fig. 1 for further details.

References

1. F. Grundahl, et al.: Astronomy & Astrophysics, 385, L14, (2002)
2. J. Hesser et al.: AJ, 87, 1470 (1982)

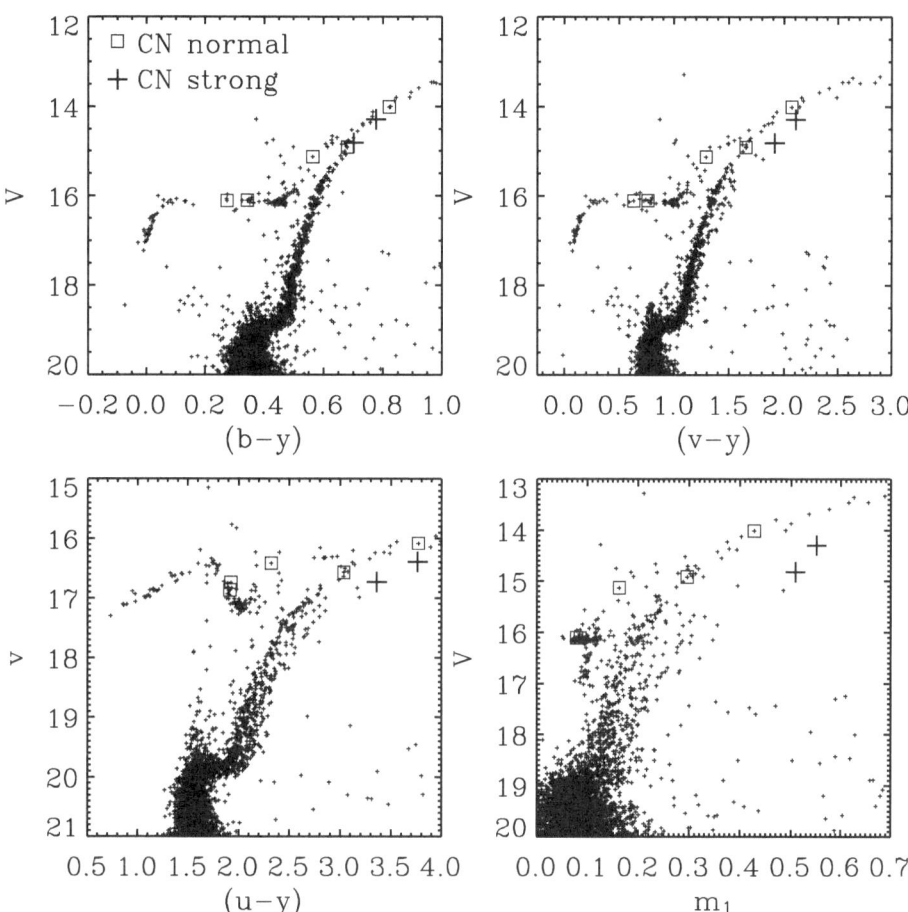

Fig. 1. Various color–magnitude diagrams for NGC 1851 obtained with *uvby* filters at the Danish 1.54m telescope on La Silla. Seven stars in our sample have previous low–resolution spectroscopy from [2] which classified them into CN strong (open squares) and CN normal (plusses) groups. Note how the CN strong stars stand out clearly from the cluster sequences when using filter combinations involving the *u* and *v* filters. We see from the lower righthand panel that the RGB stars in this cluster also show a large scatter in the m_1 index at a fixed luminosity – this is the *only* cluster in our sample of 20 which show m_1 scatter. This points to very large C variations (larger than for other clusters). Could this be related to the bimodality of the cluster horizontal branch?

Heavy Elements and Chemical Enrichment in Globular Clusters

G. James

GEPI – Observatoire de Paris, 61 Avenue de l'Observatoire, F-75014 Paris (France)

Abstract. High resolution ($R \gtrsim 40\,000$) and high S/N spectra have been acquired with UVES on the VLT-Kueyen (Paranal Observatory, ESO Chile) for several MS turnoff stars ($V \sim 17$ mag) and subgiants at the base of the RGB ($V \sim 16$ mag) in three globular clusters (NGC 6397, NGC 6752 and 47 Tuc/NGC 104) at different metallicities (respectively [Fe/H] $\simeq -2.0; -1.5; -0.7$). A sample of 25 field halo subdwarves has also been taken with equal resolution and higher S/N. These data have been used in the framework of the ESO-LP "Globular Cluster Ages, Distances and Metallicities" to determine the abundances of several heavy elements in these three clusters: Sr, Y, Ba, and Eu. These are the first abundance determinations of neutron-capture elements for such an extended sample of scarcely evolved stars. These values together with the [Ba/Eu] and [Sr/Ba] abundance ratios have been used to test the self-enrichment scenario. A comparison is done with field halo stars and other well known globular clusters in which heavy elements have already been measured in the past at least in bright giants ($V \sim 11$–12 mag). Our results show clearly that globular clusters have been uniformly enriched by r– and s–process syntheses, and that even if globular clusters present usually "abundance anomalies" for some of the light metals, most of them exhibit neutron-capture element abundance patterns that are very similar to those of field halo stars at similar metallicities.

1 Introduction

Spectroscopic observations of globular clusters (GCs) have revealed star-to-star inhomogeneities in the light metals that are not observed in field stars. These light metal "anomalies" could be interpreted with a self-pollution scenario. But what about heavier ($Z > 30$) elements? Do they also show "abundance anomalies"? Up to now, no model has been developed for the synthesis of n–capture elements in GCs, and the self-pollution models do not explain the origin of their metallicity. In 1988, Truran suggested a test for the self-enrichment scenario [4], which could possibly explain the metallicity and the heavy metal abundances in GCs: if self-enrichment occurred in GCs, even the most metal-rich clusters would show both high [α/Fe] ratios and r–process dominated heavy elements patterns, which characterize massive star ejecta as it is seen in the most metal-poor stars.

2 Results

More details about the observations and the analysis can be found in James et al. (2004a,b) [1,2]. The main results of this work can be summarized as follows:

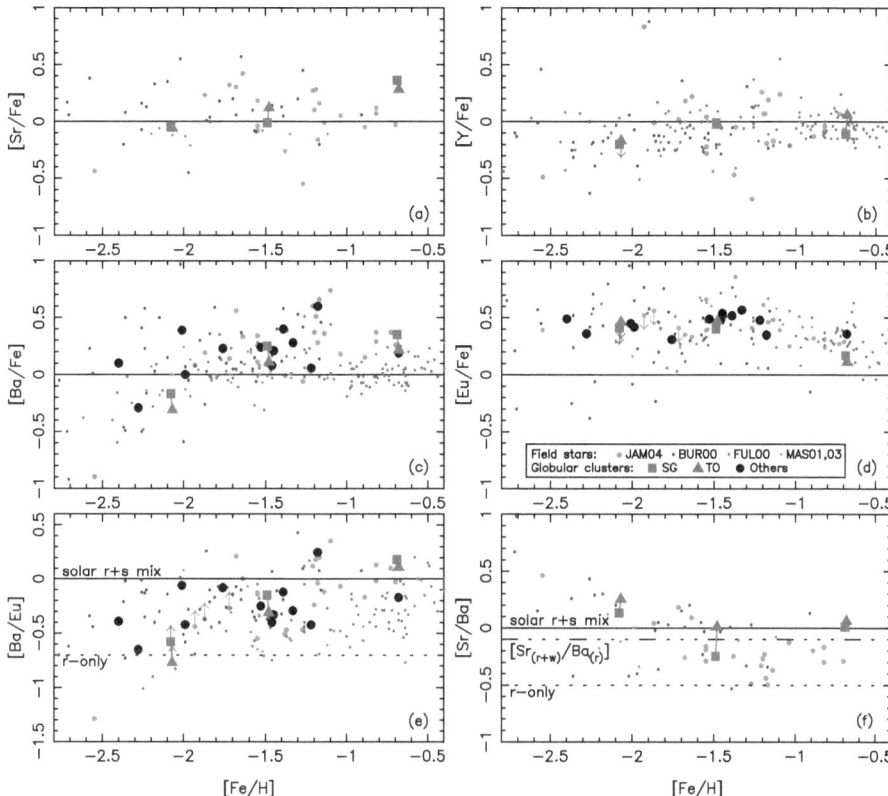

Fig. 1. *(a)* to *(d)*: [X/Fe] mean ratios for the neutron-capture elements in globular clusters and field stars. *(e)* and *(f)*: [Ba/Eu] and [Sr/Ba] ratios as a function of [Fe/H]. *SG* and *TO*: subgiants and TO stars in 47 Tuc, NGC 6752 and NGC 6397.

(1) GCs do not seem to show any (anti-)correlation in the abundance patterns of the *n*–capture elements or between heavy and light metals;

(2) almost all GCs are very homogeneous in the abundances of *n*–capture elements, independently of the evolutionary phase of the observed stars;

(3) the heavy element abundance ratios indicate a progressive chemical enrichment in *s*–process elements for all GCs similar to that of field stars. This pattern is clearly not compatible with a classical self-enrichment scenario [5,3].

References

1. G. James et al.: Astron. Astrophys. **414**, 1071 (2004a)
2. G. James et al.: Astron. Astrophys. (accepted), astro-ph/0408330 (2004b)
3. A. Thoul et al.: Astron. Astrophys. **383**, 491 (2002)
4. J. W. Truran: In *IAU Symp. 132*, p. 577 (1988)
5. J. W. Truran, J. Brown, A. Burkert: *ASP Conf. Ser. 13*, p.78 (1991)

The Effect of Metallicity on Pulsations in B-Type Stars in the Magellanic Clouds

Z. Kołaczkowski and A. Pigulski

Wrocław University Observatory, Poland

1 Introduction

Large and Small Magellanic Clouds (LMC and SMC) are less abundant in heavy elements than the Galaxy ($Z_{LMC} = 0.008$, $Z_{SMC} = 0.004$), but they have relatively well-known distance and small interstellar absorption. These galaxies are among the best objects for the study of dependencies of different astrophysical parameters on metallicity.

In the Galaxy, we know 93 β Cephei (Stankov & Handler 2004) and about 100 SPB-type stars (De Cat et al. 2004). They fall within the instability strips predicted by the theory. The κ-mechanism driving pulsations in β Cephei and SPB stars strongly depends on the abundance of the iron-group ions in the driving zone at temperatures around 2×10^5 K (Dziembowski & Pamyatnykh 1993, Dziembowski et al. 1993). Theoretical models predict that pulsations of β Cephei and SPB-type vanish for $Z = 0.01$ and $Z = 0.006$, respectively (Pamyatnykh 1999).

2 Observations and Analysis

Time-series photometry: OGLE-II photometry in 32 fields in Magellanic Clouds (Soszyński et al. 2002) supplemented, if available, by the MACHO photometry (Allsman & Axelrod 2001) for interesting stars.
Stars selected for analysis: All stars with $(V - I_C) < 0.5$ mag and brighter than 18.5 mag in V from the catalogues of Udalski et al. (1998) and Udalski et al. (2000) for the SMC and LMC, respectively. In total, about 215 000 stars were selected for analysis.
Methods of analysis: Fourier periodograms calculated in the range between 0 and 20 d^{-1}. Automatic search for periodicities with the application of prewhitening and detrending. The photometry of selected objects was examined visually.

3 Results

LMC:

- We found 92 β Cephei and 59 SPB-type stars in the LMC.
- The incidence of β Cephei stars in the LMC is about two orders of magnitude lower than in the Galaxy.

- The median period of the β Cephei sample in the LMC is much longer (0.27 d) than in the Galaxy (0.18 d).
- From the point-of-view of the theory, β Cephei-type stars in the LMC have to have larger-than-average metallicity. This is consistent with the observed periods for $Z \approx 0.0125$.
- A large sample of β Cephei stars in the LMC ($\sim$50%) show both short period(s) that can be attributed to the p-mode pulsations and about twice longer period(s), characteristic for g-mode pulsations.
- For p modes, the observed range of periods in a given star is much smaller in the LMC β Cephei-type stars than in the Galactic ones. This is consistent with the theory.
- Multiperiodic SPB stars were found in the whole range of the V magnitudes in the searched area. Unlike in the Galaxy, these stars populate the region of early B or even late O-type stars.

SMC:

- Four multiperiodic β Cephei and eleven SPB stars were found in the SMC. There are also two monoperiodic short-period variables that are good candidates for β Cephei stars in this galaxy.
- Many multiperiodic variables with periods in the range between 0.25 and 1.5 d were found. They can be identified with the known emission-line objects (Be stars), so that their light variations are of λ Eri-type.

Our analysis allows also to estimate the incidence of β Cephei stars in the LMC and SMC. They constitute of about 0.3% in the LMC, that is an order of magnitude more than in the SMC and about two orders less than in Galactic young clusters.

Acknowledgement: This work was supported by the KBN grant 2P03D 02125.

References

1. Allsman R.A., Axelrod T.S., 2001, astro-ph/0108444
2. De Cat P., Daszyńska-Daszkiewicz J., Briquet M. i in., 2004, A.S.P. Conf. Ser. 310, 195
3. Dziembowski W.A., Pamyatnykh A.A., 1993, MNRAS 262, 204
4. Dziembowski W.A., Moskalik P., Pamyatnykh A.A., 1993, MNRAS 265, 588
5. Meyssonier N., Azzopardi M., 1993, A&AS 102, 451
6. Pamyatnykh A.A., 1999, Acta Astron. 49, 119
7. Soszyński I., Udalski A., Szymański M. i in., 2002, Acta Astron. 52, 369
8. Stankov A., Handler G., 2004, in preparation
9. Udalski A., Szymański M., Kubiak M. i in., 1998, Acta Astron. 48, 147
10. Udalski A., Szymański M., Kubiak M. i in., 2000, Acta Astron. 50, 307

VLT-FLAMES Observations of a Large Sample of Bulge Clump Giants

A. Lecureur[1], V. Hill[1], A. Gómez[1], F. Royer[1], and M. Schultheis[2]

[1] Paris Observatory, France
[2] Besançon Observatory, France

Abstract. We present a progress report of our work on the galactic bulge. It is based on a sample of clump giants in Baade's window with high-resolution spectroscopic data obtained with FLAMES in the framework of the Paris Observatory GTO time. The final aim of this programme is to determine 1) the metallicity distribution, 2) the abundance ratios of α-elements, of r- and s- process elements for a statistically significant sample, 3) the mean kinematic parameters of the bulge populations and their possible correlation with metallicity.

These chemical and kinematical data will help to characterize the bulge populations and provide clues to its formation processes.

1 The Sample

The sample contains 228 stars selected from about 1400 clump-giants observed in the region BUL-SC45 of OGLE-II survey [1]. The field of view covers 25 arcmin in diameter, centered in Baade's window ($l = 0.98$ deg, $b = -3.94$ deg). The stars have V-magnitudes between 16.5 and 17.3, U, B and I photometry and positions from OGLE-II survey and near-infrared photometry from DENIS and 2MASS catalogues. Photometric data in visible and infrared bands will be used to derive effective temperatures to serve as starting point for the abundance analysis of the stars.

2 The Observations

As part of Guaranteed Time Observations programs of the Paris Observatory, the observations were carried out at the ESO/VLT with the FLAMES (Fibre Large Array Multi-Element Spectrograph) instrument. A total of 228 stars, divided into 114 brighter stars ($16.5 < V < 16.9$) and 114 fainter stars ($16.9 < V < 17.2$), has been observed with the medium-high resolution spectrograph GIRAFFE (R $\simeq 30000$) in three settings, HR13 (612-640 nm), HR14 (638-662 nm) and HR15 (660-696 nm). Only the brighter sample has been observed in the last setting due to poor weather conditions. The total exposure time was of 17h25m.

These setups have been chosen to measure abundances of iron peak, α-elements and neutron-capture elements. In parallel, a subset of 14 stars has been observed with the high resolution spectrograph UVES (R=48000) to serve as calibrators for the GIRAFFE sample.

3   Radial Velocities

Data reduction has been performed using the software BLDRS developed at the Geneva Observatory (http://girbldrs.sourceforge.netweb). The radial velocities (RV) have been computing using the Cross Correlation Technique (CCT) with the K-mask. For each star, around ten RV measurements have been obtained from the multiple exposures in the various wavelength domain and through the two sets of Medusa fibers (FLAMES has two swappable focal plates). The dispersion around the mean RV is only of about 0.4 km.s^{-1} ($\sim$1/10th of a pixel at GIRAFFE resolution), showing that the stability of the instrument and the wavelength calibration are of high quality. We could however show that there is a systematic difference of $0.4 \pm 0.1 km.s^{-1}$ between measured RV in setups H13 and H14: the measured RV depend on the observational setup used, due to possible mismatches between the mask and the clump giants spectra.

The sample has been cleaned of 5 suspected binaries and of 6 stars with a $215 < RV < 267$ km.s^{-1} making the assumption that they were members of the globular cluster NGC6528 [2]. The final sample contains 217 stars. The RV distribution is shown in Figure 1. The associated errors are smaller than 0.45 km.s^{-1}, with a mean error value of 150 m.s^{-1}.

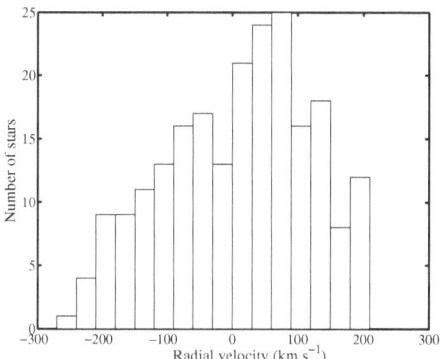

Fig. 1. Radial velocities for 217 stars

With these accurate kinematical data and the future determination of the global metallicity for each star, we plan to study the possible mixing of stellar populations.

References

1. B. Paczynski, A. Udalski, M. Szymanski, M. Kubiak, G. Pietrzynski, I. Soszynski, P. Wozniak, K. Zebrun, 1999, Acta Astronomica, 49, 319-339.
2. P. Coelho, B. Barbuy, M.-N. Perrin, T. Idiart, R. P. Schiavon, S. Ortolani, E. Bica, 2001, A&A, 376, 136-143.

B 12, a Be-Type Star with a Low Nitrogen Abundance in NGC 330

J.-K. Lee[1], D.J. Lennon[2], P.L. Dufton[1], and R.S.I. Ryans[1]

[1] Department of Pure & Applied Physics, The Queen's University of Belfast, BT7 1NN, Northern Ireland, U.K.
[2] The Isaac Newton Group of Telescopes, Apartado de Correos 321, E-38700 Santa Cruz de La Palma, Canary Islands, Spain

Abstract. The chemical composition of B 12, a Be star in the SMC cluster NGC 330, is analysed using high-resolution UVES/VLT spectra and the non-LTE model atmosphere code TLUSTY. A differential analysis relative to a SMC standard star AV 304 revealed (1) a general under-abundance of metals compared with that expected for the SMC, and (2) the lack of nitrogen enhancement. The former is attributed to the presence of a disk, and its contribution to the overall emission is estimated. Possible explanations for the lack of rotational mixing in the apparently rapidly rotating star are discussed.

1 Introduction

Many spectroscopic studies have shown strong nitrogen enrichment in all the B-type giants and supergiants in NGC 330, one of the most studied young clusters in the SMC ([2], [6]). A favoured explanation is that the atmosphere of these stars have been polluted with CN processed material through rotationally induced mixing (e.g. [5]). Here, the role of rotation on the evolution of massive stars is investigated through a detailed abundance analysis of B 12, a rapidly rotating pole-on Be star([7]; [1]) in NGC 330.

2 Data and Abundance Analysis

A striking feature of the B 12 spectrum is that the N II line is extremely weak (Fig. 1), which is peculiar considering nitrogen enrichment shown in other B-type stars in NGC 330. Be stars are generally accepted as B-type stars near the end of the main-sequence. Quantifying the atmospheric nitrogen abundance of B 12 would show to what degree, if any, the nitrogen content has been enhanced by rotational mixing. A non-LTE model atmosphere analysis of B 12 was conducted relative to a B-type SMC standard, AV 304, based on grids calculated using the TLUSTY and SYNSPEC codes (e.g. [3] & references therein).

Two noticeable results are (a) a general under-abundance of all elements relative to AV 304 and (b) no evidence for nitrogen enhancements. For example, compared with AV 304, the under-abundances range from -0.06 dex for C to -0.62 dex for O, with N, Mg and Si showing rather similar under-abundances of -0.35 to -0.45 dex. The detailed results are presented in [4].

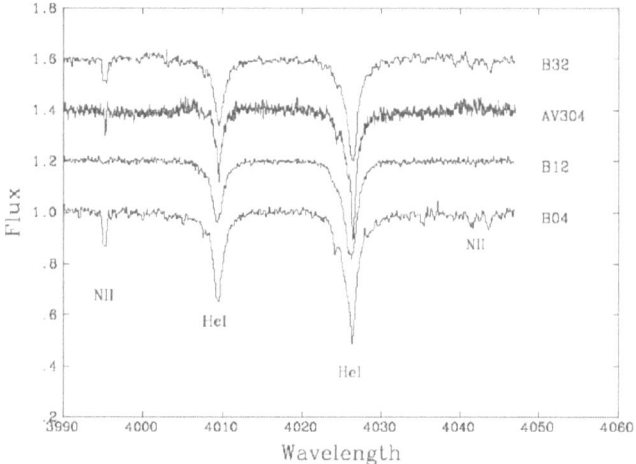

Fig. 1. A comparison of the 3990–4050 Å regions for the nitrogen rich stars B 32 and B 04 in NGC 330, the nitrogen normal star AV 304, and the Be star B 12. The He I lines at 4009 and 4026 Å are marked together with the N II lines at 3995, 4041 and 4043 Å. Note the strength of the N II line at 3994 Å in B 12, which is significantly weaker than in the other stars. The element shows under-abundances of −0.35 relative to AV 304.

3   Discussion

Two likely causes of these under-abundances are suggested by the Be nature of B 12; (i) veiling of the continuum due to a disk, and (ii) departures from standard non-rotating plane parallel models caused by rapid rotation.

The assumption of a disk contribution of 25% to the continuum makes moderate the discrepancies in abundance of B 12 with AV 304. There is a possibility that B 12 is a slow rotator which has 'recently' been spun-up. While detailed discussion can be found elsewhere [4], we speculate that our findings lend support to the idea that magnetic fields may be present in Be stars, inhibiting mixing, and that the star is also rotating at a velocity close to its critical value. This in turn leads to some residual moderate abundance anomalies, an artifact of models not including the effect of von Zeipel gravity darkening. B 12 may thus put important constraints on the current theories of the evolution of rapidly rotating massive stars.

References

1. Baade, D., Rivinius, Th., Stefl, S., Kaufer, A. A&A, **383**, 31 (2002)
2. Dufton P.L., McErlean, N.D., Lennon, D.J., Ryans, R.S.I. A&A, **353**, 311 (2000)
3. Lee, J.-K., Dufton P.L., Ryans, R.S.I., Rolleston W.J.R. A&A, in print (2004)
4. Lennon, D.J., Lee, J.-K., Dufton P.L., Ryans, R.S.I. A&A, submitted (2004)
5. Maeder, A., & Meynet, G. A&A, **373**, 555 (2001)
6. Trundle, C., Lennon, D.J., Puls, J., & Dufton, P.L. A&A, **417**, 217 (2004)
7. Robertson, J.W., A&AS, **15**, 261 (1974)

Chemical Abundances of Three Metal-Poor Globular Clusters in the Inner Halo

J.-W. Lee

Department of Astronomy and Space Science and Astrophysical Research Center for the Structure and Evolution of the Cosmos, Sejong University, 140-747, Seoul, Korea

1 Introduction

Detailed elemental abundance studies of globular clusters may provide strong constraints on the Galaxy formation picture. For example, a metallicity gradient would imply that the Galaxy formed via a slow dissipational process. A constant and enhanced [α/Fe] versus [Fe/H] relation may indicate that the globular clusters must have formed simultaneously within a couple of gigayears (Wyse & Gilmore 1988; Carney 1996), so that their proto-cluster clouds were not contaminated by SNe Ia products. The abundance ratio of r-process elements to s-process elements, such as [Ba/Eu] and [La/Eu], as a function of metallicity in globular cluster systems, may also suggest how rapidly they were polluted by the low- or intermediate-mass stars before they formed. In spite of the importance of the chemical abundance studies, high resolution spectroscopic studies of the globular clusters near the Galactic center have not been performed due to the observational limitations set by large interstellar reddening.

We performed a chemical abundance study of the old inner halo clusters NGC 6287 ($R_{GC} = 1.6$ kpc), NGC 6293 ($R_{GC} = 1.4$ kpc), and NGC 6541 ($R_{GC} = 2.2$ kpc) using the CTIO 4-meter telescope and its Cassegrain echelle spectrograph (Lee et al. 2001, Lee & Carney 2002). Our metallicity estimates for NGC 6287, NGC 6293, and NGC 6541 are [Fe/H] $= -2.00 \pm 0.05$, $-2.01 \pm$ 0.02, and -1.78 ± 0.02, respectively, and our metallicity measurements are in good agreement with previous estimates.

The mean α-element abundances of our program clusters are in good agreement with other globular clusters, confirming the previous results of Carney in 1996. However, the individual α-elements appear to follow different trends. The silicon abundances of the inner halo clusters appear to be enhanced and the titanium abundances appear to be depleted compared to the intermediate halo clusters. In particular, [Si/Ti] ratios appear to be related to the metallicities or Galactocentric distances, in the sense that [Si/Ti] ratios decrease with metallicity and Galactocentric distance. We propose that these [Si/Ti] gradients with metallicity or Galactocentric distance may be due to the different masses of the SNe II progenitors. The high [Si/Ti] ratios toward the Galactic center and at lower metallicities may be due to high-mass SNe II contributions. On the other hand, the abundances of the neutron capture elements may tell a different story. Our program clusters appear to have enhanced s-process elemental abundances,

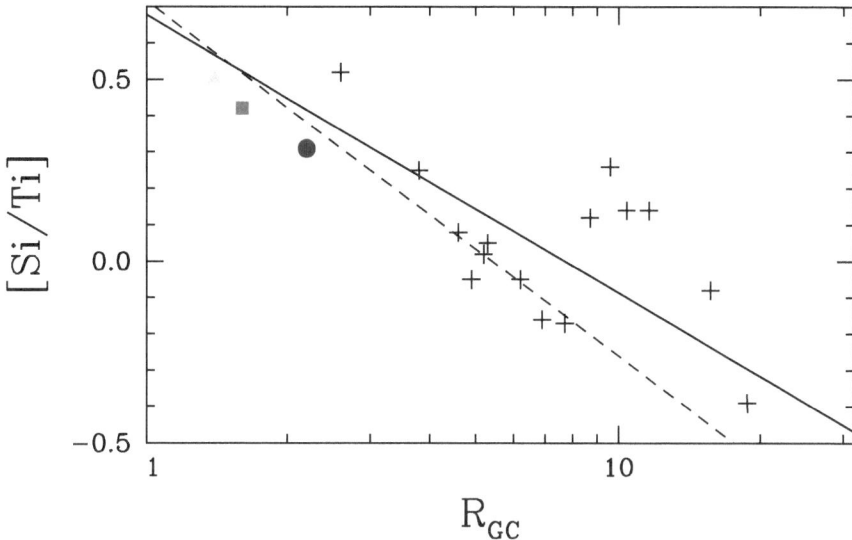

Fig. 1. [Si/Ti] as a function of R_{GC}. Crosses represent "old halo" clusters. NGC 6287 is represented by filled squares, NGC 6293 by filled triangles, and NGC 6541 by filled circles. The solid line for the bisector linear fit to the data and the dotted line represents the bisector linear fit to the clusters with $R_{GC} \leq 8$ kpc (12 clusters).

which are thought to be synthesized in the low- or intermediate-mass AGB stars produced in a longer timescale, $\approx 10^9$ yr.

References

1. B.W. Carney: PASP **108**, 900 (1996)
2. J.-W. Lee, B.W. Carney, L.K. Fullton, P.B. Stetson: AJ, **122**, 3136 (2001)
3. J.-W. Lee, B.W. Carney: AJ, **124**, 1511 (2002)
4. R.F.G. Wyse, G. Gilmore: AJ, **95**, 1404 (1988)

From C-Enhanced, Metal Poor Stars to AGB Nucleosynthesis

S. Lucatello[1], R. Gratton[1], T. Beers[2], and E. Carretta[3]

[1] INAF-Osservatorio Astronomico di Padova, Italy
[2] Michigan State University & JINA
[3] INAF-Osservatorio Astronomico di Bologna, Italy

1   C-Enhanced Metal Poor Stars

The largest to date surveys for metal poor stars (*i.e.* HK survey Beers *et al.* 1992 and HES Christlieb *et al.* 2001) find that as many as ~25% of stars with [Fe/H]≤ -2.5 dex are have [C/Fe]>1 dex (CEMP stars). High resolution studies have revealed that the C-enhancements is accompanied by different abundance patterns, s and/or r-process enrichment, but there are also cases with no n-capture elements overabundance,and with or without extraordinary α elements enhancements. The mechanisms that originate the range of phenomena observed are far from being fully understood.

2   Our Sample

In order to shed light on the different processes which contributed to the abundance pattern observed, we observed 8 CEMP stars with UVES, selected from the HK survey on the sole basis of metallicity, C enhancement and temperature. Our analysis shows that the overwhelming majority of them are CEMP stars with extreme s-process enrichment (CEMP-s). Moreover, combining our data with those present in the literature we showed that *all* CEMP-s stars are members of binary systems (Lucatello *et al.* 2004). Much like classical CH stars, they originate from shell processed material transferred from a now extinct intermediate mass companion during its AGB phase. Therefore, their peculiar composition does not originate from peculiar mixing/nucleosynthetic processes due to their low metal content. CEMP-s are thus an excellent testing ground for low metallicity AGB models.

3   Constraints on AGB Nucleosynthesis at Low Metallicity

The data available so far show no evidence for a dependence of the s-process efficiency (traced by [Pb/Ba]) from metallicity (see Figure 1, left panel). Conversely, the state-of the art models (*e.g.* Busso *et al.* 2001) predict that, as the number of seed nuclei decreases with decreasing metallicity, the path of the s-process shifts more toward the third peak (*e.g.* Pb) with respect to the second peak (*e.g.* Ba). Thus an increase of [Pb/Ba] is expected as the metallicity lowers.

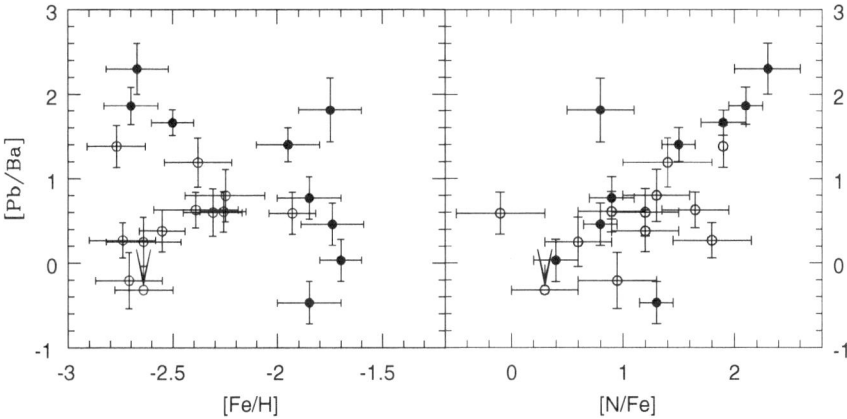

Fig. 1. [Pb/Ba] as a function of metallicity and N content for CEMP-s stars analyzed so far. Open symbols are literature data, closed symbols are program stars

On the other hand, the observations suggest the existence of a correlation between N abundance and [Pb/Ba], as shown in Figure 1, right panel, indicating that the stars in which the s-process is more efficient are those with the highest content of N. This is somewhat surprising, as the models usually consider N as a poison for the s-process, which is thus expected to be inhibited by the presence of large amount of such element. These results might indicate that there is still some crucial ingredient lacking in the low metallicity AGB models.

Another possibility might be that the nucleosynthesis of the donor star is affected by the presence of a close companion. In fact, Herwig *et al.* (2004) recently noted that the most extreme abundance patterns among CEMP-s are found among short period binaries. Hence, the comparison of the CEMP-s with models of *single* AGB star nucleosynthesis might not appropriate.

References

1. Beers, T. C., Preston, G. W., & Shectman, S. A. 1992, AJ, 103, 1987
2. Busso, M., Gallino, R., Lambert, D. L., Travaglio, C., & Smith, V. V. 2001, ApJ, 557, 802
3. Christlieb, N., Green, P. J., Wisotzki, L., & Reimers, D. 2001, A&A, 375, 366
4. Herwig, F. *et al.* 2004 in this proceedings
5. Lucatello, S., Tsangarides, S., Beers, T., Carretta, E., Gratton, R., Ryan, S., 2004, ApJ accepted

The Metallicity Dependence
of the Cepheid Period-Luminosity Relation:
Methodology and Results

M. Mottini[1], M. Romaniello[1], F. Primas[1], M. Groenewegen[2], and G. Bono[3], and P. François[4]

[1] European Southern Observatory, Karl-Schwarzschild-Strasse 2,
   D-85748 Garching b. München, Germany
[2] Instituut voor Sterrenkunde, Celestijnenlaan 200B, B-3001 Leuven, Belgium
[3] INAF - Osservatorio Astronomico di Roma, via Frascati 33,
   I-00040 Monte Porzio Catone, Italy
[4] Observatoire de Paris-Meudon, GEPI, 61 avenue de l'Observatoire, F-75014 Paris,
   France

Abstract. We present the results of an observational campaign undertaken to assess the influence of the iron content on the Cepheid Period-Luminosity relation. Our data indicate that this dependence is not well represented by a simple linear relation. Rather, the behaviour is markedly non monotonic, with the correction peaking at about solar metallicity and declining for higher and lower values of $[Fe/H]$.

Data Analysis and Results

Cepheid stars, through their Period-Luminosity relation, are one of the pillars on which the extragalactic distance scale is built. To this day, however, the debate is still open on the role played by the chemical composition on the pulsational properties of these stars, with different theoretical models and observational results leading to markedly different conclusion (e.g. [1], [2], [4], [5]). To tackle this problem we used high resolution spectra collected with UVES and FEROS .Our sample includes a total of 76 stars: 40 Galactic, 22 LMC, and 14 SMC Cepheids. As first step in our analysis of the chemical composition of Cepheids we focused on the iron content. We developed a robust analysis procedure in order to accurately determine the iron abundance. First, we carefully assembled a reliable list of 263 iron lines (both neutral and ionized) between 480 and 780 nm. We selected the oscillator strength from the VALD database and we visually inspected each line profile on the observed spectra in order to detect and eliminate those lines affected by other elemental blends. Secondly, we measured the equivalent widths of all the lines assembled as described above using the *FITLINE* routine based on Gaussian fit. Finally, our iron abundances were determined using Kurucz's *ATLAS9* model atmospheres and the *WIDTH9* code ([6]).

We have determined the iron abundances for a sub-sample of stars: 13 Galactic, 13 LMC and 12 SMC Cepheids. Our main result is summarised in the Fig.1, where we plot the V-band residuals $\delta(M_V)$ of our stars from the standard PL relation of [3] as a function of the iron abundance we have derived from the

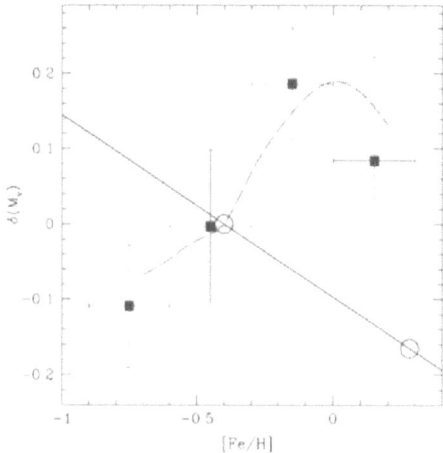

Fig. 1. Result from the subset analysed so far. The data are binned in metallicity to reflect the typical uncertainty on our determination of [Fe/H], marked by horizontal error-bars. The median value of $\delta(M_V)$ in each metallicity bin is plotted as filled squares, with the vertical error-bars representing its associated error. For comparison we also plot the empirical results from [5] in two Cepheid fields in M101 (*open circles and solid line*) and the theoretical predictions by [2] from non-linear pulsational models (*dashed line*)

FEROS and UVES spectra. This standard PL relation was derived for the LMC as a whole $(< [Fe/H] >\cong -0.4)$ and $\delta(M_V)$ is, effectively, the correction to be applied to a "universal" PL relation as a function of metallicity. A positive $\delta(M_V)$ means fainter than the standard PL relation.

Our data indicate that the stars become fainter as metallicity increases, until a plateau or turnover point is reached at about solar metallicity. Our data are incompatible with both no dependence of th PL relation on iron abundance and with the linearly decreasing behaviour often found in the literature (e.g. [5], [8]). On the other hand, non-linear theoretical models of [2] provide a fairly good description of the data. For an in-depth discussion see [7].

References

1. I. Baraffe, Y. Alibert: A&A **371**, 592 (2001)
2. G. Fiorentino, F. Caputo, M. Marconi, I. Musella: ApJ **576**, 402 (2002)
3. W.L. Freedman, B.F. Madore, B.K. Gibson et al: ApJ **553**, 47 (2001)
4. A.M. Fry, B.W. Carney: AJ **113**, 1073 (1997)
5. R.C. Kennicutt, P.B. Stetson, A. Saha, D. Kelson et al: ApJ **498**, 181 (1998)
6. R.L. Kurucz: CD-ROMs #1, 13, 18 (1993)
7. M. Romaniello, F. Primas, M. Mottini, M. Groenewegen, G.Bono, P. François: A&A submitted (2004)
8. S. Sakai, L. Ferrarese, R.C. Kennicutt, A. Saha: ApJ **608**, 42 (2004)

Chemical Abundances of Supersolar Metallicity Stars of Bulgelike Kinematics

L. Pompéia[1], B. Barbuy[1], and M. Grenon[2]

[1] Instituto Astronômico e Geofísico - USP, Rua do Matão, 1226 São Paulo, Brazil
[2] Observatoire de Genéve, Chemin des Maillettes 51, CH-1290 Sauverny, Switzerland

Abstract. In the present work we report abundance ratios for Ca, Ti, Si, Ni, Cr and V for metal-rich stars belonging to the bulge-like sample (e.g. [5]) with metallicities ranging [Fe/H] = 0.0 to +0.55 dex. The bulge-like sample contains very old stars with ages $\sim$ 10 Gyr, and highly eccentric orbits, with probable origin in the inner disk or in the bulge of the Galaxy. Previous works in the same metallicity range indicate an increasing trend with metallicity for Si, Ni and Ti, and a flat trend for Ca, Cr and V. In our sample, iron-peak group element ratios show no trend with metallicity and overlap with that of the disk, while α-elements Si, Ca and Ti show slightly underabundant behavior when compared to thin disk samples.

1 Introduction

Studies of chemical abundances of metal rich samples are still not numerous due to the paucity of stars in the super-solar metallicity range. Nevertheless, some recent works have succeeded to delineate the chemical evolution pattern of important elements, showing interesting behaviors for such domain (e.g. [2], [6]). In the present work we report preliminary abundances for super-solar metallicity bulge-like stars.

2 Chemical Abundances and Discussion

Chemical abundances are inferred from the EW of the lines. Selected lines and atomic data are from our previous paper [5], from [4] and . Stellar parameters were first inferred from Geneva photometry and Hipparcos parallaxes. Then temperatures, microturbulence velocities, gravities and metallicities were iteratively changed in order to i) obey the excitation equilibrium of the Fe I lines; ii) require that Fe I and Fe II abundances agree within 0.1-0.15 dex; and iii) require that Fe I lines with different equivalent widths (EW) give the same iron abundance.

In Fig. 1 the [iron-peak/Fe] and [α/Fe] distributions are depicted. As can be seen from these plots, the chemical distributions of the iron-peak elements for our sample overlap that of the galactic disk. [Cr/Fe] show a slightly different pattern, with smaller ratios for higher metallicities stars when compared to the disk. But such behavior is within the errors. α-elements ratios show an interesting behavior. For all the three α-elements, the bulge-like group shows smaller values than the disk samples while for the disk we see a hint of an increasing trend with metallicity, the bulge-like stars show no apparent trend with metallicity.

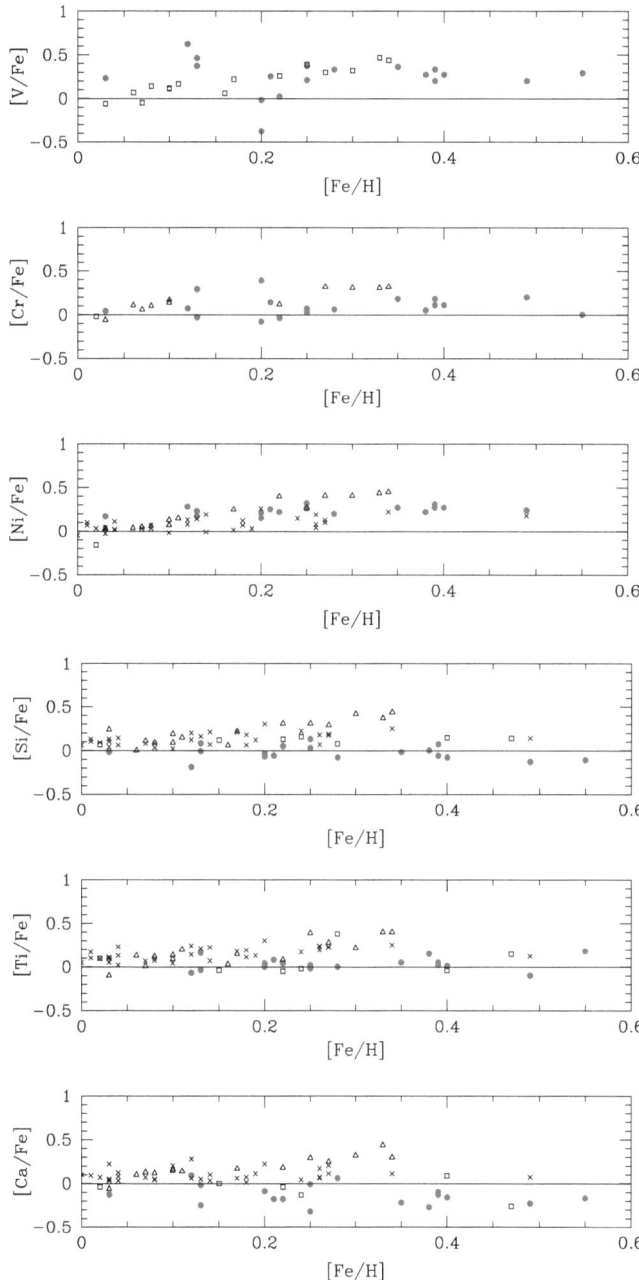

Fig. 1. Abundances distributions for iron-peak elements (upper 3 panels) and for α-elements (lower 3 panels). References for the figures are: filled circles - our sample; open squares - [2] open triangles [6]; crosses - [1].

The behavior for the super-solar domain seems to indicate a higher contribution of the SNe Ia, in the sense that the generations of super-solar metallicities have been more contaminated with SNe Ia yields than those of the disk. This is also consistent with the sub-solar values seen in our sample for the [α/Fe] ratios at [Fe/H] $\sim$ -0.1 dex (see sect. 6.4 of [5]). The present results reinforce the picture of a different identity of the present population compared to that of the galactic disk.

References

1. AllendePrieto,C.; Barklem,P.S.; Lambert,D.L.; Cunha,K. 2004, A&A 420, 183
2. Feltzing, S.; Gonzalez, G 2001, A&A 367, 235
3. Fulbright, J. P. 2000, AJ, 120, 1841
4. Melendez & Barbuy (in preparation)
5. Pompéia, L., Barbuy, B., Grenon, M. 2003, ApJ 592, 1173
6. Thóren, P. & Feltzing, S. 2000, A&A 363, 692

Abundance Anomalies in Hot Horizontal Branch Stars of the Galactic Globular Cluster NGC 1904

A. Recio-Blanco[1], D. Fabbian[2], and R.G. Gratton[3]

[1] Cassiopée UMR 6202, Observatoire de la Côte d'Azur, BP 4229,
   06304 Nice Cedex 4, France
[2] Research School of Astronomy & Astrophysics, Australian National University,
   Mount Stromlo Observatory, Cotter Road, Weston ACT 2611, Australia
[3] Osservatorio Astronomico di Padova, Vicolo dell'Osservatorio 5, 35122 Padova,
   Italy

Abstract. We present abundance measurements, based on high resolution optical spectroscopic data obtained with the UVES at VLT, for 10 stars in the blue horizontal branch (BHB) of the Galactic globular cluster NGC 1904 (M79). In agreement with previous findings for other clusters, we obtain normal abundances for stars cooler than Teff $\sim$ 11000 K, and largely anomalous abundances for hotter stars: large He depletions and overabundances of Fe, Ti, Cr, P and Mn. The abundances of Mg, Si and Ca are roughly normal, in the hot stars as well as in the cooler ones. This abundance pattern can be attributed to the onset of diffusion and to radiation pressure in the stable atmospheres of hot HB stars. A possibly related discontinuity in the stellar rotation rate seems also to occur at Teff $\sim$ 11000 K.

1 Abundance Measurements

The observations were carried out with the UVES spectrograph at VLT, covering the range 3730-5000 Åand a resolution of R = 40000. The targets were identified in the Strömgren u, y photometry by Grundahl et al. (1999) and in the Johnson U, V photometry by Momany et al. (2004, A&A, 420, 605). The selected stars populate the blue HB of NGC 1904 in the temperature range $\sim$ 8000-14000 K.

We have performed the abundance analysis using the programme WIDTH3, developed by R.G. Gratton and adapted by D. Fabbian to temperatures up to Teff 20000 K. The adopted grid of model atmospheres is the one by Kurucz (1994) . The line list was extracted from Moore et. al. (1966), Hambly et al. (1997) and Kurucz & Bell (1995). Laboratory oscillator strengths were considered whenever possible. Equivalent Widths (EW) were measured on the unidimensional, extracted spectral orders using an automatic routine within the ISA package, prepared by R. G. Gratton. But for a few exceptions, only lines with 10< EW <100 mÅ were considered.

Effective temperature values were determined from the position of the stars in the u, (u-y) colour-magnitude diagram (CMD) by Grundahl et al. (1999) and the position in a Johnson U, (U-V) CMD by Momany et al. (2004). The zero age horizontal branch models by Cassisi et al. (1999, A&AS, 134, 103) were used. A mean relation between log (Teff) and log g was derived from Behr et al. (1999) measurements of BHB stars in M13, and applied to our very similar targets in NGC 1904. In the same way, for the cooler stars, another relation between log (Teff) and the micro-turbulence velocity was determined, while we adopted $\xi = 0$ km/s for the warmer stars, as given by the analysis of the star 392, which has the best spectrum. Metallicities were obtained varying the metal abundance [A/H] of the model until it was close to the derived [Fe/H] value.

2   Results and Conclusions

Table 1 contains the abundance values derived for each of our target stars relative to the solar abundances of Grevesse & Sauval (1998 AJ, 85, 161).

Remarkable enhancements of iron and other metal species are found among the blue HB stars hotter than about 11000 K. In addition, He abundances appear below the expected solar He/H ratio, again for stars with Teff > 11000 K. These abundance anomalies in NGC 1904 are likely to be due to the same diffusion processes that were invoked by Behr et al. (1999) to explain their measurements in M13: radiative levitation of metals and gravitational settling of He in the stable non-convective atmospheres of the hotter, higher gravity stars, as hypothesized by Michaud et al. (1983). Interestingly, the onset of radiative levitation coincides with a stellar rotation discontinuity, as reported by Recio-Blanco et al. (2002). New measurements in globular clusters with different metallicities are already in course in other to unveil the possible dependences of the abundance anomalies with cluster and stellar parameters.

Table 1. Abundances relative to solar for our target stars in NGC 1904.

star	[He/H]	[Mg/H]	[Si/H]	[P/H]	[Ca/H]	[Ti/H]	[Cr/H]	[Mn/H]	[Fe/H]
209	...	-1.1±0.2	-1.3±0.2	...	-0.6±0.3	-1.3±0.2	-1.3±0.2	< -0.7	-1.48±0.20
281	...	-0.8±0.2	-1.0±0.2	...	...	-1.1±0.2	-0.9±0.3	...	-1.37±0.20
354	+0.3±0.3	-1.3±0.2	-1.1±0.2	...	-0.9±0.3	-0.7±0.2	...	...	-1.17±0.20
389	...	-1.0±0.2	-0.2±0.3	+1.9±0.3	-0.7±0.3	-0.1±0.2	...	...	+0.39±0.20
363	< -1.6	-1.3±0.2	...	...	0.0±0.3	+0.3±0.2	...	...	+0.49±0.20
392	-0.1±0.3	-1.2±0.2	-2.3±0.2	+1.3±0.3	-0.6±0.3	-0.2±0.2	-0.8±0.3	+0.8±0.2	+0.42±0.20
434	< -1.7	-1.0±0.2	-1.4±0.2	...	-0.3±0.3	+0.1±0.2	+0.8±0.5	...	+0.39±0.20
469	-1.2±0.3	-0.8±0.2	-0.9±0.3	...	-0.3±0.2	+0.4±0.2	+0.1±0.5	+1.1±0.3	+0.60±0.20
535	-1.2±0.3	-1.3±0.2	-0.3±0.2	...	-0.5±0.3	-0.4±0.2	...	...	+0.39±0.20
489	-1.2±0.3	-0.9±0.2	-0.4±0.2	+1.2±0.3	-0.1±0.3	+0.6±0.2	+1.7±0.3	...	+0.36±0.20

References

1. Behr, B.B., Cohen, J. G., McCarthy, J.K., & Djorgovski, S.G. 1999, ApJL, 517, L31
2. Grundahl, F. et al. 1999, ApJ, 524, 242

3. Michaud, G., Vauclair, G. & Vauclair, S. 1983, ApJ, 267, 256
4. Momany, Y. et al., 2004, A&A, 420, 605
5. Recio-Blanco, A., Piotto, G., Aparicio, A., & Renzini, A. 2002, ApJ, 572, L7

Spectroscopy of Blue Stragglers and Turnoff Stars in NGC 2506 with FLAMES

F. Royer[1,2] and A.E. Gómez[3]

[1] European Southern Observatory,
   D-85748 Garching bei München, Germany
[2] Observatoire de Genève,
   CH-1290 Sauverny, Switzerland
[3] GEPI – Observatoire de Paris-Meudon,
   F-92195 Meudon, France

Abstract. We present preliminary results of lithium abundances in turnoff stars in the open cluster NGC 2506. Some fifty turnoff stars and a few blue stragglers have been observed using the FLAMES facility on the VLT, during half a night from the French Guaranteed Time on this instrument.

The lithium abundance, using the Li I 6707 doublet, will be used as a signature of internal mixing to confirm whether or not internal mixing is one of the mechanisms responsible for the existence of blue straggler stars.

1 Introduction

The formation mechanisms that can explain the existence of blue stragglers are still very unclear. The different explanations involve mass-transfer in binary systems, or internal mixing.

[8] and [6] suggested that the blue stragglers were stars undergoing a mechanism of internal mixing produced by a fast rotation or an intense magnetic field. This mixing would bring additional hydrogen in the core long after the exhaustion of the fuel and would then extend the lifetime of the star on the main sequence.

We proposed to observe the open cluster NGC 2506 which contains some identified blue stragglers ([1]), and to both investigate blue stragglers and turnoff stars, in the aim of getting new hints about the mechanisms involved in the formation of blue stragglers. Lithium abundance (Li I 6707.8 Å) can be used as a signature of internal mixing ([5]), and will allow us to confirm whether or not internal mixing is involved in the formation of blue stragglers.

Moreover the proportion of turnoff stars supposed to be future blue stragglers (i.e. the subturnoff-mass blue stragglers) would give an independent estimate of the ratio found by [8]: the number of blue stragglers to the number of giant stars.

2 Observations

[1] list 12 blue stragglers for this cluster in their catalog. Reddening, distance and age of the cluster are given by [4]: $E(B-V) = 0$–0.07, $(m-M)_0 = 12.6$, $\tau = 1.5$–2.2 Gyr.

Using FLAMES we observed, 45 turnoff stars, 7 giant stars with GIRAFFE (LR2 from 396.4 to 456.7 nm, $R \sim 6400$; and H15 from 660.6 to 696.5 nm, $R \sim 19300$). At the same time, 3 stars known as blue stragglers ([1]) were observed by UVES-link: $R = 47000$, and a 200 nm wide spectral range centered around 580 nm.

The target stars are plotted in the Hertzsprung-Russel diagram of NGC 2506 in Fig. 1. We propose to detect progenitors of blue stragglers among turn-off stars, by measuring lithium abundance to diagnose whether or not internal mixing is occurring in some of these stars, which will stay longer in the main sequence while other ones will evolve as giants. For blue straggler stars, brighter than the turnoff, UVES spectra will allow a detailed abundance determination. This work

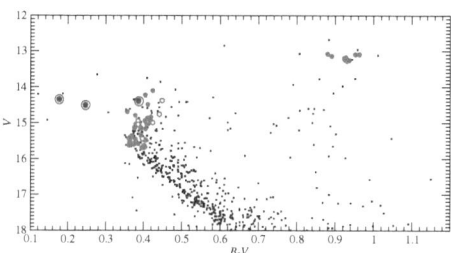

Fig. 1. HR diagram of NGC 2506: crosses are the data from [4]. The targets observed with FLAMES-GIRAFFE are indicated by circles (open circles are the outliers in the radial velocities distribution), and the blue stragglers observed with FLAMES-UVES are indicated by double circles.

is still very preliminary. Only the GIRAFFE spectra have been analyzed, and this analysis is in progress.

The radial velocities have been computed with the low resolution set-up (more spectral lines, no telluric line), using a cross-correlation technique. When excluding the seven outliers, the peak in centered at $83.0 \pm 0.4 \, \text{km s}^{-1}$ with a dispersion of $1.9 \pm 0.2 \, \text{km s}^{-1}$. Lithium abundance is being determined using Li I 6707.8 Å. We used the $B - V$ index to determined the T_{eff} ([3]), and the curve of growth from [7] to derive $N(\text{Li})$.

References

1. Ahumada, J. & Lapasset, E., ApJSS **99**, 135 (1995)
2. Gerbaldi, M., Faraggiana, R., Castelli, F., A&AS **111**, 1 (1995)
3. Hauck, B. & Künzli, M., BA **5**, 303 (1996)
4. Marconi, G., Hamilton, D., Tosi, M., Bragaglia, A., MNRAS **291**, 763 (1997)
5. Pritchet, C. J. & Glaspey, J. W., ApJ **373**, 105 (1991)
6. Saio, H. & Wheeler, J. C., ApJ **242**, 1176 (1980)
7. Soderblom, D. R., Jones, B. F., Balachandran, S., Stauffer, J. R., Duncan D. K., Fedele, S. B., Hudon, J. D., AJ **106**, 1059 (1993)
8. Wheeler, J. C., ApJ **234**, 569 (1979)

Chemical Inhomogeneities in the Stellar System ω Centauri

A. Sollima[1,2], E. Pancino[1], F.R. Ferraro[2], and M. Bellazzini[1]

[1] INAF – Osservatorio Astronomico di Bologna, Italy
[2] Dipartimento di Astronomia, Università di Bologna, Italy

Abstract. We present preliminary results of an extensive low and high-resolution ESO-VLT spectroscopic survey of Subgiant stars in the stellar system ω Centauri. Basing on infrared Ca II triplet lines we derived metallicities and radial velocities for more than 110 stars belonging to different stellar populations of the system. The most metal rich component, the SGB-a, appears to have metallicity [Fe/H] $\sim$ -0.5 . Moreover, SGB-a stars have been found to stray from the dynamical behaviour of the bulk population. Such evidence adds new puzzling questions on the formation and the chemical enrichment history of this stellar system.

Chemical Inhomogeneities in ω Centauri

As part of the Ital-FLAMES GTO we performed an extensive survey of a large sample of sub-giant stars belonging to different sub-populations of ω Cen. Using the Fibre Large Array Multi Element Spectrograph (FLAMES) mounted at UT2 of the ESO-VLT at Cerro Paranal (Chile) we obtained a series of high quality (S/N $\sim$ 50) low-resolution (R $\sim$ 8,000) spectra in the infrared spectral range 8200-9400 Å where CaII triplet lines are measurable. In particular: *(i)* 84 SGB stars have been selected in an external region ($\sim$ 10' away from the cluster centre) and *(ii)* 31 stars along the anomalous SGB (SGB-a, Ferraro et al. 2004) in the central region of the cluster. The preliminary metallicity distribution function confirms the metallicity spread observed in previous analysis performed on giant stars. SGB-a stars appear to have a metallicity higher than the bulk population, representing the extreme metal-rich extension of the stellar content of ω Cen. In order to assess the relative age of each population we fit the observed branches with theoretical isochrones having appropriate metallicities: [Fe/H] $\sim$-1.8 and [Fe/H] $\sim$-1.3 for the two main metal poor populations and [Fe/H] $\sim$-0.5 for the SGB-a. This comparison suggest: (a) an age spread of a few Gyrs between the metal poor populations and (b) that SGB-a stars are 1-2 Gyrs older than the main population, if we assume no helium enhancement between these two population. Moreover, From accurate radial velocity determinations we observed a clear trend in the velocity dispersion distribution as a function of metallicity. SGB-a stars do not follow this trend: they show a significantly larger velocity dispersion and seem to be dynamically warm (see Figure 1).

In addition, high-resolution (R $\sim$ 45,000) spectra with high signal to noise (S/N$\geq$100) in the spectral range 4870-6800 Å were obtained for 6 SGB stars

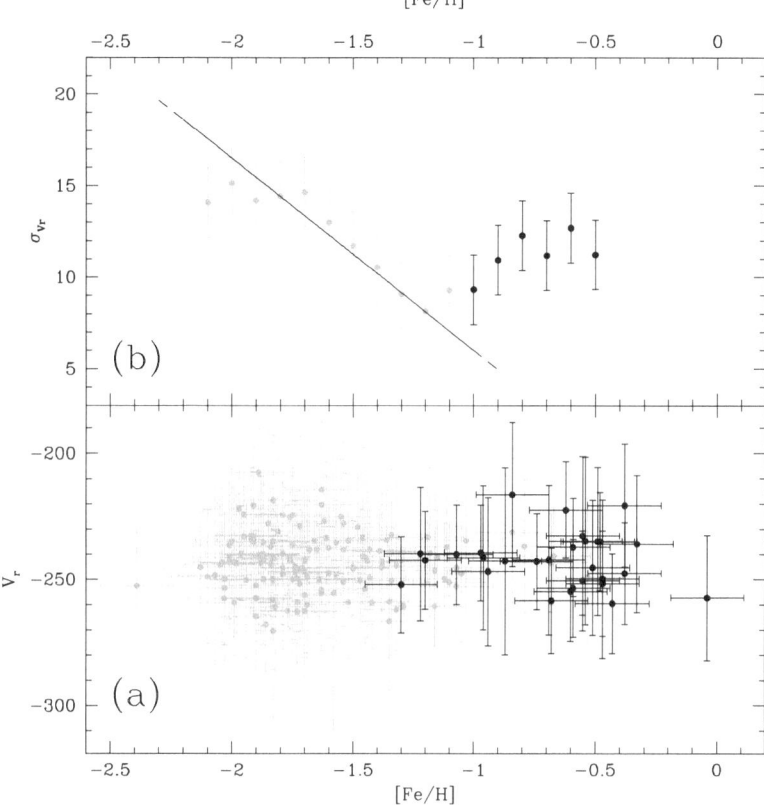

Fig. 1. Radial velocity as a function of metallicity (panel a). The velocity dispersion is plotted against metallicity in panel b. SGB-a stars are marked in both panels with black symbols.

using UVES@VLT. The spectral analysis allows us to obtain accurate determinations of metallicities and α-elements abundances. The preliminary analysis shows clear evidences of metallicity segregation along the SGB.

References

1. Ferraro, F. R., Pancino, E., Bellazzini, M., in "New Horizons in Globular Cluster Astronomy", 2003, ed. G.Piotto, G. Meylan, S.G.Djorgovski & M.Riello, ASP Conf.Series, 296, 215
2. Ferraro, F. R., Sollima, A., Pancino, E., Bellazzini, M., Origlia, L., Straniero, O., Cool, A., 2004, ApJ, 603, L81
3. Norris, J. E., Freeman, K. C., Mighell, K. J., 1996, ApJ, 462, 241
4. Suntzeff, N. B. & Kraft, R. P., 1996, AJ, 111, 1913

High Resolution IR Spectroscopy
of Bulge Globular Clusters

E. Valenti[1,2], L. Origlia[2], and R.M. Rich[3]

[1] Dip. Astronomia Universitá di Bologna, via Ranzani, 1, 40127 Bologna, Italy
[2] INAF – Osservatorio Astronomico di Bologna, via Ranzani, 1, 40127 Bologna, Italy
[3] Dept. of Physics and Astronomy, University of California, Los Angeles,
   CA 90095-1562, (USA)

Abstract. Using the NIRSPEC spectrograph at KeckII we have obtained high resolution echelle spectra in the range $1.5-1.8$ μm of bright giants in Bulge Globular Clusters.
We present our abundances of several metals like Fe, C, O and other α-elements.

The Global Project

In order to understand the history of the Bulge formation and its chemical enrichment scenario, it is of primary importance to study and to compare detailed
abundance patterns of Bulge Globular Clusters (GCs) with those of Bulge field
population. Many of these bulge clusters suffer of huge foreground extinction
as to largely preclude optical studies of any kind, particularly at high spectral
resolution. Within this framework we started a long-term project devoted to
study the abundance patterns of bulge cluster & field populations in the near-IR
spectral domain.

We have obtained near-IR high resolution spectra, with the NIRSPEC (see [5])
spectrograph at KeckII, of several giants in 6 Bulge GCs (namely, NGC 6342,
NGC 6528, NGC 6553, Terzan 4, Terzan 5 and Liller 1). The raw two dimensional
spectra have been processed using the NIRSPEC-IDL pipeline, while we have
removed atmospheric features by using a reference, featureless O-star spectrum.
Stellar temperatures are both estimated from optical/IR colors and molecular
lines, gravity from theoretical evolutionary tracks, according to the location of
the stars on the Red Giant Branch (RGB). The abundances analysis has been
performed by using full spectra synthesis technique (see [7], [8]) and equivalent
width measurements of representative lines: C and O from the CO ($\Delta v=3$) and
OH ($\Delta v=2$) molecular lines, respectively, and Fe, Mg, Si, Ca, Ti from neutral
atomic lines.

Table 1 lists our abundances of Fe, C, O and α-elements for the observed
clusters. An overall excess of α-elements is shared by all the clusters up to solar
metallicity, consistent with SNeII being responsible for the gas enrichment. Our
findings are also in good agreement both with all previous high resolution studies
(see [2], [3], [6]) and with recent abundance determinations for field stars in the
Galactic Bulge (see [10]).

As can be seen in Table 1, all the program clusters also show very low $^{12}C/^{13}C$
isotopic ratios, which cannot be explained by first-dredge up alone, but they

required some extra-mixing mechanisms due to the *cool bottom processing* during the evolution along the RGB (see [1]).

The growing scenario for the Bulge is pointing toward an early formation and a rapid chemical enrichment, as for the Galactic Halo but possibly at higher star formation rate.

Table 1. The observed cluster sample: reddening from [5], metallicities, abundance ratios and heliocentric radial velocities from [7], [8], [9]

Cluster	E(B-V)	[Fe/H]	[α/Fe]	[C/Fe]	$^{12}C/^{13}C$	v_r
NGC 6342	0.46	-0.60	+0.34	-0.34	≤ 5	+114
NGC 6528	0.56	-0.17	+0.33	-0.35	≤ 8	+210
NGC 6553	0.75	-0.30	+0.30	-0.30	≤ 5	-7
Terzan 4	2.35	-1.60	+0.50	-0.30	≤ 5	-53
Terzan 5	2.37	-0.21	+0.30	-0.30	≤ 10	-94
Liller 1	0.00	-0.30	+0.30	-0.30	≤ 5	+63

References

1. A. I. Boothroyd & I. J. Sackmann, 1999, AJ, 96, 588
2. E. Carretta, J. Cohen, R. G. Gratton & B. Behr, 2001, AJ, 122, 1469
3. J. Cohen, R. G. Gratton, B. Behr & E. Carretta, 1999, ApJ, 523, 739
4. W. E. Harris, 1996, AJ, 112, 1487
5. I. McLean, et al. 1998, Proc., SPIE, 3354, 566
6. J. Melendez, B. Barbuy, E. Bica, M. Zoccali, S. Ortolani, A. Renzini & V. Hill, 2003, A&A, 411, 417
7. L. Origlia, R. M. Rich & S. Castro, 2002, AJ, 123, 1559 A&A, 321, 859
8. L. Origlia & R. M. Rich, 2004, AJ, 127, 3422
9. L. Origlia, E. Valenti & R. M. Rich, 2004, MNRAS, astro-ph/0410519, in press
10. R. M. Rich & A. McWilliam, 2000, SPIE, 4005, 150

Part III

Tracing Mixing in Stars

Pre-Main-Sequence Lithium Depletion

R.D. Jeffries

Astrophysics Group, School of Chemistry and Physics, Keele University, Staffordshire, ST5 5BG, UK

Abstract. In this review I briefly discuss the theory of pre-main-sequence (PMS) Li depletion in low-mass ($0.075 < M < 1.2\,M_\odot$) stars and highlight those uncertain parameters which lead to substantial differences in model predictions. I then summarise observations of PMS stars in very young open clusters, clusters that have just reached the ZAMS and briefly highlight recent developments in the observation of Li in very low-mass PMS stars.

1 Introduction

During pre-main-sequence (PMS) evolution, Li is burned at relatively low temperatures (2.5–3.0×10^6 K) and, in low-mass stars ($< 1.2\,M_\odot$), convective mixing can rapidly bring Li-depleted material to the photosphere. For this reason, photospheric Li abundance measurements provide one of the few methods of probing stellar interiors and are a sensitive test of PMS evolutionary models. Understanding PMS Li depletion also offers a route to estimating the ages of young stars and of course is a pre-requisite for quantifying any subsequent main-sequence Li depletion (see Randich 2005, these proceedings).

2 Models of PMS Li Depletion

2.1 Very Low-Mass Stars

PMS stars with $M < 0.35\,M_\odot$ have a simple structure – they are fully convective balls of gas all the way to the ZAMS. As the star contracts along its Hayashi track the core heats up, but the temperature gradient stays very close to adiabatic except in the surface layers. Li begins to burn in p, α reactions when the core temperature, T_c reaches $\simeq 3 \times 10^6$ K and, because the reaction is so temperature sensitive ($\propto T_c^{16-19}$ at typical PMS densities) and convective mixing so very rapid, all the Li is burned in a small fraction of the Kelvin-Helmholtz timescale (see Fig. 1).

The age at which Li depletion occurs increases with decreasing mass (and Li-burning temperatures are never reached for $M < 0.06\,M_\odot$). As luminosity, $L \propto M^2$ for PMS stars, the luminosity at which complete Li depletion takes place is therefore a sensitive function of age between about 10 and 200 Myr [6]. This relationship depends little on ingredients of the PMS models such as the treatments of convection and interior radiative opacities because the stars are fully convective. The extreme temperature dependence means nuclear physics uncertainties play little role, and there is only a small dependence on the kind of atmosphere assumed as a boundary condition, or the adopted equation of state. Indeed, whilst the chosen form of the atmosphere (grey or non-grey) changes the $T_{\rm eff}$ at which Li is burned, it hardly affects the luminosity. Ages determined from the luminosity at the "Li depletion boundary" (LDB) vary by only 10 per cent between different models and even analytical treatments [7].

2.2   Higher Mass Stars

Li depletion is *much* more complex in higher mass stars. They have lower central densities and as T_c rises during PMS contraction, the opacity falls sufficiently for the temperature gradient to become sub-adiabatic. A radiative core forms which pushes outward to include a rapidly increasing fraction of the stellar mass. For $M < 1\,M_\odot$ there is small window of opportunity to burn some Li before the radiative core develops (at $\simeq 2\,{\rm Myr}$ for $1\,M_\odot$). For $M < 0.6\,M_\odot$ all the Li is burned in this way (see Fig. 1). For higher mass stars the radiative core develops before Li burning is complete and the temperature at the base of the convective envelope, T_{bcz}, decreases. In the absence of convective mixing, Li-depleted material cannot get to the photosphere, so once T_{bcz} drops below the Li-burning threshold, photospheric Li-depletion ceases. Photospheric Li depletion begins at about 2 Myr in a $1\,M_\odot$ star and should terminate at about 15 Myr. This window shifts towards older ages in lower mass stars. However, the overall amount of

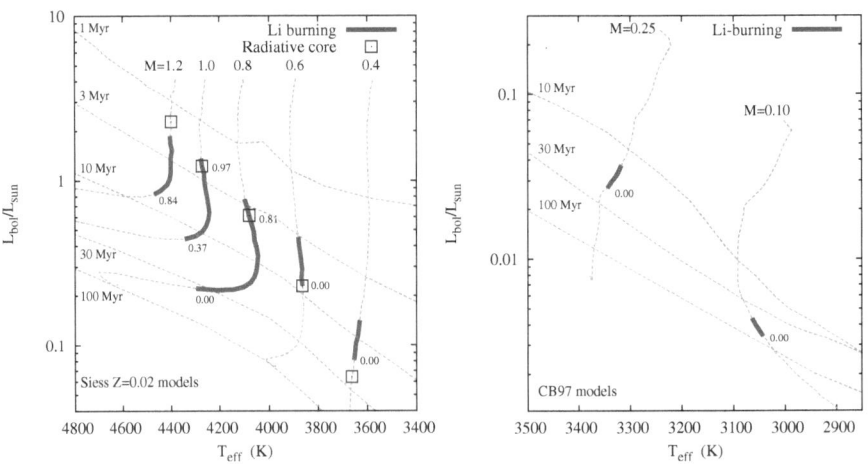

Fig. 1. Evolutionary tracks (labelled in $M_\odot$) and isochrones (in Myr) for low-mass stars taken from two models [8,31]. The epochs of photospheric Li depletion (and hence Li-burning in the core of a fully convective star or at the convection zone base otherwise) and the development of a radiative core are indicated. The numbers to the right of the tracks indicate the fraction of photospheric Li remaining at the point where the radiative core develops and at the end of Li burning.

Li-depletion is extremely sensitive to mass (and other model parameters – see below). There should be relatively little depletion in solar mass stars compared with lower-mass stars (see Fig. 2).

The exact amount of Li depletion expected is exquisitely dependent on a number of model details. The reason is that whilst Li depletion is occurring, even with a radiative core, the overall temperature gradient in the stars is still very close to adiabatic (see Fig. 2 in [26]). It takes only a small perturbation to this gradient to change the time at which the radiative core develops, the position of the convection zone base and hence T_{bcz}. As a result large changes in Li depletion predictions can result from relatively minor perturbations in model parameters. Similarly, because photospheric Li-depletion arises from rapid Li burning in a very thin region above the convection zone base, a model grid with temporal and spatial resolution merely sufficient to model the structure of the star may be an order of magnitude to coarse to accurately predict Li depletion [26].

Convective efficiency is a crucial model parameter. If convection is efficient then T_{bcz} is higher (at a given mass) and hence stays above the Li-burning threshold for longer, resulting in much more photospheric Li depletion [10]. A typical approach to modelling convection is to use mixing length theory with the mixing length set by requiring a model to reproduce the solar structure (revealed by helioseismology) at the age of the Sun. It is not clear that this approach is valid. The mixing length may vary with time, depending on evolutionary stage, surface gravity or effective temperature. Adopting alternate convection theories, such as the full spectrum of turbulence models which have more efficient convection in the deep layers, results in orders of magnitude more PMS Li depletion at the same mass (see [9] and Fig. 2).

Opacity effects are also important. This can refer to differences in the treatment of interior opacities or to the effects of uncertain stellar compositions on the opacities. An increase in opacity makes temperature gradients larger, keeps the star convective for longer, raises T_{bcz} once the radiative core develops and so leads to enhanced Li depletion. Opacity is increased by an increase in overall metallicity or a decrease in the Helium abundance. Changes of only 0.1 dex in metallicity can lead to an order of magnitude change in Li depletion (e.g. see Fig. 2 of [37]).

Other factors, such as the adopted equation of state or chosen treatment of the atmospheric boundary conditions have some effect on Li-depletion predictions, but are much less significant.

3 Observations

There have been more measurements of Li in stars than any other chemical element. The vast majority have been derived from high resolution spectra of the strong Li I 6708Å resonance doublet. Only a fraction of the observational material can be reviewed here. The reader is referred to some other reviews for a more complete picture [16,25].

3.1   The Initial Li Abundance

Theory doesn't tell us what initial Li a star has, only what depletion it suffers. An accurate estimate of the initial Li abundance is therefore a pre-requisite before observations and models can be compared. The Sun is a unique exception, where we know the present abundance, $A(\text{Li}) = 1.1 \pm 0.1$ (where $A(\text{Li}) = \log[N(\text{Li})/N(\text{H})] + 12$) and the initial abundance of $A(\text{Li}) = 3.34$ is obtained from meteorites. For recently born stars, the initial Li abundance is estimated from photospheric measurements in young T-Tauri stars, or from the hotter F stars of slightly older clusters, where *theory suggests* that no Li depletion can yet have taken place. Results vary from $3.0 < A(\text{Li}) < 3.4$, somewhat dependent on assumed atmospheres, NLTE corrections and T_{eff} scales [23,33]. It is of course quite possible that the initial Li, like Fe abundances in the solar neighbourhood, shows some cosmic scatter. Present observations certainly cannot rule this out, leading to about a ± 0.2 dex systematic uncertainty when comparing observations with Li depletion predictions.

3.2   ZAMS Clusters

Clusters that are old enough for stars to have reached the ZAMS empirically show us the results of PMS Li depletion. The canonical dataset is that for the Pleiades (Fig. 2, [32]). With an age of 120 Myr, all stars with $M > 0.5\,M_\odot$ have reached the ZAMS. Assuming an initial $A(\text{Li})$ of 3.2, then there seems to have been little PMS Li depletion among F-stars, ≤ 0.2 dex in G stars and then a strongly increasing level of Li depletion with decreasing mass. There is also evidence for a *scatter* in Li abundances that develops for $T_{\text{eff}} < 5300$ K and probably continues to $T_{\text{eff}} \simeq 4000$ K, where Li becomes undetectable [13].

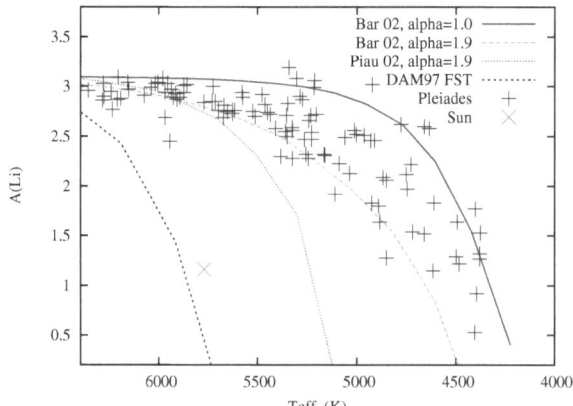

Fig. 2. Measured Li abundances for the Sun and the Pleiades [32] compared with a variety of models [5,9,26]. The majority of the differences are due to the convective efficiency used by the models

Similar results are now available for a number of clusters with ages 50-200 Myr (e.g. [2,14,18,22,29]).

The difference in the Li abundances in the G-stars of the Pleiades and the Sun, combined with the probable similarities in their overall chemical composition tell us that PMS Li depletion cannot be the whole story. Another mechanism, additional to convective mixing, must be responsible for Li depletion whilst solar-type stars are on the main-sequence. Recent PMS models that have their convective treatments tuned to match the structure of the Sun reproduce the mass dependence of Li depletion, but deplete too much Li compared with the Pleiades, and can even explain the solar $A(Li)$ in the case of full spectrum turbulence models [9]. The over-depletion with respect to the Pleiades gets worse at lower masses. Better fits to the Pleiades data are achieved with PMS models that feature relatively inefficient convection with smaller mixing lengths.

3.3 An Li Abundance Scatter?

The apparent scatter among Li abundances in K-type and lower mass stars of the Pleiades and other young clusters is intriguing. It is either telling us something about the physics of mixing and Li-burning inside PMS stars or it is telling us something about the atmospheres of these stars such that we cannot properly estimate their Li abundances. Clues include: the T_{eff} at which the scatter develops, which coincides with those stars that did most of their Li depletion in a fully convective state; and the strong correlation between apparent Li abundance and rotation rate for the K-type stars, such that fast rotators appear to have high $A(Li)$, whereas slower rotators can have either higher or lower $A(Li)$ than average. This correlation may be weaker or absent in the lower mass stars [13,20]. Efforts to understand the apparent Li abundance scatter divide into those that propose a physical mechanism for the Li abundance scatter (i.e. that assume the scatter is real) and those that assume the scatter is not real and instead suggest that the strength of the Li I 6708Å feature does not reliably yield true Li abundances.

Rotationally Driven Mixing: Non-convective mixing can take place in radiative regions, driven by angular momentum loss (AML), and causes additional Li depletion. Fast rotating ZAMS stars have suffered little AML and so would have the highest Li abundances. Slow rotators may have undergone little AML (if they started out with less angular momentum), or lots (if they remained magnetically coupled to a circumstellar disc for an extended period) and so could have a range of Li abundances. Problems with this persuasive picture are that *additional* PMS Li depletion is predicted, widening the disagreement between solar-tuned models and ZAMS clusters and that very little scatter is actually produced in theoretical models even with a realistic range of initial angular momenta [27].

Structural Effects of Rotation: Rapid rotation in a fully convective star decreases the core temperature, but actually increases T_{bcz} once a radiative core has developed. The net effect on Li depletion seems to be rather small and cannot explain the dispersion of Li abundances seen among the slow rotating ZAMS stars [24].

Composition Variations: Li depletion is sensitive to interior opacities, which themselves depend on the stellar composition. Small star-to-star variations might cause an Li abundance scatter, which would grow towards lower masses. However, current limits on metallicity variations in the Pleiades (and other clusters) seem too small for this to be the dominant explanation of any scatter [38]. In addition, the correlation of Li-depletion with rotation is unexplained.

Accretion: Li abundances can be altered in two ways by accretion. During PMS Li depletion the additional mass will lead to increased Li depletion at a given T_{eff} when the star reaches the ZAMS [26]. If accretion occurs after Li-burning has ceased then the convective zone is enriched with Li. Too much accretion is required to be compatible with observations of disks around PMS stars unless the accreted material is H/He-deficient. But then accretion of sufficient H/He depleted material to explain the Li abundance scatter would also lead to (for instance) Fe abundance anomalies of order 0.2-0.3 dex – much higher than allowed by current observational constraints [38].

Magnetic Fields: Low-mass PMS stars are known to be magnetically active. B-fields in the convection zone can provide additional support, raise the adiabatic temperature gradient, hasten the onset of a radiative core and hence decrease Li depletion. Magnetic activity *may* be correlated with rotation in PMS stars at the critical ages of 2-20 Myr but this remains to be established. Basic models including B-fields in the convection zone have now been developed [11,37], suggesting this mechanism could inhibit Li depletion by orders of magnitude!

Atmospheric effects: The atmospheres of PMS stars are doubtless more complicated than the 1-d, homogeneous models usually used to estimate their Li abundances. Starspots and plages complicate the interpretation and could lead to a scatter in the strength of Li I spectral features at a given abundance [3,12]. The 6708Å line is also formed high in the atmosphere and is susceptible to NLTE effects and possible overionisation from an overlying chromosphere [36]. It is telling that the analogous K I resonance line mimics the behaviour of the Li I line, despite there being no possibility of significant K abundance variations [21,28,32]. Varying activity levels could at least be responsible for some of the *apparent* Li abundance scatter. Arguing against this are that very little time variability is seen in the strength of the Li I 6708Å line, despite magnetic/chromospheric activity being quite variable in cool ZAMS stars [15]. In addition, measurements of the weak Li I 6104Å feature, which is probably less susceptible to details of the model atmosphere, have implied a scatter in Li abundances at least as large as that derived from the resonance line [12].

3.4   The Metallicity Dependence of PMS Li Depletion

PMS Li depletion is supposed to be very sensitive to overall metallicity. Groups of ZAMS clusters with similar ages but differing metallicities can be used to test this prediction. The results are surprising. Metallicity variations of 0.1-0.2 dex appear to make no difference to PMS Li depletion [2,17]. An explanation might be that whilst [Fe/H] (what is usually measured as a proxy for metallicity) varies, other elements which are important for interior opacities, especially O, Si, Mg, might vary in the opposite direction to compensate. Quite small differences of 0.1–0.2 dex in [O/Fe] would be required [26], but these differences are still uncomfortably high compared with the spread in [O/Fe] measured for field dwarfs [30]. In addition, it would require a cosmic conspiracy of some proportions to ensure that the half dozen ZAMS clusters investigated so far, all had similar interior opacities. Careful and consistent multi-element abundance determinations are required for these clusters to definitively address the issue.

An interesting aside to this discussion concerns the composition mix assumed in the theoretical models. Recent measurements have suggested that the solar O abundance might be 0.2 dex lower than previously believed [1]. A change of this size in the model compositions could lead to significantly less PMS Li depletion among solar-type stars, reducing the discrepancy between the Li depletion predicted by solar-tuned convective models and the ZAMS cluster data.

3.5 Very Low-Mass Stars

Whilst problems remain in the modelling and interpretation of $0.6 < M < 1.2 M_\odot$ stars, the situation is more favourable in lower mass objects that are always fully convective. In agreement with theory, observations of four young clusters (Pleiades, Alpha Per, IC 2391 and NGC 2547) have now found the sharply defined LDB, where the original undepleted Li abundance is seen in the coolest objects and which marks the age-dependent point at which cores are still too cool to burn Li [4,19,34,35]. Because the LDB is a model-insensitive chronometer, these LDB ages can be used to test the physics which goes into isochronal ages determined from higher-mass stars. The conclusions are that LDB ages are 50 per cent older than nuclear turn-off ages without convective core overshoot, but in reasonable agreement with isochronal ages defined by the descent of low-mass stars to the ZAMS.

4 Summary

The study of PMS Li depletion divides into two regimes. For very low mass stars $0.075 < M < 0.35 M_\odot$, the few extant observations are fully in agreement with available theoretical predictions. Furthermore there is little variation in the predictions of different models and little dependence on uncertain physical processes or parameters. However, models of PMS Li depletion for higher mass stars ($0.35 < M < 1.2 M_\odot$ in which a radiative core develops, make wildly varying (by orders of magnitude) quantitative predictions for Li depletion. None of these models satisfactorily explain all aspects of the data, particularly the presence of an apparent scatter in Li abundances at the end of the PMS phase and the lack of any sensitivity of Li-depletion to stellar metallicity. The model dependence does at least give hope that some aspects of PMS evolution may ultimately be tightly constrained by Li abundance measurements. As an example the current data-model comparisons suggest that PMS convective efficiency is lower than suggested by tuning models to produce the Sun, particularly among cooler stars.

References

1. M. Asplund, N. Grevesse, A. Jacques Sauval: 'The solar chemical composition'. In: *Cosmic abundances as records of stellar evolution and nucleosynthesis*, ed. by F.N. Bash, T.G. Barnes (ASP San Francisco 2005), in press
2. D. Barrado y Navascués, C.P. Deliyannis, J.R. Stauffer: ApJ **549**, 452 (2001)
3. D. Barrado y Navascués, R.J. García-López, G. Severino, M.T. Gomez: A&A **371**, 652 (2001)
4. D. Barrado y Navascués, J.R. Stauffer, R. Jayawardhana: ApJ **614**, 386 (2004)
5. I. Baraffe, G. Chabrier, F. Allard, P.H. Hauschildt: A&A **382**, 563 (2002)
6. L. Bildsten, E.F. Brown, C.D. Matzner, G. Ushomirsky: ApJ **482**, 442 (1997)
7. C.J. Burke, M.H. Pinsonneault, A. Sills: ApJ **604**, 272 (2004)
8. G. Chabrier, I. Baraffe: A&A **327**, 1039 (1997)
9. F. D'Antona, I. Mazzitelli: Mem. Soc. Astr. It. **68**, 807 (1997)
10. F. D'Antona, J. Montalbán: A&A **412** 213 (2003)
11. F. D'Antona, P. Ventura, I. Mazzitelli: ApJ **543**, L77 (2000)
12. A. Ford, R.D. Jeffries, B. Smalley: A&A **391**, 253 (2002)
13. R.J. García-Lopez, R. Rebolo, E.L. Martín: A&A **282**, 518 (1994)
14. M. Hünsch, S. Randich, M. Hempel, C. Weidner, J.H.M.M. Schmitt: A&A **418**, 539 (2004)
15. R.D. Jeffries: MNRAS **304**, 821 (1999)
16. R.D. Jeffries: 'Lithium depletion in open clusters'. In: *Stellar Clusters and Associations: Convection, Rotation, and Dynamos, ASP Proceedings Vol. 198*, ed. by R. Pallavicini, G. Micela, and S. Sciortino (ASP, San Francisco 2000) p.245
17. R.D. Jeffries, D.J. James: ApJ **511**, 218 (1999)
18. R.D. Jeffries, D.J. James, M.R. Thurston: MNRAS **300**, 550 (1998)
19. R.D. Jeffries, J.M. Oliveira: MNRAS, in press (2005)
20. B.F. Jones, M. Shetrone, D. Fischer, D.R. Soderblom: AJ **112**, 186 (1996)
21. J.R. King, A. Krishnamurthi, M.H. Pinsonneault: AJ **119**, 859 (2000)
22. E.L. Martín, D. Montes: A&A **318**, 805 (1997)
23. E.L. Martín, R. Rebolo, A. Magazzù, Ya.V. Pavlenko: A&A **282**, 503 (1994)
24. L.T.S. Mendes, F. D'Antona, I. Mazzitelli: A&A **341**, 174 (1997)
25. R. Pallavicini, S. Randich, J.R. Stauffer, S.C. Balachandran: 'Lithium in young open clusters'. In: *The Light Elements and their Evolution, Proc. IAU Symp. 198*, ed. by L. da Silva, R. de Medeiros, M Spite (2000) p.350
26. L. Piau, S. Turck-Chièze: ApJ **566**, 419 (2002)
27. M.H. Pinsonneault, T.P. Walker, G. Steigman, V.K. Narayanan: ApJ **527**, 180 (1999)
28. S. Randich: A&A **377**, 512 (2001)
29. S. Randich, R. Pallavicini, G. Meola, J.R. Stauffer, S.C. Balachandran: A&A **372**, 862 (2001)
30. B.E. Reddy, J. Tomkin, D.L. Lambert, C. Allende Prieto: MNRAS **340**, 304 (2003)
31. L. Siess, E. Dufour, M. Forestini: A&A **358**, 593 (2000)
32. D.R. Soderblom, B.F. Jones, S. Balachandran, J.R. Stauffer, D.K. Duncan, S.B. Fedele, J.D. Hudon: AJ **106**, 1059 (1993)
33. D.R. Soderblom, J.R. King, L. Siess, B.F. Jones, D. Fischer: AJ **118**, 1301 (1999)
34. J.R. Stauffer, G. Schultz, J.D. Kirkpatrick: ApJ **499**, L199 (1998)
35. J.R. Stauffer et al.: ApJ **527**, 219 (1999)
36. R. Stuik, J.H.M.J. Bruls, R.J. Rutten: A&A **322**, 911 (1997)
37. P. Ventura, A. Zeppieri, I. Mazzitelli, F. D'Antona: A&A **331**, 1011 (1998)
38. B.S. Wilden, B.F. Jones, D.N.C. Lin, D.R. Soderblom: AJ, **124**, 2799 (2002)

Li, Be and B Destruction in Astrophysical Environments: Indirect Cross Section Measurements

R.G. Pizzone[1], C. Spitaleri[1,2], L. Lamia[1,2], S. Cherubini[3], M. La Cognata[1,2], A. Musumarra[1,2], M.G. Pellegriti[1,2], S. Romano[1,2], and A. Tumino[1,2]

[1] INFN - Laboratori Nazionali del Sud, Catania, Italy
[2] Dipartimento di Metodologie Fisiche e Chimiche per l'Ingegneria, Università di Catania, Catania, Italy
[3] Ruhr Universität Bochum, Bochum, Germany

1 Introduction

Big efforts have been devoted in the last years to the study of light elements abundances. Definitively their importance is strongly related to cosmology as well as to stellar structure and evolution. In fact hints on the primordial nucleosynthesis can be achieved from Li, Be and B primordial abundances. Moreover these studies can be a precious tool for testing and understanding the inner stellar structure, especially for what regards the mixing processes in stellar envelopes [1].

In this framework the different nuclear processes which produce or destroy Li, Be and B must be studied in details and an accurate knowledge of the involved nuclear cross sections are necessary. In particular we will focus our attention on one of the main destruction channels for these elements in stellar environments, the (p, α) reactions.

Such cross sections must be measured at energies typical of the different astrophysical scenarios, i.e. at much lower energies with respect to the Coulomb barrier of the interacting particles. This leads to big difficulties in the experimental measurements since the considered cross sections are very low (of the order of picobarn). In order to determine reaction rates within the astrophysical energy range extrapolations from higher energies are usually performed [2].

Even though cross section measurement within the Gamow window were performed in some cases (e.g. the LUNA experiment in Gran Sasso INFN laboratories) an "unexpected" problem arises at such low energies: the electron screening effect. It should be stressed that it is necessary to measure the bare nucleus for any astrophysical application since stellar screening can be quite different from the laboratory one. Thus even when cross sections are measured at very low energies in laboratory, extrapolation must be performed for determining the bare nucleus one.

Among a number of indirect methods which have been suggested for nuclear astrophysics applications the Trojan Horse Method (THM) is particularly suitable in the case of (p,α), (n,α),(n,p) reactions. In the following we will not discuss in details the method, so we refer to [3] for further details.

2   Experimental Results

The THM has been recently applied to several reactions whose cross section is crucial for the study of light element abundance in stellar environments. In particular the reactions $^6Li(p,\alpha)^3He$, $^7Li(p,\alpha)^4He$, $^9Be(p,\alpha)^6Li$ and $^{11}B(p,\alpha)^8Be$ were studied and the corresponding bare nucleus cross sections were measured. An exhaustive discussion of the experimental results is reported in references [4–7] respectively.

In table 1 the results for the bare nucleus S(E)-factor at E=0 are reported together with the results obtained from the direct measurements. The agreement between the two data-sets is good for each case and only for the $^{11}B(p,\alpha)^8Be$ a discrepancy between THM and direct data , not yet fully understood, shows up.

Table 1. Bare nucleus astrophysical S(E)-factor at zero energy for the reactions discussed in the text.

reaction	S(E=0) MeV·b (THM)	S(E=0) MeV·b (direct)	ref.
$^6Li(p,\alpha)^3He$	3.00± 0.19	2.97	[4]
$^7Li(p,\alpha)^4He$	0.055 ± 0.003	0.058	[5]
$^{11}B(p,\alpha_0)^8Be$	0.41 ± 0.09	2.1	[6]

In the $^6Li(p,\alpha)^3He$ and $^7Li(p,\alpha)^4He$ cases, THM results lead to unchanged astrophysical implications [8] with respect to the most commonly adopted reactions rate compilation (e.g. NACRE). Moreover, for an improved study of boron destruction the $^{11}B(p,\alpha_1)^8Be$ cross section will be studied in the next future.

References

1. A.M. Boesgaard et al., APJ, 492, 727, (1998);
2. C. Rolfs & R. Rodney: *Cauldrons in the Cosmos*, Univ. of Chicago press (1988);
3. C. Spitaleri et al., Phys. Rev. C **60**, 055802 (1999);
4. A. Tumino et al., Phys. Rev. C **67**, 065803 (2003);
5. M. Lattuada et al., Ap. J. **562**, 1076 (2001);
6. C. Spitaleri et al., Phys. Rev. C **69**, 055806 (2004);
7. L. Lamia et al., INFN-LNS activity report (2003);
8. R.G. Pizzone et al. Astron. & Astroph. 398, 423(2003).

Mixing on the Main Sequence:
Lithium and Beryllium in Old Open Clusters

S. Randich

INAF – Osservatorio Astrofisico di Arcetri, Largo E. Fermi, 5, I-50125 Firenze, Italy

Abstract. Lithium surveys of very young clusters indicate that solar–type stars undergo very little (if any) Li depletion during the pre-main sequence phases (e.g., [8]); on the other hand, according to standard models of stellar evolution (those including convection only), solar–type stars should not destroy Li while on the main sequence (MS), since the base of their convective zones does not extend deep enough to reach the layers where Li has been burnt. Yet, the Sun has a very low Li abundance, a factor of ~ 100 lower than the meteoritic value, which is representative of the original solar Li content. This evidence of MS Li depletion has motivated theoreticians to introduce non-standard processes in the models (see [4], [15] and references therein); at the same time, a large amount of observational effort has been devoted to Li (and, to some extent, to Be) surveys with the goal of placing empirical constraints on proposed mechanisms. In this paper I will review the current empirical scenario, focusing on the results from old open clusters (ages ≥ 600 Myr).

1 Introduction

As first emphasized by [25], measuring Li in clusters older than the Hyades is a key tool to investigate MS Li depletion and its timescales. Now, more than 30 years later, Li surveys have been carried out for several old clusters including NGC 752, M 67, and the very old NGC 188 ([22], [9], [18] and references therein).

Due to their different burning temperatures (~ 2.5 and 3.5 MK, respectively), the simultaneous determination of Li and Be in the same star allows one to trace mixing to different depths and eventually to put more stringent constraints on internal mixing mechanisms (e.g., [7]). However, measurements of Be are by far more challenging than for Li, since the Be II resonance doublet ($\lambda\lambda = 3130.420$, 3131.064 Å) lies close to the atmospheric UV cutoff; until a few years ago, only field stars and the brightest Hyades had been observed at Be. Now, thanks to the large collecting areas of 8-m class telescopes and the high near-UV efficiencies of the HIRES and UVES spectrographs, additional high quality Be data for the Hyades ([1]) have been obtained, together with Be measurements for Praesepe ([2]), and even for the older, considerably more distant IC 4651 and M 67 ([17]).

In the next sections I will present the main results from these recent Li and Be datasets. In particular, I will address the following questions: **1.** What are the timescales of Li depletion?; **2.** Is the Sun typical? **3.** What parameter drives Li depletion at old ages? **4.** Do stars deplete Be at the same rate as Li? Whereas I will mostly focus on stars with masses close to solar, I mention in passing that light element depletion is also strongly dependent on stellar mass.

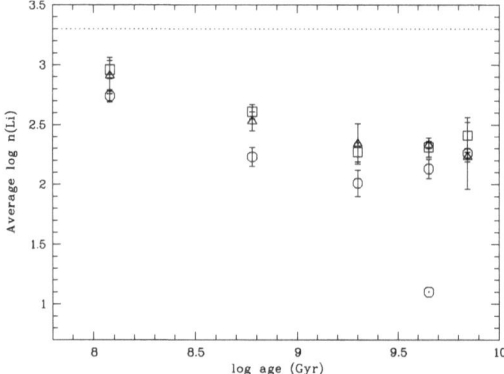

Fig. 1. Mean Li abundance as a function of age. Li abundances are in the usual notation log n(Li)= N(Li)/N(H)+12. Different symbols indicate stars in different mass bins, namely 1 ± 0.02 M$_\odot$ (circles), 1.05 ± 0.02 M$_\odot$ (squares) and 1.1 ± 0.02 M$_\odot$ (triangles). The Sun is also shown. The horizontal line denotes the initial log n(Li). The following clusters have been considered: 120 Myr: Pleiades; 600 Myr: Hyades; 2 Gyr: IC 4651, NGC 3680, NGC 752; 4.5 Gyr: M 67 (only the upper envelope of the Li vs. T$_{\rm eff}$ distribution –see text); 7 Gyr: NGC 188. The cluster samples have all been analysed with the same method. Error bars correspond to 1σ deviations from the mean.

2   Time Scales of Li Depletion

In Fig. 1 I show the average Li abundance as a function of age for stars in three mass bins. Masses have been estimated from effective temperatures using the isochrones of [3]. 1, 1.05 and 1.1 M$_\odot$ correspond to $\sim$ 5650, 5800, and 5950 K at 120 Gyr and to $\sim$ 5750, 5860, and 5960 K at 7 Gyr.

The figure indicates that the evolution of Li from $\sim$ 100 Myr to 6–8 Gyr is similar for the three mass ranges. Log n(Li) values decline at an almost constant rate with log age up to $\sim$ 2 Gyr, where Li depletion virtually stops. Considering the age interval between the Pleiades and the 2 Gyr clusters, stars in the three mass bins undergo a factor of $\sim 4-5$ depletion, resulting in a depletion timescale of the order of $\sim$ 1–1.3 Gyr; after 2 Gyr the timescales become virtually infinite, since no additional depletion seems to occur and Li abundances converge toward a plateau. The exact value of the plateau abundance depends on the mass, but it is surprisingly similar (within a factor of two) to the Spite' s plateau for Pop. II stars ([24]; see also [20]). Whilst this point may be a coincidence –and I will not draw any conclusion from it–, this coincidence is very intriguing and worth of further investigation.

The position of the Sun in the diagram clearly indicates that it is over depleted in Li with respect to stars with similar age and even older (e.g., NGC 188). This result indicates that, as far as Li is concerned, the Sun is not typical, and, more in general, that Li depletion is not a tight function of mass and age. Depending on some additional parameter the non-standard mixing process becomes either ineffective at ages greater than $\sim$ 2 Gyr, or extremely fast.

3   The Star-To-Star Scatter in M 67

The Sun is not an unique exception to the Li vs. age relationship. A large star-to-star scatter in Li is observed both among old field stars ([11], [13]) and, most surprisingly, for virtually identical members of solar age cluster M 67 ([9] and references therein); a fraction ($\sim 40\%$) of stars in this cluster have a Li abundance comparable to the Sun, while the remaining fraction (the upper envelope of M 67) have a factor of about 10 higher Li. No dispersion is observed in the two Gyr old clusters IC 4651, NGC 3680 and NGC 752 ([16] [22]) nor in the old NGC 188 ([18]). Whereas the lack of a dispersion in the intermediate age clusters (and the Hyades) might suggest that the spread develops at ages older than ~ 2 Gyr, the results for NGC 188 instead imply that the appearance of a dispersion depends more on the characteristics of the cluster than on age.

Non membership and/or binarity are not the reasons for the scatter M 67, since both low and high Li stars are confirmed radial velocity members and the spread is still present when considering single stars only. Under the very reasonable assumption that cluster stars were all born with the same Li content, the scatter is therefore intrinsic and due to different amounts of Li depletion.

The originally proposed hypothesis that there might have been more than an episode of star formation within M 67 and that, accordingly, Li-rich/poor cluster stars might represent the young/old population ([10]) can be excluded, since we now know that Li at old ages is not necessarily low (see Fig. 1). The scatter in M 67 indeed reinforces the conclusion that at least one further parameter besides age and mass drives Li depletion, the possible additional parameters being the presence of planets, chemical composition ([10], [14]) and rotation and/or rotational history ([9]).

As discussed by [21], the Li (and Be) distributions of planet-host stars and stars without a planet are very similar; thus the possibility that the presence of a planetary companion might affect Li evolution in M 67 members appears unlikely. As for chemical composition, it does affect stellar opacities and thus internal structure and mixing processes (both standard and non standard ones). For example [14], have shown that even a small (0.05 dex) variation in CNO abundances changes the amount of Li depletion and that a scatter in heavy element abundances could explain the observed scatter in M 67. On the other hand, models including slow rotationally driven mixing (more specifically mixing due to angular momentum loss -AML- and transport -e.g, [5]) predict that stars with different initial rotational velocities undergo different amounts of mixing during the MS. Li–poor M 67 members (and, more in general, Li–poor old stars) would be stars that started the zero age main sequence (ZAMS) as fast rotators and that have then suffered a larger amount of AML and thus Li depletion.

[10] already ruled out that different [Fe/H] values within M 67 could be the reason for the scatter in Li, since they did not find any dichotomy in metallicity between Li-rich and Li-poor stars. [19] confirmed this result and extended it to other elements (mostly α-elements), although they did nor perform CNO measurements that are instead crucial for this issue.

On the other hand, a direct observational proof in support of the AML scenario is not possible, since old stars in M 67 have by now converged to rather low rotational rates and we do not have information on the original rotational velocities. Nevertheless, support (or lack thereof) to the scenario of Li depletion due to AML can be found using at least two different empirical tests; namely, *i)* additional observations of Li in large samples of old cluster stars; *ii)* measurements of Be.

4   Statistics: Additional Old Clusters

If the scatter in Li is due to a dispersion in the initial rotational properties, one would expect that a dispersion is observed in all old clusters, unless they had a different initial distribution of rotation; the latter hypothesis is rather unlikely, due to the fact that very similar distributions of rotational properties are currently derived for young clusters. As mentioned above (see also [23]), neither the three 2 Gyr clusters, nor NGC 188 show a significant star-to-star scatter. Although in all the four clusters fewer stars than in M 67 have been observed, both [18] and [22] provide convincing arguments that the lack of a scatter in these clusters is not due to low number statistics.

FLAMES+ Giraffe on VLT/UT2 is an ideal instrument to increase the statistics of old cluster stars with Li measurements. Two rather large observing projects have been/are being carried out that address the issue of MS Li depletion (plus other topics): the first programme has been discussed by [12]. The second one is a normal GO programme (PI: S. Randich) that includes Giraffe observations of MS stars in six intermediate–age and old open clusters. The two projects together will result in a huge database of Li (and heavy element) abundances for almost 1000 MS stars in nine clusters which will allow a detailed investigation of the origin of the dispersion in Li and, more in general, of mixing mechanisms during the MS. Preliminary analysis of the old cluster Collinder 261 (see again [12]) and of the 2 Gyr metal-rich NGC 6253 indicates that both clusters do show a dispersion in Li abundances.

5   Beryllium

Current models including mixing driven by AML make precise predictions on the Be vs. Li depletion at different ages ([5]). More specifically, according to these models, some Be depletion is expected for solar–type stars at the age of M 67 (while virtually no depletion is predicted at 600 Myr) and Be should correlate with Li; therefore, similar stars with different Li abundances should also differ in their Be content.

In Fig. 2 log n(Be) is plotted as a function of log n(Li) for the Hyades, IC 4651, and M 67. We recently obtained UVES spectra of a new sample of F– and G–type stars in M 67. Be abundances for these stars are shown in the figure together with those of the sample of [17]. As found and discussed by [1], the figure shows a tight correlation between Be and Li abundances for Hyades stars

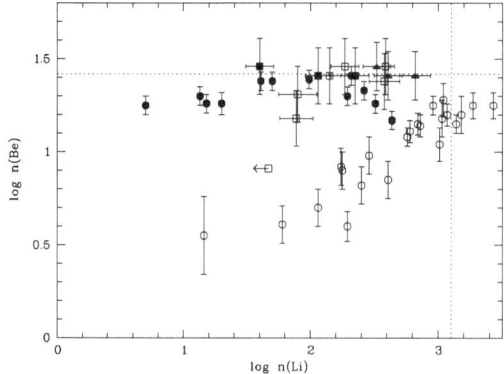

Fig. 2. Be vs. Li abundances for the Hyades (circles –[1]), IC 4651 (triangles –[17]) and M 67 (squares –[17] and new UVES data). Filled and open symbols denote stars with masses below and above 1.1 $M_\odot$, respectively. The horizontal and vertical lines indicate initial Be and Li abundances. Log n(Be) derived by different authors have been put on the same scale.

more massive than 1.1 $M_\odot$ (or warmer than $\sim$ 6000 K). Such a correlation is most evident for stars in the so–called Li–dip and is also observed in Praesepe ([2]) and among field stars ([6]). The M 67 sample includes very few stars more massive than 1.1 $M_\odot$: whereas a Be vs. Li correlation might be present for part of them, others show different amount of Li, but, within the errors, the same Be content. Most important, stars less massive than 1.1 $M_\odot$ in all the three clusters do not show any Be depletion, having Be abundances consistent with the initial value. In particular Li–poor stars in M 67 have the same Be content as Li–rich stars. In other words, neither a dispersion in Be, corresponding to the dispersion in Li, nor a Be vs. Li correlation are present. Stars more massive than 1.1 $M_\odot$ deplete Be faster than lower mass stars and the mixing mechanism responsible for their Li depletion extends deep enough for Be depletion to occur. This is instead not the case for lower mass stars.

6 Conclusions

Whereas we still do not know how the Sun depleted its original Li, observations of old clusters allow one to provide empirical answers to the questions outlined in Sect. 1.

Solar–type stars start destroying Li soon after their arrival on the ZAMS. The difference in the average Li between the Pleiades and the 2 Gyr clusters implies depletion timescales of the order of 1 Gyr. After 2 Gyr they become extremely long for part of the stars (Li depletion virtually stops) or very short for another fraction. The Sun belongs to the second category and, as such, it is not representative of all stars with the same mass, age, and metallicity. Li depletion is clearly not a monotonic function of age and at least one additional parameter besides age and mass seems to drive Li depletion. [Fe/H] and/or chemical composition are most likely not the main parameters responsible for differences in Li depletion in otherwise similar stars, although additional precise determinations of heavy element abundances (in particular CNO) in old clusters are clearly needed. The role of rotation and, more in general of slow mixing driven by AML, are still not empirically confirmed. Both Li–rich and Li–poor solar–type stars in M 67 do not destroy Be and no Be–Li correlation is present for these stars, at variance with current predictions of models including slow rotationally driven mixing. The large dataset of Li abundances for old cluster stars that is coming on line thanks to FLAMES will hopefully provide further insights on the mechanism(s) driving MS Li depletion.

Acknowledgments

This work is supported by a grant COFIN 2002027319 003.

References

1. A.M. Boesgaard, J. King: ApJ **565**, 587 (2002)
2. A.M. Boesgaard, E. Armengaud, J. King: ApJ **605**, 864 (2004)
3. V. Castellani, S. degl'Innocenti, P. Prada Moroni, V. Tordiglione: MNRAS **334**, 193 (2002)
4. C. Charbonnel: 'Rotation and Internal Gravity Waves in Low-Mass Stars'. In: This volume.
5. C.P. Deliyannis, M.H. Pinsonneault: ApJ **498**, L147 (1997)
6. C.P. Deliyannis, et al.: ApJ **498**, L147 (1998)
7. C.P. Deliyannis: ASP Conf. Series **198**, 235 (2000)
8. R.J. Jeffries: 'Pre-main-sequence lithium depletion'. In: This volume.
9. B.F. Jones, et al.: AJ **117**, 330 (1999)
10. R.J. García López, R. Rebolo, J.E. Beckman: PASP **100**, 1489 (1988)
11. R. Pallavicini, M. Cerruti-Sola, D.K. Duncan: A&A **174**, 116 (1987)
12. R. Pallavicini, et al.: 'Multi-object spectroscopy of open clusters with FLAMES: preliminary GTO results'. In: This volume.
13. L. Pasquini, Q. Liu, R. Pallavicini: A&A **287**, 191 (1994)
14. L. Piau, S. Randich, F. Palla: A&A **408**, 1037 (2003)
15. M.H. Pinsonneault: 'Stirring the pot'. In: This volume.
16. S. Randich, L. Pasquini, R. Pallavicini.: A&A **356**, L25 (2000)
17. S. Randich, F. Primas, L. Pasquini, R. Pallavicini: A&A **387**, 222 (2002)
18. S. Randich, P. Sestito, R. Pallavicini: A&A **399**, 133 (2003)
19. S. Randich, et al.: A&A, in preparation (2005)
20. S.G. Ryan: 'The halo lithium plateau: outstanding issues'. In: This volume.
21. N. Santos: 'On the chemical abundances of stars with giant planets'. In: This volume.
22. P. Sestito, S. Randich, R. Pallavicini: A&A **426**, 809 (2004)
23. P. Sestito, S. Randich, R. Pallavicini: 'Lithium and metallicity in the intermediate age open cluster NGC 752'. In: This volume.
24. F. Spite, M. Spite: A&A **115**, 357 (1982)
25. R.R. Zappalá: ApJ **172**, 57 (1972)

Lithium and Metallicity
in the Intermediate Age Open Cluster NGC 752

P. Sestito[1], S. Randich[2], and R. Pallavicini[3]

[1] Dipartimento di Astronomia, Università di Firenze, Largo E. Fermi 5,
    I-50125 Firenze, Italy
[2] INAF/Osservatorio Astrofisico di Arcetri, Largo E. Fermi 5, I-50125 Firenze, Italy
[3] INAF/Osservatorio Astronomico di Palermo, Piazza del Parlamento 1,
    I-90134 Palermo, Italy

Abstract. We have determined Li abundances ($\log n(\mathrm{Li})$) and metallicity ([Fe/H]) in the $\sim$2 Gyr old open cluster NGC 752. The cluster turned out to have a nearly solar Fe content, at variance with previous reports of sub-solar metallicity. The Li distribution vs. effective temperature (T_{eff}) of NGC 752 is very similar to those of IC 4651 and NGC 3680, which have similar age but different [Fe/H]. Moreover, similarly to the other two clusters, NGC 752 does not show a Li scatter as large as that observed in the solar age cluster M 67. In general, the Li vs. T_{eff} distribution does not appear to depend significantly on metallicity, as shown by the comparison of NGC 752 with IC 4651 and NGC 3680; however, a weak dependence on metallicity might be present when comparing the three clusters in the [$\log n(\mathrm{Li})$, mass] plane.

Li and [Fe/H] in NGC 752

Lithium is a very good tracer of mixing processes in stars, but several discrepancies exist between observations and model predictions (see [3]; [8]). Among the most puzzling results, we mention the apparent lack of Li–metallicity dependence (which on the contrary is predicted by models) and the large spread observed in the solar-age solar-metallicity open cluster M 67. We investigated the $\sim$2 Gyr old cluster NGC 752 in order to address these issues; a description of the sample and the analysis of Li (following [13]) and Fe are reported in [11]. We found [Fe/H] = $+0.01 \pm 0.04$ for the cluster, while previous authors reported a sub-solar Fe content (e.g. [2]).

Figure 1a shows the Li distribution of NGC 752 as a function of T_{eff}, compared with other samples: two clusters (data from [9]) with similar age but different metallicity (IC 4651, [Fe/H] = $+0.10$, [7]; NGC 3680, [Fe/H] = -0.17, [6]) and M 67 (data from [4]). NGC 752 is characterized by a rather tight Li vs. T_{eff} distribution, nearly identical to that of the other $\sim$2 Gyr old clusters but at variance with M 67 which shows a large Li spread over the whole T_{eff} range.

From these observational results we draw two main conclusions: *(i)* the lack of Li spread in NGC 752 reinforces the conclusion that M 67 might represent a peculiar sample, since it is the only old open cluster (out of 5) for which a large dispersion in Li abundances has been ascertained until recently (but see below). We mention that also NGC 188, older than M 67, does not show a Li spread ([10],

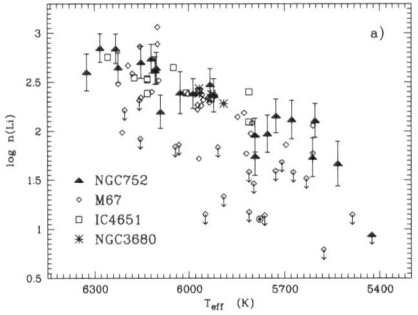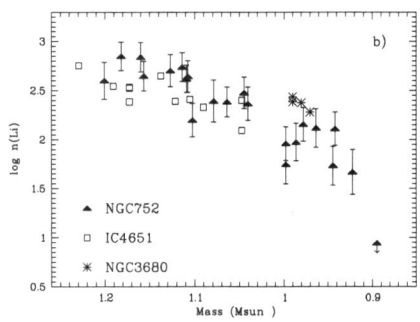

Fig. 1. Li distribution of NGC 752: a) comparison with IC 4651, NGC 3680, M 67 in the $\log n(\mathrm{Li})$ vs. T_{eff} plane; b) comparison with IC 4651 and NGC 3680 in the $\log n(\mathrm{Li})$ vs. mass plane. Stellar masses have been obtained using the isochrones of [1].

[5]). The possible presence of a Li scatter has been reported recently for the ~ 8 Gyr old cluster Cr 261 [5] and for the ~ 3 Gyr old NGC 6253 [8], both observed with FLAMES at the VLT. This makes the general picture of old open clusters even more confusing. Observations of additional samples are clearly required. *(ii)* The metallicity seems not to affect Li depletion, at least when studying Li evolution as a function of T_{eff} (Fig. 1a). However, if we follow the evolution of Li depletion as a function of mass, we find a hint of a possible dependence on [Fe/H], as shown in Fig. 1b; the three Li distributions of NGC 752, NGC 3680 and IC 4651 are in fact slightly shifted systematically one with respect to the other. This result, although intriguing, is based however on a small sample of stars and therefore requires further investigation [12]; in addition, we note that theoretical models predict a much larger effect of the Fe content on Li abundances.

References

1. V. Castellani, S. degl'Innocenti, P.g. Prada Moroni, V. Tordiglione: MNRAS **334**, 193 (2002)
2. S.A. Daniel, D.W. Latham, R.D. Mathieu, B.A. Twarog: PASP **106**, 281 (1994)
3. R.D. Jeffries: 'Pre main sequence mixing: lithium in young open clusters'. This Vol.
4. B.F. Jones, D. Fischer, D.R. Soderblom: AJ **117**, 330 (1999)
5. R. Pallavicini: 'Multi-object spectroscopy of open clusters with FLAMES'. This Vol.
6. L. Pasquini, S. Randich, R. Pallavicini: A&A **374**, 1017 (2001)
7. L. Pasquini, S. Randich, M. Zoccali et al.: A&A **424**, 951 (2004)
8. S. Randich: 'Mixing on the main sequence: lithium and beryllium in old open clusters'. This Vol.
9. S. Randich, L. Pasquini, R. Pallavicini: A&A **356**, L25 (2000)
10. S. Randich, P. Sestito, R. Pallavicini: A&A **339**, 133 (2003)
11. P. Sestito, S. Randich, R. Pallavicini: A&A in press, astro-ph/0407305 (2004)
12. P. Sestito, & S. Randich: 'Li abundance and mixing mechanisms in open clusters'. In: *Cool stars, stellar systems and the Sun 13*, ed. by F. Favata (ESA SP, 2005)
13. D.R. Soderblom, B.F. Jones, S. Balachandran, et al.: AJ **106**, 1059 (1993)

Multi-Object Spectroscopy of Open Clusters with FLAMES: Preliminary GTO Results

R. Pallavicini[1], P. Spanò[2], L. Prisinzano[1], S. Randich[3], and P. Sestito[4]

[1] INAF/Osservatorio Astronomico di Palermo, Piazza del Parlamento 1, I-90134 Palermo, Italy (pallavic@astropa.unipa.it)
[2] DISFA, Università di Palermo and INAF/Osservatorio di Palermo, Italy
[3] INAF/Osservatorio Astrofisico di Arcetri, Largo Fermi 5, I-50125 Firenze, Italy
[4] Dipartimento di Astronomia e Scienza dello Spazio, Università di Firenze, Italy

Abstract. We report on preliminary results of VLT/FLAMES observations of the old open clusters NGC 2506, Mel 66 and Cr 261, obtained as part of our Guaranteed Time on this instrument. We focus in particular on the very old cluster Cr 261, one of the oldest open clusters in the Galaxy. We compare the derived Li abundances with those of other old clusters, and we discuss briefly Li depletion on the main-sequence from the age of the Hyades to $\sim$8 Gyr.

1 The GTO Program on Open Clusters

With the advent of the multi-object spectrograph FLAMES at the VLT [13], the study of open clusters has received a great boost due both to the larger telescope area and to the multiplexing capability offered by the new instrument. Not only it is now possible to reach at intermediate resolution (R$\sim$20,000) fainter objects and more distant clusters, but it is also possible to obtain in one shot spectra of $\sim$130 objects, with an increase in observational efficiency of several hundreds with respect to single-object spectrographs on 4m-class telescopes. Open clusters are among the most suitable targets for observations with FLAMES since they provide homogeneous samples of stars of the same age and chemical composition over projected areas in the sky typically smaller, except for the closest clusters, than the FLAMES field-of-view (25 arcmin diameter).

In return to the Italian participation in the development of FLAMES, the Ital-FLAMES Consortium obtained 11 guaranteed nights that were devoted to various science programs. 3 of these nights were used for observations of open clusters of different ages and metallicities (P.I. R. Pallavicini) with the purpose of investigating Li abundances, metallicities and membership with GIRAFFE, and Fe and α-element abundances with the fiber link to UVES. Our sample includes young clusters such as the Orion Nebula Cluster, NGC 6530 (a few Myr in age) and Blanco 1 ($\sim$100 Myr), all observed also in X-rays with *Chandra* and/or XMM-*Newton*, as well as the old clusters NGC 2506, Mel 66 and Cr 261 (all older than the Hyades). Here we focus on the old clusters which offer a unique opportunity for investigating Li depletion and non-standard mixing mechanisms in dwarf stars during their main-sequence (MS) evolution.

2   Li in Old Clusters

The scientific motivations of our study were reviewed recently by several authors (e.g. [1]; [6]; [20]; [12]). In brief, observations of clusters of different ages and metallicities, carried out mostly with 4m-class telescopes and, to a much less extent so far, with HIRES at Keck and UVES at the VLT, have shown that solar-type stars suffer Li depletion on the MS, contrary to expectations of standard models, i.e. models that assume convection as the only mixing mechanism on the MS. Moreover, while the Hyades (age $\sim$600 Myr) and most clusters older than the Hyades observed so far (e.g. [22]; [18]; [19]; [21]) show a tight Li vs. $T_{\rm eff}$ relationship, with little or no scatter, a large Li spread is observed in the solar-age cluster M 67 ([15]; [7]) as well as in field stars [14]. In particular, the low Li abundance of the present Sun, which is two orders of magnitude lower than the Li abundance of T-Tauri stars and of the meteorites, remains totally unexplained. This feature, together with the presence of a Li spread in M 67 and of Li depletion on the MS, with little or no observed dependence on metallicity ([6]; [21]), indicates the occurrence of non-standard mixing mechanisms (e.g. rotational mixing, diffusion, mass loss, gravitational waves) in addition to convection (e.g. [17]; [1]; and references therein).

As part of our GTO program on FLAMES, we have observed the open clusters NGC 2506 (age $\sim$2 Gyr, [9]), Mel 66 ($\sim$4 Gyr, [8]) and Cr 261 ($\sim$8 Gyr, [4]; [2]). The observations were carried out in May and Dec 2003 (for Cr 261 and Mel 66, respectively) and Feb 2004 (for NGC 2506). Astrometry and photometry were taken from EIS imaging observations [11] for NGC 2506 and Mel 66 and from dedicated WFI observations at the ESO/MPA 2.2m telescope for Cr 261. The data, which comprise more than 100 objects for each cluster, were obtained with the GIRAFFE spectrograph in the MEDUSA mode (R$\sim$20,000) using the grating setting HR15 centered at 679.7 nm (i.e. in the Li region). 7 bright members of each cluster were observed at a higher resolution (R$\sim$50,000) with the fiber-link to UVES. The GIRAFFE spectra were reduced with the standard pipeline and are now being analysed to derive radial velocities, Li abundances, and metallicities for turn-off (TO) and lower MS stars. Preliminary results on Li abundances in Cr 261 are presented here. Results for the other clusters will be reported elsewhere (Spanò et al., in preparation).

3   The Old Open Cluster Cr 261

Cr 261 is one of the oldest open clusters in the Galaxy, with an estimated age ranging from $\sim$7 to $\sim$11 Gyr ([5]; [16]; [4]). The metallicity, derived from both low-resolution and high-resolution spectra, is slightly subsolar ([Fe/H]=-0.2; [3]; [2]) and the reddening is highly uncertain, with quoted $E(B-V)$ values ranging from 0.22 to 0.34 ([5]; [10]; [4]). Proper motion and/or radial velocity memberships are not known, and the contamination by field stars is high. However, the TO and MS are clearly visible in the color-magnitude diagrams with the TO at V$\simeq$16.7, $(B-V)\simeq$0.85, $(V-I)\simeq$0.95 (cf. [4]). We have observed with

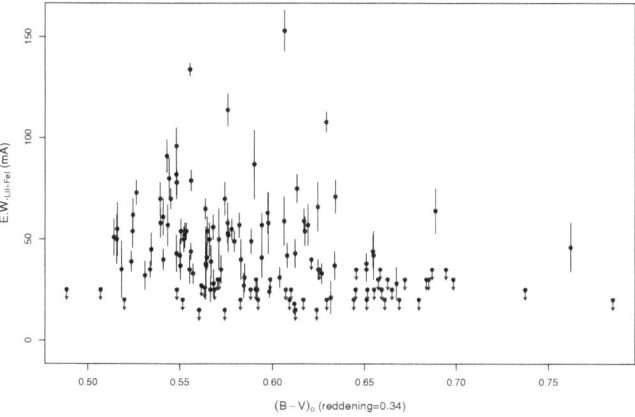

Fig. 1. Li equivalent widths vs. dereddened $(B - V)_o$ color for MS stars of Cr 261.

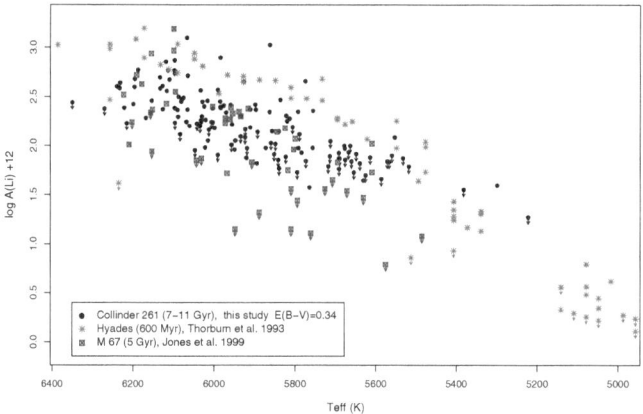

Fig. 2. Li abundances vs. T_{eff} for MS stars in Cr 261 (filled circles), compared to the Hyades (asterisks, [22]) and M 67 (squares, [7]). A reddening $E(B - V)=0.34$ has been assumed.

FLAMES/GIRAFFE stars of magnitudes V~16.5 - 18.5 that have photometry consistent with cluster membership and we have confirmed (or not) the membership by means of radial velocity measurements. This has given us a sample of 138 likely cluster members (out of 220), with a maximum of 9 possible spurious objects.

The Li equivalent widths show a clear scatter as a function of $(B - V)_o$ (Fig. 1). The resulting scatter in Li abundances (Fig. 2) is nearly as large as the one in M 67 except for the fact that the upper limits in Cr 261 are significantly higher than in the latter cluster. The derived maximum Li abundances for solar-type stars are comparable to those measured in the very old ($\sim$6-8 Gyr) cluster NGC 188 [19] and in the upper envelope of the Li distribution of M 67, but are affected by a large systematic uncertainty due to the poorly determined reddening of the cluster. Comparison of the Li vs. T_{eff} distribution of Cr 261 with those of the Hyades and M 67 suggests a high value of the reddening ($E(B-V) \sim 0.34$) close to the highest value found in previous studies. A high-value of the reddening is also required to bring photometric and spectroscopic temperatures into agreement (cf. [2]).

These results, as well as those for NGC 2506 and Mel 66, will be compared with those being acquired in our on-going GO observations with FLAMES (P.I. S. Randich) in order to obtain a better understanding of Li depletion on the MS from the age of the Hyades to $\sim$8 Gyr.

References

1. C.P. Deliyannis, M.H. Pinsonneault, C. Charbonnel: 'Sinks of light elements in stars – Part I'. In: *The Light Elements and their Evolution*, ed. by L. Da Silva, M. Spite, J.R. De Medeiros (ASP, San Francisco 2000) pp. 61–73
2. E.D. Friel, H.R. Jacobson, E. Barrett, L. Fullton, S.C. Balachandran, C.A. Pilachowski: AJ **126**, 2372 (2003)
3. E.D. Friel, K.A. Janes, M. Tavarez, J. Scott, R. Katsanis, J. Lotz, L. Hong, N. Miller: AJ **124**, 2693 (2002)
4. E. Gozzoli, M. Tosi, G. Marconi, A. Bragaglia: MNRAS **283**, 66 (1996)
5. K.A. Janes, R.L. Phelps: AJ **108**, 1773 (1994)
6. R.D. Jeffries: 'Lithium depletion in open clusters'. In: *Stellar Clusters and Associations: Convection, Rotation, and Dynamos*, ed. by R. Pallavicini, G. Micela, S. Sciortino (ASP, San Francisco 2000) pp. 245–254
7. B.F. Jones, D. Fischer, D.R. Soderblom: AJ **117**, 330 (1999)
8. M. Kassis, K.A. Janes, E.D. Friel, R.L. Phelps: AJ **113**, 1723 (1997)
9. G. Marconi, D. Hamilton, M. Tosi, A. Bragaglia: MNRAS **291**, 763 (1997)
10. B. Mazur, W. Krzeminski, J. Kaluzny: MNRAS **273**, 59 (1995)
11. Y. Momany, et al.: A&A **379**, 436 (2001)
12. R.Pallavicini, S. Randich, P. Sestito: 'Lithium abundances in intermediate age and old clusters'. In: *13th Cambridge Workshop on Cool Stars, Stellar Systems, and the Sun*, ed. by F. Favata et al. (ESA, Special Publication), in press (2004)
13. L. Pasquini, et al.: Proc. SPIE **4841**, 1683 (2003)
14. L. Pasquini, Q. Liu, R. Pallavicini: A&A **287**, 191 (1994)
15. L. Pasquini, S. Randich, R. Pallavicini: A&A **325**, 535 (1997)
16. R.L. Phelps, K.A. Janes, K.A. Montgomery: AJ **107**, 1079 (1994)
17. M.H. Pinsonneault: ARA&A **323**, 86 (1997)
18. S. Randich, L. Pasquini, R. Pallavicini: A&A **356**, L25 (2000)
19. S. Randich, P. Sestito, R. Pallavicini: A&A **339**, 133 (2003)
20. S. Randich: 'Mixing on the main sequence: lithium and beryllium in old open clusters'. This volume.
21. P. Sestito, S. Randich, R. Pallavicini: A&A, in press, astro–ph/0407305 (2004)
22. J.A. Thorburn, L.M. Hobbs, C.P. Deliyannis, M.H. Pinsonneault: ApJ **415**, 150 (1993)

The Halo Lithium Plateau: Outstanding Issues

S.G. Ryan[1] and L. Elliott[2]

[1] Centre for Earth, Planetary, Space and Astronomical Research,
Department of Physics and Astronomy, The Open University, Walton Hall,
Milton Keynes, MK7 6AA, United Kingdom; s.g.ryan@open.ac.uk

[2] Department of Mathematics and Statistics, School of Mathematical Sciences,
Monash University, Clayton, Vic. 3800, Australia; Lisa.Elliott@sci.monash.edu.au

Abstract. We examine outstanding issues in the analysis and interpretation of the halo Li plateau. We show that the majority of very Li-poor halo Li-plateau stars (5 out of 8) have high projected rotation velocities $v\sin i$ between 4.7 and 10.4 km s^{-1}. Such stars have very different evolutionary histories to Li-normal plateau stars, and hence cannot be included in studies of Li depletion by normal halo dwarfs. Uncertainties in the effective temperature scale for metal-poor stars continue to challenge the analysis of Li.

1 Introduction

From observations of 11 main-sequence stars belonging to the Galactic halo, Spite & Spite [27] concluded that the lithium abundance was essentially independent of metallicity for halo stars hotter than 5600 K, and inferred that the Li abundance was "hardly altered" from the Big Bang. Two decades of work has followed, increasing the number of stars observed and the range of metallicity that they span, in an effort to establish the primordial Li abundance more securely.

The primordial Li abundance was sought primarily because of its ability to constrain the baryon to photon ratio in the Universe, or equivalently the baryon contribution to the critical density. In this way, Li was able to complement estimates from ^{4}He, the primordial abundance of which varied only slightly with baryon density. Li also made up for the fact that the other primordial isotopes, ^{2}H (i.e. D) and ^{3}He, were at that time difficult to observe and/or interpret. During the late 1990's, however, measurements of D in damped Lyman alpha systems (high column-density gas believed to be related to galaxy discs) provided more reliable constraints on the baryon density than Li could do (e.g. [19]). Even more recently, the baryon density has been inferred from the angular power spectrum of the cosmic microwave background radiation, for example from the WMAP measurements [26]. We consider the role of Li plateau observations post WMAP.

2 Interpretation of the Li Observations

The difficulties in inferring the primordial Li abundance from halo star observations can be separated into two broad categories:

- establishing the current abundance of Li in the stars being analysed, and
- determining the degree to which the current Li abundance of a star deviates from the primordial value due to the production of Li in the interstellar gas from which the star formed, or the destruction of Li in the star subsequent to its formation.

The first of these difficulties arises primarily because of uncertainty in the temperature structure of the stars whose spectra are observed, measured, and interpreted to obtain Li abundance estimates. The problem can in turn be divided into three aspects:

- the uncertain effective temperature scale of metal-poor stars
  Differences as large as 150 to 200 K are frequently noted between the effective temperature scales adopted by different spectroscopists, and such differences can easily include dependences on metallicity and colour (e.g. see calibrations considered by [25], and the final section of the present paper). A change of 100 K corresponds to 0.08 dex in abundance.
- the uncertain temperature gradient in the outer layers of the atmosphere
  Even if two models have the same effective temperature, they may possess different temperature gradients. Since the Li absorption line forms at slightly shallower depths than the range over which the continuum forms, a difference in the temperature gradients in two models means that gas of different temperatures is involved in line formation in each case, resulting in different line strengths. Differences as large as 0.1-0.15 dex have been found from model dependent differences (e.g. [20].)
- the inappropriateness of 1D model atmospheres
  A 1D model atmosphere seeks to represent the variation in the key physical quantities as a function of distance only along a single line of sight. Recently, 3D model atmospheres have been computed for a handful of stars, and allow radiation transfer to be computed along a greater number of lines of sight into a dynamic medium. One triumph of such models is that the microturbulent velocity which must be included in 1D radiative transfer computations to account for non-thermal motions is no longer required in the dynamic 3D atmosphere. The Li abundances derived using 3D LTE models were lower than those from 1D models [3], though NLTE corrections applied to 3D models come close to reproducing 1D LTE results [2]. These early results point the way to a future of more realistic spectral analysis in which 3D modelling becomes the norm.

The second difficulty is to relate the observed abundance to the primordial one [5]. The question in 1985 was whether the primordial value was close to the abundance measured in young Population I stars, implying that Li had been depleted by an order of magnitude in halo stars, or whether Galactic chemical evolution had provided a source of Li that raised the Galactic abundance to the Population I value from a primordial level close to that observed by Spite & Spite. The evidence relied on arguments over the likely mass- (effective temperature-) and metallicity-dependence of depletion mechanisms, and whether a large degree of depletion could in the end bring about a relatively uniform plateau abundance in a range of stars. A slight dependence on effective temperature and metallicity has been claimed [20,25], but not without challenge [7,14]. The elimination of the trends would require the stellar temperatures to be revised by more than the estimated uncertainties, but not by *much* more than the estimated range of uncertainties if we have been unlucky and all errors are pushed to their extremes. However, the recent analysis of very high S/N spectra [4] likewise finds a metallicity dependence. The preliminary results of the VLT "First Stars" study [6] foreshadow the forthcoming availability of results to emerge from the full analysis of that data set. The existence of slight metallicity and temperature dependences of the plateau Li abundances, if not due to temperature errors such as those described above, indicate that some Galactic production and some stellar destruction of Li has occurred. It therefore becomes crucial to understand the extent of both processes. The fact that dwarfs down to [Fe/H] $= -3.7$ have been measured, and that there are few Galactic objects with lower metallicities, mean that very little Galactic production could have occurred by the time the halo stars formed. That is, the amount by which the halo plateau value exceeds the primordial one is quite small, $0.11^{+0.07}_{-0.09}$ dex [21]. Unfortunately, the size of the other term, i.e. the degree to which halo plateau stars may have destroyed some of the Li they formed with, is less well constrained. That term will be examined in the following section.

3   Li Destruction in Main-Sequence Halo Stars

Simple, sometimes called "standard", stellar evolution models in which the outer parts of the envelope mix according to standard convective prescriptions predict essentially no depletion of Li by plateau stars over their lifetimes [9]. However, such models are rather unsatisfactory at explaining some important characteristics of Population I stars, including the depletion of Li and other light elements [10]. Models which invoke more complex, but consequently less well constrained, rotationally-induced mixing seem to reproduce Population I observations better and hence offer another mechanism by which halo plateau stars might destroy Li from a higher initial value to arrive at the observed plateau value [16–18]. The test of these models is whether a range of depletion factors has existed (depending on the initial angular momentum of each protostar), and hence whether a range of plateau Li abundances results. The range of abundances in the sample of [25], once GCE is allowed for, is very small, having a standard deviation $\sigma = 0.031$ dex, entirely consistent with the observational uncertainties. Ryan et al. [25] argue that the intrinsic spread in plateau Li abundances at a given metallicity is very small, and they rule out rotationally-induced depletion of Li by more than 0.1 dex. However, the interpretation of the sample depends crucially on whether one is justified in excluding halo stars which have extremely low Li abundances. Eight such stars are known, having Li abundances that are at least 0.5 dex below those of plateau stars. (Only upper limits can be quoted for their

Li abundances.) Pinsonneault et al. [18] argue is contrast that depletion up to 0.2 dex could still be consistent with the existence of some Li depleted stars.

To determine whether the Li-poor stars have to be considered along with the plateau stars, or alternatively whether their evolutionary histories are not representative of normal plateau stars, we undertook a more detailed investigation of the Li-poor objects. Norris et al. [15] [24] showed that of four such stars examined, two had element compositions indistinguishable from those of Li-preserving plateau stars of the same metallicity. The other two showed tantalising evidence of unusual abundances of neutron-capture elements: G186-26 showed an excess of Sr, Y and Ba relative to Fe, while G139-8 showed a very high [Ba/Sr] ratio, though both [Sr/Fe] and [Ba/Fe] were sub-solar. A study of four further members of the class [13,11] showed no other element anomalies apart from an elevated Na abundance in one. Clearly there is no widely occurring chemical anomaly in this group of stars other than Li, though the presence of high [Ba/Sr] ratios in G139-8 and G186-26 are suggestive of the involvement of more-massive stars in the evolution of the Li-poor stars or the clouds from which they formed.

The concentration of Li-poor stars around the main sequence turnoff led [22] to speculate that Li-poor stars might be related to blue stragglers. Specifically, they proposed that the same mass transfer processes that produce field blue stragglers should also produce sub-turnoff-mass objects which would be indistinguishable from normal stars except for the absence of Li. The discovery that three out of four Li-poor stars had rotation velocities in the range $v\sin i = 5.5$ to 7.6 km s^{-1}, while Li-normal stars had undetectable rotation generally below 3 km s^{-1}, led [23] to conclude that angular momentum transfer in the blue-straggler forming process was responsible for these stars having been spun up to higher than normal rotation speeds. Our new study takes in the remaining four known Li-poor stars, and improves the resolution and S/N of the data on two of those already measured. It shows that five of the eight Li-poor stars have rotation velocities in excess of 4 km s^{-1} [12]. Four of these five stars are confirmed binaries, which indicates that mass transfer did not result in the complete merger of the components, consistent with small transferred masses as inferred by [23]. The results are given in Table 1 along with previous measurements of similar stars. Allowing for the possibility that some of these stars may be viewed at low inclination, i.e. almost pole-on, we cannot rule out the possibility that most are rotating, with $\langle v \rangle \sim 6$ km s^{-1}.

We conclude that the Li-poor stars definitely have different evolutionary histories to Li-normal plateau stars. A mass-transfer mechanism may explain the origin of these objects, but irrespective of whether this is the correct explanation, the Li-poor objects cannot be included in studies of Li depletion mechanisms that affect normal single stars. Consequently we conclude that the small (zero?) intrinsic spread in plateau Li abundances inferred by [25] is representative of normal halo stars, and thus signifies at most a small depletion in Li, < 0.1 dex by the models of [17].

Table 1. Projected rotation velocities of Li-poor stars

Star	$v\sin i$/km s^{-1} [23]	$v\sin i$/km s^{-1} [12]
Wolf 550 = G66-30	5.5 ± 0.6	5.6 ± 0.3
G202-65	8.3 ± 0.4	8.6 ± 0.2
BD+51°1817	7.6 ± 0.3	
HD 97916		10.4 ± 0.2
G122-69		4.7 ± 0.5
BD−31°19466	<2.2	
G139-8		<1.9
G186-26		<2.5

4   Li in the Post-WMAP Era

If the baryon density of the Universe is well constrained from WMAP observations, what purpose do Li observations in plateau stars now serve? The apparent lack of agreement between WMAP and the halo star Li observations could be due to imperfect modelling of big bang nucleosynthesis of Li, for example if the ^{7}Be(d,p)2^{4}He rate is wrong [8]. Alternatively, if we are to reconcile the observed Li abundance with the value currently inferred from big bang nucleosynthesis calculations, then we have a new constraint on the degree to which it has been processed in halo stars. We can use this measurement of the abundance decrement, in concert with precise measurements of the metallicity- and mass- (temperature-) dependence of surviving Li abundances, to learn much about the way stars' outer envelopes are processed during their lifetimes.

The optimistic sentiments of the penultimate paragraph presume, of course, that our abundance analyses are correct, and that we have not erred by a large factor in inferring the present Li abundances of plateau stars. An underestimate of the more important systematic uncertainties could alter the picture. A very recent paper ([14]; their Figure 2) has suggested an upward revision of the effective temperatures of [25] i.e. for stars with [Fe/H] < −2, by typically 200 to 400 K. The resulting Li abundances are marginally consistent with WMAP when overshooting model atmospheres are used. However, temperatures of metal-poor stars based on Hα line profiles appear not to be so much higher than those of [25], perhaps only ∼120 K higher [4]. Additionally, analysing the sample of [25], [1] derived spectroscopic effective temperatures on average 80 K *cooler* than those of [25], though they were concerned that the excitation temperatures were not well constrained because of the small number of Fe I lines and the low range of excitation potentials represented in the data. WMAP may have reported, but the interpretation of the Li spectra demands continued effort.

Acknowledgments

SGR is pleased to acknowledge fruitful discussions and collaborative studies with T. C. Beers, C. P. Deliyannis, A. Ford, J. E. Norris, and M. H. Pinsonneault on issues related to Li, many of which are reflected in this paper.

References

1. Arnone, A., Ryan, S. G., Argast, D., Norris, J. E. & Beers, T. C. 2004, A&A in press
2. Asplund, M., Carlsson, M. & Botnen, A. V. 2003, A&A, 399, L31
3. Asplund, M., Nordlund, Å, Trampedach, R., & Stein, R. F. 1999, A&A, 346, L17
4. Asplund 2005, this volume
5. Boesgaard, A. M. & Steigman, G. 1985, ARAA, 23, 319
6. Bonifacio, P. et al. 2003, IAU JD, 15, 39
7. Bonifacio, P. & Molaro, P. 1997, MNRAS, 285, 847
8. Coc, A., Vangioni-Flam, E., Descouvemont, P., Adahchour, A. & Angulo, C. 2004, ApJ, 600, 544
9. Deliyannis, C. P., Demarque, P., & Kawaler, S. D. 1990, ApJS, 73, 21
10. Deliyannis, C. P., Boesgaard, A. M., Stephens, A., King, J. R., Vogt, S. S., & Keane, M. J. 1998, ApJ, 498, L147
11. Elliott, L., Ford, A., Ryan, S. G., & Gregory, S. 2005, in prep.
12. Elliott, L. & Ryan, S. G. 2005, in prep.
13. Ford, A., Elliott, L., & Ryan, S. G. 2004, Origin and Evolution of the Elements, A. McWilliam & M. Rauch (eds), Carnegie Observatories Astrophysics Series 4 (http://www.ociw.edu/ociw/symposia/series/symposium4/proceedings.html)
14. Melendez, J. & Ramirez, I. 2004, astro-ph/0409383
15. Norris, J. E., Ryan, S. G., Beers, T. C. & Deliyannis, C. P. 1997, ApJ, 485, 370
16. Pinsonneault, M. H., Deliyannis, C. P., and Demarque, P. 1992, ApJS, 78, 179
17. Pinsonneault, M. H., Walker, T. P., Steigman, G., & Narayanan, V. K. 1999, ApJ, 527, 180
18. Pinsonneault, M. H., Steigman, G., Walker, T. P., & Narayanan, V. K. 2002, ApJ, 574, 398
19. O'Meara, J. M., Tytler, D., Kirkman, D., Suzuki, N., Prochaska, J. X., Lubin, D. & Wolfe, A. M. 2001, ApJ, 552, 718
20. Ryan, S. G., Beers, T. C., Deliyannis, C. P. & Thorburn, J. A. 1996, ApJ, 458, 543
21. Ryan, S. G., Beers, T. C., Olive, K. A., Fields, B. D., & Norris, J. E. 2000, ApJ, 530, L57
22. Ryan, S. G., Beers, T. C., Kajino, T., & Rosolankova, K. 2001, ApJ, 547, 231
23. Ryan, S. G., Gregory, S., Kolb, U., Beers, T. C., & Kajino, T. 2002, ApJ, 571, 501
24. Ryan, S. G., Norris, J. E., & Beers, T. C. 1998, ApJ, 506, 892
25. Ryan, S. G., Norris, J. E., & Beers, T. C. 1999, ApJ, 523, 654
26. Spergel et al. 2003, ApJS, 148, 175
27. Spite, F. & Spite, M. 1982, A&A, 115, 357

Lithium and Beryllium in Globular Cluster Stars

L. Pasquini

European Southern Observatory, Karl-Schwarzschild-Strasse 2,
D-85748 Garching bei München, Germany

Abstract. The observations of light elements (Lithium and Beryllium) in Globular Cluster (GC) stars are reviewed. Light element observations in GC are very powerful tracers of mixing processes in the stellar interior and shed new light on the GC formation history.

1 Introduction

Globular clusters are quite distant and their turnoff (TO) stars are intrinsically relatively faint. Following the advent of state-of-the-art instrumentation in 4m class telescopes, the first Li observations were carried on in GC stars, while with the advent of 8m class telescopes a quality jump occurred: high quality spectra can now be obtained for the TO stars of the closest clusters, comparable to that available for field stars. In spite of this advancement, only a handful of published refereed papers have been devoted to the study of Li in globular cluster stars, and only one to beryllium. Based upon the wealth of information made available as a result of this data, I will present new findings concerning stellar mixing, primordial Li production and GC formation.

2 Turnoff (or Close To) Stars

Turnoff (or just above it) stars have been observed in only three GC: NGC6397, 47 Tuc and M92. Early investigations concentrated on the relevance of the GC observations for the interpretation of the Spite plateau (Spite and Spite 1982): in NGC6397 Molaro and Pasquini (1994) and Pasquini and Molaro (1996) found that the Li abundance of three TO stars was compatible with the Pop II plateau, and no evidence for Li variations were observed. Similar results were found in three stars of 47 Tuc, although some Li differences could not be excluded in that sample by the limited S/N ratio of these observations (Pasquini and Molaro 1997). A quite different result emerged from the observations of faint subgiants in M92 (Boesgaard et al. 1998, Deliyannis et al. 1995): where it was claimed that the stars observed showed evidence of Li variations. Some of the M92 stars also showed anomalous Mg and Na abundances, as if the gas were previously processed by Mg-Al and Ne-Na cycles, but this cycling was considered incompatible with the high level of Li observed, and internal mixing was proposed as an alternative explanation. However, these M92 results have been challenged by Bonifacio (2002), who showed that a different treatment of the data lead to abundance values compatible with a Li constant value and with the plateau.

If recent UVES observations of NGC6397 have, on the one side, confirmed that in this cluster Li is constant and at the plateau level (Thevenin et al. 2001, Bonifacio et al. 2002), they have, on the other side, introduced new, interesting areas for investigation: given that the stars in NGC6397 show low oxygen abundance, indicating a probable CNO processing of the gas, how is it possible that the much fragile Li has not been depleted from its primordial value? Bonifacio et al. discuss this point extensively, and the only possible solution at present is that, as proposed by Ventura et al. 2002, Li was destroyed and then re-created in AGB stars envelope and later given back to the ISM via Comeron-Fowler mechanism and stellar winds. The almost perfect coincidence of the Li abundance in NGC6397 stars with the plateau, however, requires an improbable fine-tuning for this cluster.

3   Evolved Stars

Most studies concentrated on the presence of super Li rich giants in GC. But these Li-rich giants have been quite elusive: only a few have been found despite extensive searches which may have included hundreds of stars (Kraft et al. 1999, Smith et al. 1999, Pilachowski et al. 2000). Since Li-rich giants are treated in another contribution (De la Reza, these proceedings), I will just summarize these searches by saying that Li-rich giants are not common among GC.

A different approach has been followed by Castilho et al. 2000 and Grundahl et al. 2002, who observed stars along the whole RGB (including subgiants) of NGC6397 and NGC6752 respectively. For instance the work of Grundahl et al. clearly shows that, while in stars fainter than the RG bump, Li is clearly detected, in stars brighter than the RG bump, only Li upper limits can be found. This is a clean, beautiful proof that extra mixing occurs in this evolutionary phase, in agreement with indications found in field stars (see e.g. Charbonnel and do Nascimento 1998). A similar result is shown in Figure 1, where the Li eq. widths vs. magnitude diagram is shown for the M71 stars of Ramirez and Cohen 2002: stars brighter than the bump show only Li upper limits.

4   New Results in TO Stars

In the framework of a large program on GC abundances (Gratton et al. 2001), we started to analyze the Li spectra of TO stars of NGC6752. In this cluster, not only the classical anticorrelations are observed among giants, but Na-O anticorrelation has also been observed among TO stars, the same objects which are now analyzed for Li. Figure 2 shows the Li equivalent width vs. Na abundance for the observed stars. Since the stars all have a similar effective temperature, the direct comparison of the Li equivalent widths should closely mimic the behaviour of Li abundances. A clear Li-Na anticorrelation is observed. To the best of our knowledge, this is the first time that such an anticorrelation has been

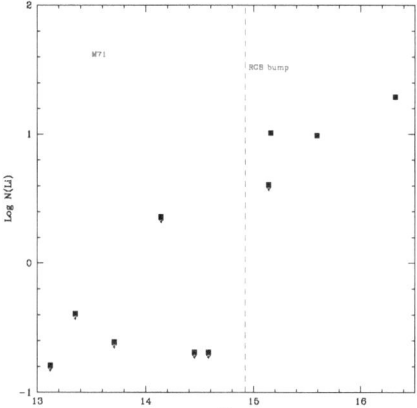

Fig. 1. Li equivalent widths vs. magnitude for the M71 stars of Ramirez and Cohen (2002). The RGB-bump magnitude is indicated with a dashed vertical line. Stars brighter (at the left of the line) than the bump show Li upper limits only.

seen and that systematic Li variations in Pop II 'plateau' stars are observed. In fact, TO stars of NGC6752 have metallicity, T_{eff} and gravity typical of plateau stars. It is natural to interpret these Li variations in the framework of chemical anticorrelations in Globular Clusters. Ventura et al. 2002 have predicted the Li content of NGC6752 stars in the framework of intermediate mass AGB pollution hypothesis. They predict the existence of the anticorrelation and that some Li must be observed even in the most polluted stars. Pending a quantitative analysis of the abundance, we can only say that the observations agree qualitatively very well with these predictions.

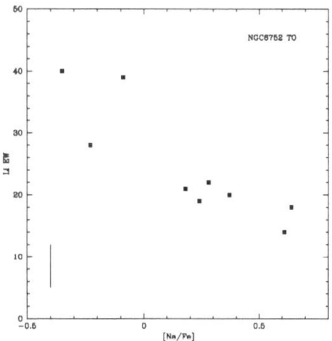

Fig. 2. Li equivalent widths vs. Na abundance for TO stars in NGC6752, showing a clear Li-Na anticorrelation (Pasquini et al. 2005, in preparation).

5   Beryllium

Be is a very interesting element, produced by spallation of galactic cosmic rays. The only two usable lines are in the extreme UV (313 nm), in a crowded spectral region, and the stellar radiation is heavily absorbed by the earth's atmosphere, so that their observations are challenging in faint stars. Only very recently (Pasquini et al. 2004) the first Be observations became available, in 2 TO stars of the nearby NGC6397.

As shown in Figure 3, the Be lines are clearly detected and an abundance of Be/H=-12.35 is measured in both stars. In addition, the stars have very different Oxygen contents and are both enriched in nitrogen with respect to field stars. Besides the use of Be as a cosmochronometer, the relevance of this finding stems from the information that Be may bring about the cluster formation mechanism. As we have seen in Section 2, for NGC6397 an ad-hoc hypothesis is needed to produce a Li abundance as high as the primordial one through AGB production and mass losses. However, Be would be destroyed during the AGB phase but could not be produced with the same mechanism. Its presence (and its abundance, which is that expected for field stars of the same iron content) is therefore clear evidence that the material which formed the stars we observe now must have had this composition before the stars were born, and that this gas must have been exposed to the GCR for a few hundred million years.

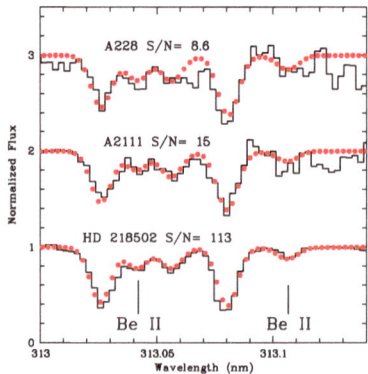

Fig. 3. Beryllium spectra for the two NGC6397 stars observed by Pasquini et al. 2004. The spectrum of a comparison field star is also added. Best fit models (red dots) are overimposed.

References

1. A.M. Boesgaard et al. ApJ 493, 906 (1998)
2. P. Bonifacio: A&A 395, 515 (2002)
3. P. Bonifacio, L. Pasquini et al: A&A 390, 91 (2002)

4. B. Castilho et al.: A&A 361, 92 (2000)
5. C. Charbonnel, J.D. Do Nascimento, A&A 336, 915 (1998)
6. C.P. Deliyannis, A.M. Boesgaard, J.R. King: ApJ 452, L13 (1995)
7. R. Gratton, P. Bonifacio et al.: A&A 369, 87 (2001)
8. N. Grundahl et al.: A&A 385, L14 (2002)
9. R.P. Kraft et al.: ApJ 118, L53 (1999)
10. P. Molaro, L. Pasquini: A&A 281, L77 (1994)
11. L. Pasquini, P. Molaro: A&A 307, 761 (1996)
12. L. Pasquini, P. Molaro: A&A 322, 109 (1997)
13. L. Pasquini, P. Bonifacio, S. Randich, D. Galli, R. Gratton: A&A 426, 651 (2004)
14. C. Pilachowski et al.: AJ 119, 2895 (2000)
15. S.V.Ramirez, J.G Cohen : AJ 123, 3277 (2002)
16. V.V. Smith, M.D. Shetrone, M.J. Keane: ApJ 512, 1006 (1999)
17. F. Spite, M. Spite: A&A 115, 357 (1982)
18. F. Thevenin, et al.: A&A 373, 905 (2001)
19. P. Ventura, F. D'Antona, I. Mazzitelli: A&A 393, 215 (2002)

Lithium Rich Red Giant Branch Stars

R. de la Reza

Observatorio Nacional/MCT - Rio de Janeiro/ Brazil

Abstract. Lithium rich K giant stars are found at the luminosity bump of the first ascending red giant branch. The discovery of these stars has given a recent impulse to advances in the theory of stellar evolution. Important connections appeared between thermonuclear processes, rapid mixing, surface activity, rotation and mass loss episodes. How could such lithium surface enrichment be produced? Two concurring scenarios offer completely different explanations: an internal stellar origin and an external one in which planets are engulfed. We will discuss the present state of these matters.

1 Introduction

Following the standard stellar evolution theory, a large part of the initial epidermical lithium in low mass stars of the first ascending red giant branch (RGB) must be diluted after the first dredge-up. Nevertheless, some RGB stars present very large Li abundances (some with even supermeteoritic values), while some show much larger Li depletions than those predicted by the theory. These unexpected results have been the source of inspiration for several theoretical approaches trying to explain the observed abundances, that involve various internal mechanisms such as rotation, mixing, surface activity, mass loss and even engulfing of external bodies. The main question is the following: Is the Li excess a peculiarity resulting of external pollution? Or is it the result of a normal evolutionary internal shortliving process during the RGB? In any case, this Li rich phenomenon has been called by Melo et al. (see these proceedings) as "one of the most exciting puzzles in stellar astrophysics". Here we will discuss what the state of art of these matters is.

2 Status of The Observed Properties

2.1 The Galactic Populations

In the Galactic disk the Li rich K giants appear not to present any peculiar behavior that can be distinguished from the common Li poor giants, as spatial distribution and velocities are concerned. Also, the metal abundances of these stars are distributed between [Fe/H]$= -1.0$ and $+0.3$ with a maximum at solar metallicity (Drake 1998). Among the disk stars, a very large difference is found as far as stellar rotation is concerned. Whereas 1 - 2% of the common low rotation giants ($v \sin i \sim 2$ to 3 km/s) are Li rich, this frequency is near 50% for the high rotating K giants ($v \sin i \geq 8$ km s^{-1}, Drake et al. 2002). Among the globular clusters this frequency falls to roughly less than 1% (Pilachowski et al. 2000).

2.2   The Lithium – Mass Loss Connection

One remarkable property was discovered by Gregorio-Hetem et al. (1992, 1993) indicating that Li rich giants were optical counterparts of IRAS sources. de la Reza et al. (1996, 1997) proposed that these far-IR excesses are due to sudden mass loss events provoked by short episodes of high Li stellar surface enrichment. The complete ejection of a circumstellar shell (CS) is described by a closed loop in the IRAS color-color diagram in which the common Li poor giants are concentrated in an initial box with no far-IR excesses. All Li rich and Li poor giants presenting far-IR excesses, distributed in the diagram, are considered to be related to their ejected CS. Because the initial CS ejection is synchronously connected to the Li enrichment event at the beginning of the loop in the above mentioned Li poor box, it is possible to measure the durations of the "Li cycles" by means of the different positions of stars in the diagram. These Li cycles mean the periods between the initial Li enrichment and its depletion.

It is interesting to note that all stars without exception are Li rich in the initial part of this cycle, characterizing an initial Li enrichment for all stars (see Figs. 7a and 7b in Drake et al. 2002). An important test of this scenario could be the observation of separated CS around RGB giants as is the case of the more evolved carbon asymptotic giant stars.

3   Two Mechanisms to Explain the Li Phenomenon

Two completely different scenarios attempt to explain the presence of large Li abundances among the RGB stars. One is the result of an external contamination (pollution) produced by the engulfing of near giant planets or brown dwarfs companions. The second one is the result of an internal action known as the Cameron-Fowler ^{7}Be mechanism. Here, we will make a brief discussion of both.

3.1   The External Process Scenario

If an external body is engulfed, it can enrich the star with the original interstellar medium abundances of ^{6}Li, ^{7}Li, ^{9}Be and ^{10,11}B (written here in increasing order of hardness to be destroyed by thermonuclear reactions). This mechanism is then supposed to produce stellar enrichment of these elements up to the maximum meteoritic value. Also, the engulfing star will suffer a rotational increase due to the gain of the planet momentum and a thermal expansion phenomenon due to the penetration of the body provoking mass loss phenomena (Siess & Livio 1999). An extension to this scenario has been proposed by Denissenkov & Weiss (2000) in order to explain supermeteoritic Li abundance values, via a combination of stellar rotation and activation of the ^{7}Be mechanism at the base of the convective layer produced by the penetration of the external body.

These two studies attempt to explain the Li enrichment - mass loss connection scenario discussed above. However, the mixed scenario of Denissenkov & Weiss produces a much delayed mass loss event, apparently not compatible with the rapid mass loss event suggested by the IRAS properties. In any case, in this external scenario excesses of ^{6}Li, ^{9}Be and specially ^{10,11}B must be present in the ^{7}Li enriched giants. Now, let us discuss the status of these elemental abundances in Li-rich giants. ^{6}Li seems to be absent in the Li rich K giants indicating, due to its fragility, no recent engulfing events. A practically absent ^{9}Be has been shown among Li K giants by Castilho et al. (1999). Recent measurements with UVES at the VLT by Melo et al. (see these proceedings) resulted in a complete absence of this isotope in a larger sample of Li rich K giants. These authors have also estimated that an almost unrealistic number of several tenths of Jupiter type planets or hundreds of Earth types planets are necessary to produce larger Li and Be enrichments. These numbers can be lower if hydrogen deficient planets are at play.

Recent Hubble observations of boron in a sample of Li-rich K giants (de la Reza, Smith & Cunha in preparation) have apparently not detected any signal of a substantial presence of this element. In some cases, the large rotation and the presence of strong UV emission lines at the 2497 Å region difficult the measure of the real B abundances. In conclusion, it appears that there is no indication of the presence of other than ^{7}Li light elements in the Li rich K giants.

We can also mention further arguments not in favor of the ingestion scenario: *a) Enrichment during the RGB*. If engulfing exists, as must really be the case at least for the case of hot jupiters, the light metal enrichment must occur at any time during the RGB evolution. This in fact is not observed. Li K giants exist only at luminosity bump of the RGB or higher. *b) Metallicity*. If we accept that most stars with planets are metal rich (see Santos in these proceedings) we would expect that giant stars Li enriched by planets will also be metal rich and this is not the case. Li rich K disk giants have in general solar metallicities.

3.2 The Internal Process Scenarios

Based on the Cameron-Fowler ^{7}Be mechanism as a way to produce internal ^{7}Li, these scenarios differ mainly in the mechanisms transporting ^{7}Be recently produced in the H-burning shell, to the base of the convective layer. When no rotation is considered, Sackmann & Boothroyd (1999) proposed an ad-hoc conveyor belt mechanism (cool bottom process) producing this extra mixing. This process produces a continuous Li enrichment. An almost episodical event of very high Li surface enrichment appears only for metal deficient giants. Differently, Fekel (1988) first proposed that stellar rotation must be the prime cause of the Li surface enrichment. Since then, various researchers have followed this direction. Rotation based scenarios involve different approaches to increase the rotation diffusion coefficient (D_R). Whereas for Denissenkov & Weiss (2000) an initial low D_R is increased by the engulfing of a planet (see Weiss & Denissenkov in these proceedings), for Charbonnel & Balachandran (2000) the supermeteoritic Li abundances could be produced only by internal structural changes. For another general approach in this respect see Chaname et al. in these proceedings. Because Li rich giants are found at the RGB bump, "the Li rich phase acts as a precursor of extra mixing produced by rotation". The presence of a "Li-flash"

(Palacios et al. 2001, also in these proceedings) produces a thermal instability enabling potentially, by the increase of temperature, to produce a global luminosity increase. This Li flash can then activate a meridional circulation process. Because a ^{13}C surface increase is predicted when ^{7}Li is decreasing after its initial enhancement, an interesting possibility appears with the possibility to measure this time delay by measuring ^{12}C/^{13}C ratios for different far-IR giant stars.

4 Conclusion

Even if engulfing processes can really exist, this phenomenon appears not to be the cause of Li enhancements in K giants. The internal processes must then be the most plausible explanations.

I would like to thank Natalia Drake and Katia Cunha for reading the manuscript of this work.

References

1. B.V. Castilho, F. Spite, B. Barbuy, M. Spite, J.R. de Medeiros, J. Gregorio-Hetem: A&A 345, 249 (1999)
2. C. Charbonnel, S.C. Balachandran: A&A 359, 563 (2000)
3. R. de la Reza, N.A. Drake, L. da Silva: ApJ 456, L115 (1996)
4. R. de la Reza, N.A. Drake, L. da Silva, C.A.O. Torres, E.L. Martin: ApJ 482, L77 (1997)
5. P.A. Denissenkov, A. Weiss: A&A 358, L49 (2000)
6. N.A. Drake: PhD Thesis, Observatório Nacional (1998)
7. N.A. Drake, R. de la Reza, L. da Silva, D.L. Lambert: AJ 123, 2703 (2002)
8. F.C. Fekel: in ESA, A Decade of UV Astronomy with the IUE Satellite, vol.1, 331 (1988)
9. J. Gregorio-Hetem, J.R.D. Lépine, G.R. Quast, C.A.O. Torres, R. de la Reza: AJ 103, 549 (1992)
10. J. Gregorio-Hetem, B.V. Castilho, B. Barbuy: A&A 268, L25 (1993)
11. A. Palacios, C. Charbonnel, M. Forestini: A&A, 375, L9 (2001)
12. C.A. Pilachowski, C. Sneden, R.P. Kraft, D. Harmer, D. Willmarth: AJ, 119, 2895 (2000)
13. I.-J. Sackmann, A.I. Boothroyd: ApJ 510, 217 (1999)
14. L. Siess, M. Livio: ApJ 308, 1133 (1999)

Evidence of Mixing
in Extremely Metal-Poor Giants

M. Spite[1], R. Cayrel[1], F. Spite[1], V. Hill[1], P. François[1], E. Depagne[2],
B. Barbuy[3], P. Bonifacio[4], P. Molaro[4], F. Primas[5], T. Beers[6], B. Plez[7],
B. Nordström[8,9], and J. Andersen[8]

[1] GEPI, Observatoire de Paris-Meudon, F-92125 Meudon Cedex, France
[2] European Southern Observatory(ESO), 3107 Alonso de Cordova, Vitacura,
    Casilla 19001, Santiago 19, Chile
[3] IAG, Universidade de Sao Paulo, Depto. de Astronomia, rua do Matao 1226,
    Sao Paulo 05508-900, Brazil
[4] Osservatorio Astronomico di Trieste, INAF, Via G.B. Tiepolo 11, I-34131 Trieste,
    Italy
[5] European Southern Observatory, Karl Schwarzschild-Str. 2,
    D-85748 Garching bei München, Germany
[6] Department of Physics & Astronomy and JINA: Joint Institute for Nuclear
    Astrophysics, Michigan State University, East Lansing, MI 48824, USA
[7] GRAAL, Université de Montpellier II, F-34095 Montpellier Cedex 05, France
[8] Astronomical Observatory, NBIfAFG, Juliane Maries Vej 30,
    DK-2100 Copenhagen, Denmark
[9] Lund Observatory, Box 43, SE-221 00 Lund, Sweden

Abstract. Thirty five extremely metal-poor (EMP) giants (22 of them with a metallicity [Fe/H] $\leq$ -3.0) have been observed with UVES at the VLT. In these stars the
abundance of lithium carbon and nitrogen and also the $^{12}C/^{13}C$ ratio have been measured. It is shown that, in about half of them, the carbon abundance and the lithium
abundance are very low, on the contrary the nitrogen abundance is high. Moreover
the value of the $^{12}C/^{13}C$ ratio is close to the equilibrium value of the CN cycle. All
these phenomena imply a mixing between the surface of the star and the H burning
layers. We also study the influence of the mixing on the oxygen, sodium and aluminum
abundance.

1   Introduction

In the frame of the ESO Large Program "First Stars, first nucleosynthesis" [1] [2]
33 extremely metal-poor giants not classified as "carbon-rich stars", have been
observed at the VLT with the high resolution spectrograph UVES. The spectral
coverage is almost complete from 330 to 900nm and covers in particular, the NH
band at 336nm.

Abundances of the elements from C to Zn have been measured and it has
been shown that for all the elements "X" lighter than Zn, the spread of the
ratios [X/Fe] or [X/Mg] relative to [Fe/H] or [Mg/H] is generally very small and
$\leq$ 0.15 dex (see: [1] [2]).

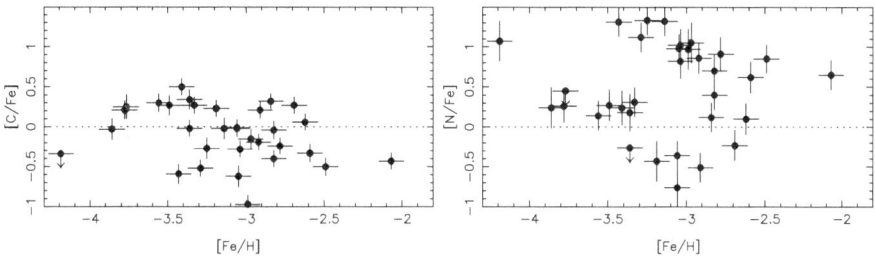

Fig. 1. The spread of the relations [C/Fe] and [N/Fe] vs [Fe/H] is very large: $\sigma = 0.35$ and 0.57dex

However there are at least two exceptions: when [C/Fe] and [N/Fe] are plotted versus [Fe/H] (or [C/Mg] and [N/Mg] versus [Mg/H]) the spread is very large (Fig. 1 and [6]) suggesting a mixing with the H burning layer.

Similar anomalies have been found by Gratton et al. [4] and Sneden and al. [5] in metal poor red giants in the metallicity range $-2 \leq [\text{Fe/H}] \leq -1$ and have been explained by an "extra" mixing (not predicted by the standard theory) between the surface and the H-burning shell. It is thus interesting to check whether this extra-mixing is also efficient in the EMP giants and can explain the very large spreads in Fig. 1.

2 Evidence of Mixing in Extremely Metal-Poor Giants

Gratton et al. [4] found that abundances of Li, C, N, and the ratio $^{12}\text{C}/^{13}\text{C}$, change abruptly at a given luminosity very similar to that of the RGB bump.

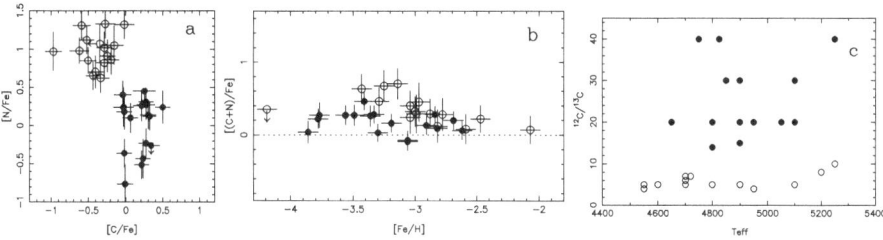

Fig. 2. a-) When [N/Fe] is plotted versus [C/Fe] two groups of stars appear clearly: in a first group (**unmixed stars, filled circles**) the C abundance is high but the the nitrogen abundance is low, and in the second group (**mixed stars, open circles**) the carbon abundance is low but the N abundance is ten times higher than in the first group. **b-)**On the other hand, the sum of C and N abundances is almost the same in the two groups of stars. **c-)** In the "mixed stars" $^{12}\text{C}/^{13}\text{C}$ is close to the equilibrium value of the CN cycle

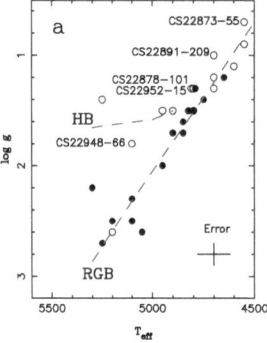
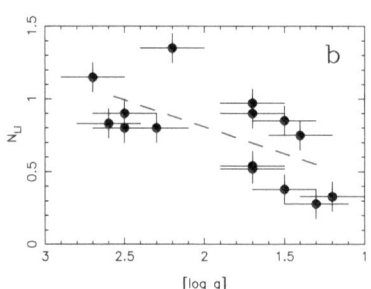

Fig. 3. a-) The unmixed stars (filled circles) belong mainly to the low RGB and the mixed stars (open circles) are on the horizontal branch or above. **b-)** In the "unmixed" giants the abundance of lithium decreases as the star evolves and log g decreases.

The absolute luminosities of our EMP giants are unknown but if we plot [N/Fe] versus [C/Fe] (Fig. 2a) two different populations of stars appear clearly: a first one with [N/Fe]> 0.5 and [C/Fe]< 0.0 (called unmixed stars) and a second one with [N/Fe]< 0.5 and [C/Fe]> 0.0 (mixed stars). The sum of C and N abundances is almost the same in the two groups of stars (Fig. 2b). Three "mixed stars" seem to have a slightly higher value of the C+N abundance (CS22948-66, CS22878-101, CS22952-15), they could belong to the AGB (see also the position of these stars in Fig. 3a).

From the measurement of the ^{13}CH and ^{12}CH bands at 430nm, it has been possible to measure also the ratio ^{12}C/^{13}C in the two groups of stars. The result of these measurements for each group of stars is given in Fig. 2c. It is clear that in the stars with a high value of the ratio [N/Fe] (mixed stars) the abundance of ^{13}C is high, and the ^{12}C/^{13}C is close to 5.

Since the absolute luminosities of the stars are unknown, log g has been taken as a first order indicator of the luminosity, and we could show that generally speaking the unmixed stars belong to the low RGB and the mixed stars are on the Horizontal Branch or above (Fig. 3-a).

We have also measured the lithium abundances in the samples of unmixed and mixed stars [6]. When low mass stars, such as those in our sample, evolve through the red giant branch, the degree of dilution of the lithium increases as the convective zone penetrates deeper and thus we expect a decline of the lithium abundance. In the mixed stars the lithium has never been detected, the upper limit of the lithium abundance is log N(Li) < 0.0, on the contrary in all the "unmixed stars" but one, the lithium line is visible and log N(Li) is > 0.20. In these stars as expected, the lithium abundance decreases when the gravity decreases (Fig. 3-b).

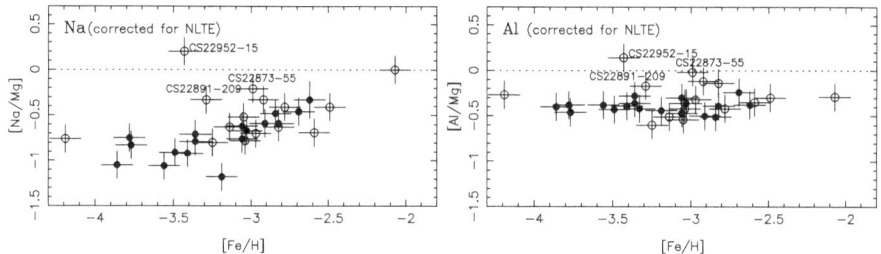

Fig. 4. All the Na-rich giants are mixed stars (but several mixed stars are not Na-rich. All the giants which are very Na-rich are also Al-rich.

3 Influence of Mixing on O, Na, Al

When [O/Fe] is plotted versus [Fe/H] in the interval $-3.5 \leq$ [Fe/H] ≤ -2.5, no difference appears between the mixed and the unmixed stars. The mean value of the oxygen abundance is not affected by mixing, its value (without 1D-3D correction) is [O/Fe]=+0.7 in both groups (see [6]).

If a deep mixing takes place in the mixed stars, as in some globular cluster stars, we could expect that the products of the Ne-Na and Mg-Al cycles be brought to the surface and that the abundances of Na and Al increase and thus also the ratios [Na/Mg] or [Al/Mg]. This deep mixing has not been observed in the moderately metal-poor field giants [4]. In Fig. 4 we have plotted [Na/Mg] and [Al/Mg] as a function of [Fe/H]. All the Na-rich stars are mixed stars. The mixed stars strongly enriched in Na are also enriched in Al. Let us note that we do not observe any anticorrelation between the sodium and the oxygen abundances (at contrast with globular clusters). From the position of the Na-rich stars in the HR diagram (Fig. 3a) it could be possible that these stars be AGB stars. However they do not present any significant excess of the heavy elements like Sr or Ba [3].

References

1. Cayrel R., Depagne E., Spite M., Hill V., Spite F., François P., Plez B., Beers T., Primas F., Andersen J., Barbuy B., Bonifacio P., Molaro P., Nordström B., 2004, A&A 416, 1117 Paper V
2. Cayrel R., Spite M., et al. 2005, ESO Astrophysics Symposia "Chemical Abundances and Mixing in Stars in the Milky Way and its Satellites", S. Randich and L. Pasquini eds. (this book).
3. François P. et al. 2005, ESO Astrophysics Symposia "Chemical Abundances and Mixing in Stars in the Milky Way and its Satellites", S. Randich and L. Pasquini eds. (this book).
4. Gratton R.G., Sneden C., Carretta E., and Bragaglia A., 2000, A&A 354, 169
5. Sneden C., Pilachowski C.A., VandenBerg D.A., 1986, ApJ 311, 826
6. Spite M., Cayrel R., Plez B., Hill V., Spite F., Depagne E., François P., Bonifacio P., Barbuy B., Beers T., Andersen J., Molaro P., Nordström B., Primas F., 2004, A&A in press Paper VI (Astro-ph 0409536).

Mixing and CNO Abundances in M Supergiants

S.C. Balachandran[1], J.S. Carr[2], and K.A. Venn[3]

[1] Department of Astronomy, University of Maryland, College Park MD 20742
[2] Naval Research Laboratory, Code 7213, Washington DC 20375
[3] Department of Physics and Astronomy, Macalester College, 1600 Grand Avenue, Saint Paul MN 55105

Abstract. We present initial results from our study of mixing in M Supergiants. C, N and O abundances are measured in five stars. N/C and N/O ratios indicate extensive mixing in excess of the standard models and in support of the rotational models.

Mixing in M Supergiants

Massive stars play an important role in numerous astrophysical contexts that range from the understanding of starburst environments to the chemical evolution in the early Universe. It is therefore crucial that their evolution be fully and consistently understood. A variety of observations of hot stars reveal discrepancies with the standard evolutionary models (see [1] for review): He and N excesses have been observed in O and B main sequence stars and large depletions of B accompanied by N enhancements are seen in B stars and A-F supergiants [2,3,4,5]. All of these suggest the presence of excess-mixing, and have led to the development of a new generation of evolutionary models which incorporate rotation (full reviews in [1],[6],[7]).

The rotation models predict significant effects on the properties and the evolution of the massive stars. They alter the ratio of red to blue supergiants and hence the nature of SNII progenitors; they affect the properties, formation and evolution of Wolf-Rayet stars; they result in the enrichment of He and C in the ISM while the abundance of O decreases; they produce higher He and α-element yields from SNII via larger He cores. Many of these effects are metallicity dependent. With such far ranging impact, the effects of rotation and mass loss on the evolution of massive stars should be thoroughly understood.

In order to constrain the highly uncertain physics of rotational mixing in the models [1], we have initiated a project to measure C, N and O abundances in M supergiants. There are several advantages to studying the cool supergiants: the first dredge-up is complete and the surface abundances reveal a fuller picture of CNO processing in the stellar interior than the hotter supergiants are able to provide; abundance measurements are more robust because non-LTE effects are absent in the molecular infrared transitions; and C, N and O can be measured from the principal molecules in which they reside.

High resolution infrared spectra of CO, OH and CN lines were obtained using Phoenix on Gemini South. The results for 5 stars are shown in Figure 1 along the abundances for the Galactic Center M2Ia supergiant IRS7 and α Ori [8]. Only one of the five stars has N/C and N/O ratios similar to α Ori and appears to conform to the no-rotation model. All of the remaining stars show much larger N/C and N/O ratios indicative of extensive mixing perhaps driven by rotation. While two stars in our sample have N/C ratios near the value of IRS 7, none equal the Galactic Center supergiant in N/O. Our full sample of 30 stars will allow us to map mixing trends with luminosity (mass) and provide stringent constraints for the models.

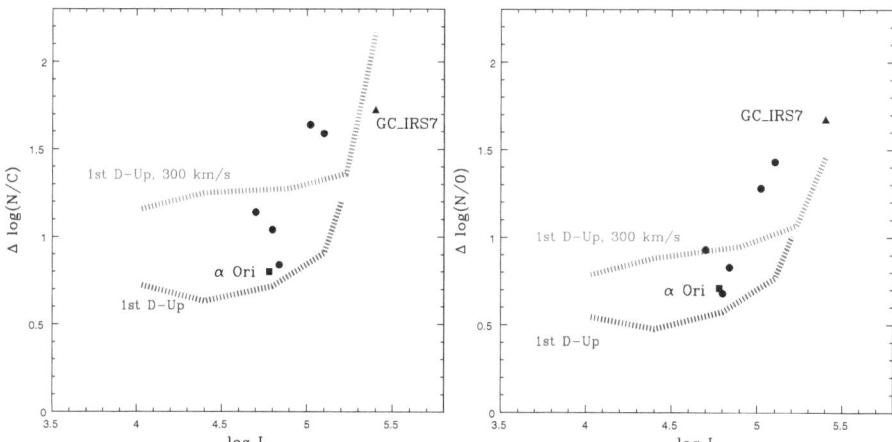

Fig. 1. N/C and N/O ratios are shown as a function of luminosity relative to the initial solar values. The lower hatched line in each plot is the standard model prediction and the upper hatched line is the predicted value for an initial rotational velocity of 300 km s^{-1} [6]. Our measurements show that the low ratios seen in αOri are not commonly seen in supergiants. Instead the ratios indicate extensive mixing as predicted by the rotation models.

References

1. A. Maeder, & G. Meynet, ARAA, **38**, 143 (2000).
2. K. A. Venn, ASP. Conf. Proc. **270**, 71 (2002)
3. N. D. McErlean, D. J. Lennon, & P. L. Dufton A&A, **349**, 553 (1999)
4. A. Herrero, R. P. Kudritzki, J. M. Vilchez, D. Kunze, K. Butler, & S. Haser, A&A, **261**, 209 (1992)
5. S. J. Smartt, D. J. Lennon, R. P. Kudritzki, F. Rosales, R. S. I. Ryans, N. Wright, A&A, **391**, 979
6. G. Meynet, & A. Maeder, A&A, **361**, 101 (2000)
7. A. Heger, & N. Langer, ApJ, **544**, 1016 (2000)
8. J. S. Carr, K. Sellgren, & S. C. Balachandran, ApJ, **530**, 307 (2000)

GIRAFFE Observations in 47 TUC: Lithium Abundances Along the Red Giant Branch

A. Lèbre[1], G. Jasniewicz[1], P. de Laverny[2], F. Thévenin[2], and C. Charbonnel[3], and A. Recio-Blanco[2]

[1] GRAAL, Université Montpellier II, Montpellier, France
[2] Cassiopée UMR6202, OCA, Observatoire de Nice-Cote d'Azur, France
[3] Observatoire de Genève, Switzerland

Abstract. From VLT/UT2 GIRAFFE GTO, we performed a lithium abundance survey along the red giant branch of the metal-rich globular cluster 47 TUC (NGC 104), in order to investigate the efficiency of extra mixing occurring at the RGB bump.

1   GIRAFFE Observations

Powerful multi-object spectroscopic tools (FLAMES-GIRAFFE at ESO) can now shed a new light on chemical abundances of faint objects. In globular cluster stars, a combination of internal mixing and external pollution is thought to play a part in the pattern of their abundance anomalies (Gratton et al., 2001). With GIRAFFE at VLT/UT2, we have sampled the H.R. diagram of 47 TUC looking for the abundance evolution of Li to provide strong evidence on the nature and efficiency of the mixing process in low mass stars. With a Color Magnitude Diagram (CMD) of 47 TUC we made from (V,V-I) photometry and astrometry (Momany, priv. comm.), we have selected stars in different regions above the turn-off, on the Red Giant Branch (RGB) and the Asymptotic Giant Branch, on the Horizontal Branch (HB) and in the Post-AGB area, with a special coverage of the RGB Bump area (V=14.58 +/- 0.03, Riello et al. 2003). In October 2003, about 230 evolved stars have been observed with GIRAFFE in medusa mode on two fibre positioning plates (bright & faint stars of our large sample) at three high resolution (R=15 000) set-ups (627.3nm, 679.7nm and 875.7nm).

2   Li and Mixing Processes Along the RGB

We have first explored the Li line region (670.7nm) for faint stars (bottom of the RGB, the RGB Bump area and the HB, see Fig.1). Data reduction has been done with the GIRAFFE DRS on 123 spectra (signal to noise ratio around 70). The $log\ g$ have been estimated with a fit of the mean RGB and HB sequences of 47 TUC to a theoretical isochrone (Bertelli et al. 1994), and $Teff$ have been estimated from the (V-I) color-temperature relations (Houdashelt et al., 2000). With [Fe/H]= - 0.65 for 47 TUC, synthetic spectra have been computed from MARCS models ($log\ g$=3.0 ; $Teff$= 4800 K to 5100 K). Li abundances (A_Li>0 dex, Li dots in Fig.1) have been derived for 41 stars (with uncertainties of about 0.2dex).

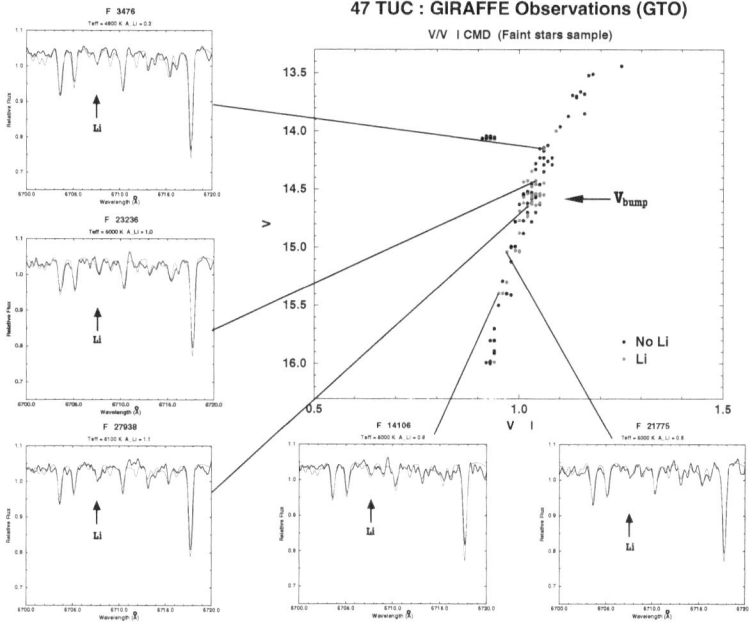

Fig. 1. CMD of our faint stars sample and selected (observed and synthetic) spectra.

When stars leave the main sequence, lithium is diluted during the 1st dredge-up on the subgiant branch. Besides, observations in various environments have shown that low mass stars undergo an additional mixing event that further modifies the surface abundances. Both observational and theoretical evidences indicate that this process occurs at the so-called RGB bump in the luminosity function (Charbonnel 1994, Charbonnel et al. 1998). Up to now, most of the Li data in globular clusters stars concern the tip of the RGB (Pilachowski et al., 2000) and very few bump RGB stars have been investigated (Grundahl et al. 2002). The present observations indicate very clearly that above the RGB bump Li is indeed strongly depleted in 47 Tuc stars (see Fig.1), confirming the occurrence of the additional mixing event in such an environment. Additional clues will soon come from the determination of the $^{12}C/^{13}C$ in our sample stars.

References

1. G. Bertelli, A. Bressand, C. Chiosi et al.: A&AS **106**, 275 (1994)
2. C. Charbonnel: A&A **282**, 811 (1994)
3. C. Charbonnel, J. Brown, G. Wallerstein: A&A **332**, 204 (1998)
4. R.G. Gratton, P. Bonifacio, A. Bragaglia et al.: A&A **369**, 87 (2001)
5. F. Grundahl, M. Briley, P.E. Nissen et al.: A&A **385**, L14 (2002)
6. M.L. Houdashelt, R.A. Bell, A.V. Sweigart: AJ **119**, 1448 (2000)
7. C.A. Pilachowski, C. Sneden, R.P. Kraft et al.: AJ **119**, 2895 (2000)
8. M. Riello, S. Cassisi, G. Piotto et al.: A&A **410**, 553 (2003)

Is an Engulfment Event at the Origin of the Li-Rich Giants?

C.H.F. Melo[1], P. de Laverny[2], N.C. Santos[3,4], G. Israelian[5], S. Randich[6], J.D.Jr. do Nascimento[7], and J.R. De Medeiros[7]

[1] European Southern Observatory, Casilla 19001, Santiago 19, Chile
[2] Observatoire de la Côte d'Azur, Département Cassiopée, UMR 6202, BP 4229, 06304 Nice, France
[3] Centro de Astronomia e Astrofíca da Universidade de Lisboa, Observatório Astronómico de Lisboa, Tapada da Ajuda, 1349-018 Lisboa, Portugal
[4] Observatoire de Genève, 51 ch. des Maillettes, 1290 Sauverny, Switzerland
[5] Instituto de Astrofísica de Canarias, 38205 La Laguna, Tenerife, Spain
[6] INAF/Osservatorio Astrofisico di Arcetri, Largo E. Fermi 5, 50125 Firenze, Italy
[7] Departamento de Física, Universidade Federal do Rio Grande do Norte, 59072-970 Natal, RN., Brazil

Abstract. We have derived beryllium abundances for all Li-rich giant stars known in the literature and visible from the southern hemisphere and 10 other Li-normal giants, aimed to investigate the origin of the Lithium in the Li-rich giants. In particular, we test the predictions of the engulfment scenario proposed by Siess & Livio (1999) where the engulfment of a brown dwarf or one or more giant planets would lead to a simultaneous enrichment of ^{7}Li and ^{9}Be. We show that regardless their nature, none of the stars observed in this work was found to have beryllium. Using simple dilution arguments we show that the engulfment of an external object as the *sole* source of Li enrichment is ruled out by the the Li and Be abundance data. The present results favor the idea that Li has been produced in the interior of the stars and brought up to the surface by a Cameron-Fowler mechanism.

1 Introduction

The discovery of lithium excesses in a few low-mass giants in recent years represents one of the most exciting puzzle for stellar astrophysics. Essentially such stars show lithium content significantly larger than the values predicted in the framework of standard stellar evolution, some of them possessing surface lithium content approaching the present interstellar medium value of $A(Li) \sim 3.0$ or even higher. In spite of an increasing number of studies with a variety of propositions, the root-cause of these highly abnormal abundances of Li in low-mass red giant stars remains unknown, adding a new critical question for these undoubtedly very complex physical systems. Some explanations are related to internal processes, such as a fresh lithium synthesis, or a preservation of the initial lithium content, whereas other explanations are based on external processes as the contamination of the stellar external layers by debris of nova ejecta or the engulfment of brown dwarfs or planets by the giant star (see de la Reza's contribution on this volume for a review.)

2 Constraints from Beryllium Abundances

According to Siess & Livio ([3]), if the Li excess found in the Li-rich giants is related to the engulfment of an external body, we should expect to observe an enrichment of other light elements such as ^{6}Li, ^{9}Be and ^{11}B, although in a lower degree. Among these 3 elements, an eventual enrichment of ^{9}Be would be the easiest one to be detected after the ^{7}Li. Thus, in order to test the engulfment hypothesis as the origin of Li-rich giants, we have carried out spectroscopic observations with UVES spectrograph attached to Kueyen VLT UT2, at Paranal Observatory, ESO, Chile aimed to measure the ^{9}Be of the Li-rich giants. The detailed analysis and the results are presented in Melo et al. ([1]). The main results of Melo et al. ([1]) are summarized as follow: i) No Be was found in any of the observed stars, regardless their nature Li-rich or Li-normal (see Table 1); ii) The presence of Li and the absence of Be strongly suggests that the Li enrichment is probably result of dredge-up of fresh Li produced in the interior of these stars by the extra-mixing mechanism proposed by Sackmann & Boothroyd ([2]). iii) Using simple dilution arguments, we show that the accreted mass necessary to produce the Li-rich enrichment observed in the Li-rich giants would also produce a Be enrichment detectable by our observations.

Table 1. Atmospheric parameters and derived abundances for the program stars. Asterisks indicate guess values for the gravity or the metallicity. For stars with very large rotational velocities, we were unable to derive abundances of O, Mn and Tm.

HD	$T_{\mathrm{eff}}(K)$	$\log g$	$M/M_\odot$	[Fe/H]	$V_{\mathrm{sini}}(\mathrm{km\,s}^{-1})$	[Be/Fe]	$A(Li)$
Li-rich giants:							
787	3990	1.3	1.9	+0.00	1.9	<-5	1.8
19745	4730	2.0(*)		+0.10	1.0	<-5	4.1
30238	3990	1.4	1.9	+0.00	1.0	<-5	0.8
39853	3920	1.1	1.6	-0.40	1.0	<-5	2.8
95799	4800	2.0(*)		-0.11	1.0	<-5	3.2
176588	3800	2.0(*)		+0.00	1.0	<-5	1.1
183492	4720	2.6	2.0	+0.00	1.0	<-5	2.0
217352 too large $V \sin i$							
219025 too large $V \sin i$							
Li-normal stars:							
360	4750	2.7	2.1	-0.20	1.0	<-5	0.2
1522	4400	1.8	3.2	+0.20	1.0	<-5	< 0.0
4128	4760	2.4	3.2	+0.15	1.0	<-5	< 0.2
5437	3950	1.3	1.4	-0.20	1.0	<-5	0.1
61772	3880	1.0	1.9	+0.10	1.0	<-5	-0.5:
61935	4760	2.7	2.4	+0.00	1.0	<-5	0.2
95272	4620	2.4	2.1	+0.00	1.0	<-5	< 0.0
105707	4190	1.4	3.2	+0.10	1.0	<-5	0.8
126271	4410	2.3	1.4	+0.00	1.0	<-5	< -0.4
220321	4630	2.4	2.1	-0.30	1.0	<-5	< -0.2

References

1. C. Melo, et al.: A&A, accepted (2005)
2. I.-J. Sackmann, A. I. Boothroyd: ApJ **510**, 217 (1999)
3. L. Siess, M. Livio: MNRAS **308**, 1133 (1999)

Part IV

Local Group Galaxies

Abundances as Tracers of the Formation and Evolution of (Dwarf) Galaxies

E. Tolstoy

Kapteyn Institute, University of Groningen, 9700 AV Groningen, the Netherlands

Abstract. This aims to be an overview of what detailed observations of individual stars in nearby dwarf galaxies may teach us about galaxy evolution. This includes some early results from the DART (Dwarf Abundances and Radial velocity Team) Large Programme at ESO. This project has used 2.2m/WFI and VLT/FLAMES to obtain spectra of large samples of individual stars in nearby dwarf spheroidal galaxies and determine accurate abundances *and* kinematics. These results can be used to trace the formation and evolution of nearby galaxies from the earliest times to the present.

1 Introduction

Dwarf galaxies are the most numerous type of galaxy we know of and they are commonly assumed to be if not the actual building blocks of larger galaxies then they most closely resemble them. The Local Group contains ∼36 dwarf galaxies out of a total of ∼42 members covering a large range of properties [1], and including more than one example of most if not all the known classes of dwarf galaxy. There are nucleated dwarfs (e.g., NGC205, NGC185); extremely low surface brightness dwarfs (e.g., Sextans, Ursa Minor); interacting dwarfs (e.g., Sagittarius); star bursting dwarfs (e.g., IC10, Sextans A); isolated dwarfs (e.g., Tucana, Cetus). They fall predominantly into two classes - those with gas which are still forming stars and those which appear not to have gas and are not presently forming stars.

The abundance patterns of individual stars of different ages and environments enable us to unlock the evolutionary history of galaxies. Many physical characteristics of a galaxy may change over time, such as shape and colour, however the metal content and abundance ratios of stellar atmospheres are not so easy to tamper with. Stars retain the chemical imprint of the interstellar gas out of which they formed, and metals can only increase with time. This method to study galaxy evolution has been elegantly named *Chemical Tagging* [2].

There have been a number of detailed abundance studies of stars in nearby galaxies which cast ever more serious doubt on the premise that the galaxies we see today are in any way related to galactic building blocks (e.g., [3], [4], [5]). The [α/Fe] ratios of stars in dwarf spheroidal (dSph) galaxies are generally lower than similar metallicity Galactic stars. There is marginal overlap in the [α/Fe] ratios between dSph stars and Galactic halo stars but this similarity does not extend to other element ratios where, for example, a significant over abundance in [Ba/Y] is typically observed in dSph stars compared to Galactic stars (see Venn, this volume). The stars in larger galaxies, such as the LMC and Sagittarius are also chemically distinct from the majority of the Galactic stars (see Hill, this volume; Bonifacio, this volume; McWilliam, this volume). This makes a merging hypothesis difficult to explain any component of our Galaxy, unless the merging were to occur predominantly at very early times. These observations can thus be interpreted in two ways - either to say that dwarf galaxies are not building blocks of larger galaxies, which begs the questions - what are they then? and how do they avoid too much merging with larger galaxies? or we can say that perhaps the whole idea of hierarchical structure formation as it currently stands needs some serious revision because dwarf galaxies really don't fit the picture. Both interpretations have merit and neither can be ruled out.

2   Initial Results from DART: Sculptor Dwarf Spheroidal

The DART large programme at ESO made v_{hel} and [Fe/H] measurements from FLAMES spectroscopy of 401 red giant branch (RGB) stars in the Sculptor (Scl) dSph [6]. The relatively high signal/noise, S/N ($\approx$ 10-20 per pixel) resulted in both accurate metallicities ($\approx$ 0.1 dex from internal errors) and radial velocities ($\approx$ ±2 km/s). This is the first time that a large sample of accurate velocities *and* metallicities have been measured in a dwarf galaxy.

Scl is a close companion of the Milky Way, at a distance of 72 ±5 kpc [7], with a low total (dynamical) mass, $(1.4 \pm 0.6) \times 10^7 M_\odot$ [8], and modest luminosity, $M_V = -10.7 \pm 0.5$, and central surface brightness, $\Sigma_{0,V} = 23.5 \pm 0.5$ mag/arcsec^2 [9] with no HI gas [10]. CMD analysis, including the oldest Main Sequence turnoffs, has determined that this galaxy is predominantly old and that the entire star formation history can have lasted only a few Gyr [11].

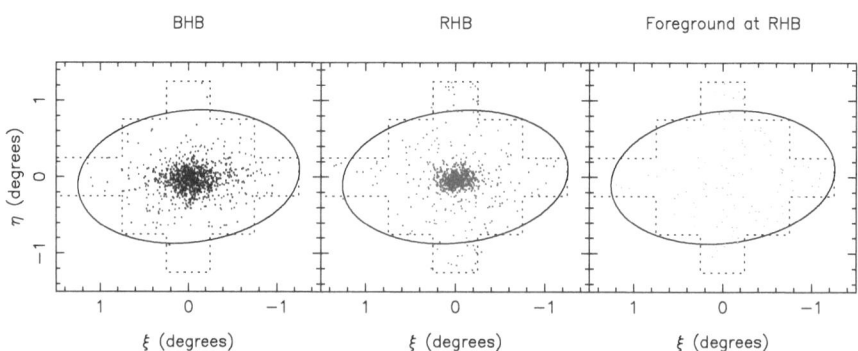

Fig. 1. The distribution of Horizontal Branch stars from WFI imaging of the Scl dSph showing the different spatial distributions of BHB and RHB as selected from a M_v, V−I Colour-Magnitude Diagram [6]. Also shown, to illustrate the foreground contamination in the RHB distribution, are a CMD-selected sample of foreground stars to match the RHB contamination density. The ellipse is the tidal radius of Scl.

Imaging: Previous studies already suggested that the spatial distribution of the Horizontal Branch stars in Scl shows signs of a gradient, with the red horizontal branch stars (RHB) being more centrally concentrated than the blue horizontal branch (BHB) stars [12], [13]. The DART WFI imaging data extends beyond the nominal tidal radius and with an average 5-σ limiting magnitude of V=23.5 and I=22.5 also probes well below the Horizontal Branch. This has enabled us to unequivocally demonstrate that the BHB and RHB stars have markedly different spatial distributions (see Fig. 1).

The different spatial occupancy of the two populations, taking into account the foreground contamination present in the RHB sample, is striking and provides strong evidence that we are seeing two distinct components. The characteristics of the BHB and RHB are also consistent with different ages (e.g., ≤ 2 Gyr), or different metallicities (Δ[Fe/H]$\sim$0.7 dex), from theoretical modelling of globular cluster Horizontal Branches [14].

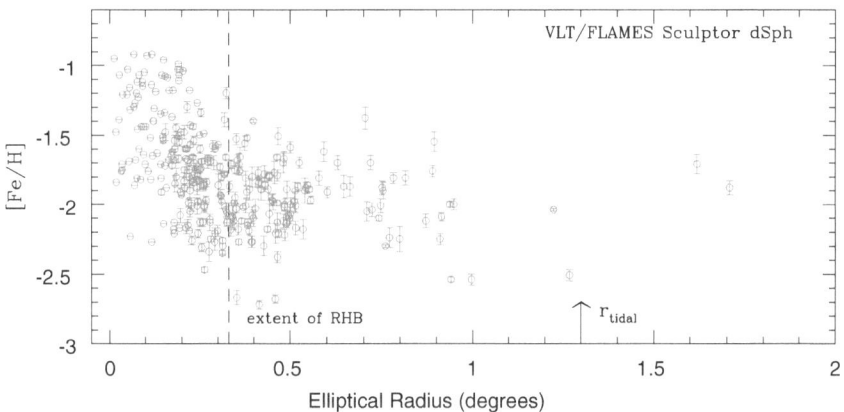

Fig. 2. VLT/FLAMES [Fe/H] measurements for 308 potential members of Sculptor dSph versus elliptical radius. Also marked, is the extent of the RHB distribution (as a dashed line) and the tidal radius, r_{tidal}.

Spectroscopy: The WFI images were used to select samples of stars on the RGB in Scl to take spectra in the Ca II triplet region with VLT/FLAMES. This resulted in radial velocity measurements and metallicity estimates for more than 400 stars in Scl over the fields outlined in Fig. 1, of which 300 have a high membership probability.

In Fig. 2 we show the distribution of [Fe/H] as a function of elliptical radius (the equivalent distance along the semi-major axis from the centre of Scl) for those RGB stars which were determined to have a high probability of membership. A well-defined metallicity gradient is apparent with a similar scale size to the RHB versus BHB spatial distributions.

In the central region of Sculptor we have high resolution spectra providing direct abundance measurements for numerous elements (Hill et al., in prep). In Fig. 3 we show the preliminary results for the α-elements (Ca, Mg & Ti) compared with similar observations of stars in our Galaxy. It is clear that the overall distribution of [α/Fe] versus [Fe/H] in Scl does not match our Galaxy, except for a small number of the most metal poor stars in Scl which overlap with Galactic halo stars.

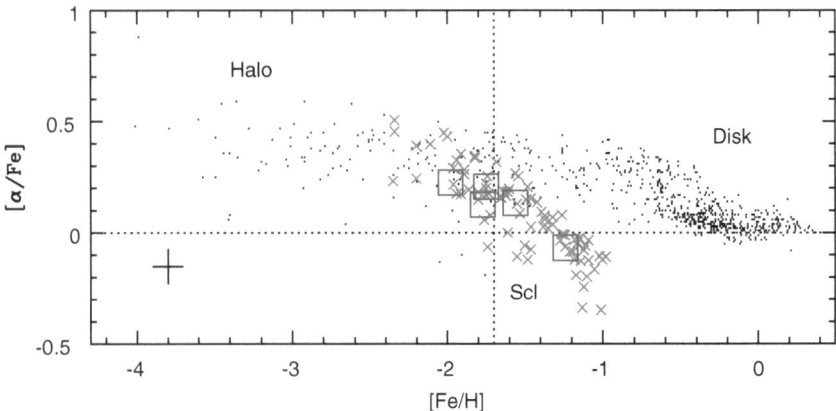

Fig. 3. The α-abundance (average of Ca, Mg and Ti) for stars in our Galaxy compared to those in Scl. The VLT/FLAMES high resolution measurements of 92 members in the central field are shown as crosses (from Hill et al., in prep). The Galactic stars come from standard literature sources (see [5] for references). The 5 open squares are UVES measurements of individual stars in Scl [3].

3   Two Stellar Components

Our FLAMES results clearly show that Scl contains two distinct stellar components with different spatial, kinematic and abundance properties [6]. There appears to be a metal-rich, $-0.9 > [Fe/H] > -1.7$, and a metal-poor, $-1.7 > [Fe/H] > -2.8$ component. The metal-rich component is more centrally concentrated than the metal poor, and on average appears to have a lower velocity dispersion, $\sigma_{metal-rich} = 7 \pm 1$ km/s, whereas $\sigma_{metal-poor} = 11 \pm 1$ km/s (see Battaglia, this volume).

There are indications that the presence of two populations is a common feature of dSph galaxies. Our preliminary analysis of HB stars, v_{hel} and [Fe/H] measurements in the other galaxies in our sample (Fornax and Sextans dSph; Battaglia et al., in prep) also shows very similar characteristics to Scl, especially in the most metal poor component. Pure radial velocity studies [15], [16] have also considered the possibility that kinematically distinct components exist in Ursa Minor, Draco and Sextans dSph galaxies.

What mechanism could create two ancient stellar components in a small dwarf spheroidal galaxy? A simple possibility is that the formation of these dSph galaxies began with an initial burst of star formation, resulting in a stellar population with a mean [Fe/H] ≤ -2. Subsequent supernovae explosions from this initial episode could have been sufficient to cause gas (and metal) loss such that star formation was inhibited until the remaining gas could sink deeper into the centre [17]. Thus the subsequent generation(s) of stars would inhabit a region closer to the centre of the galaxy, and have a higher average metallicity and different kinematics. Another possible cause is external influences, such as minor mergers, or accretion of additional gas. It might also be that events surrounding the epoch of reionisation influenced the evolution of these small galaxies [18] and resulted in the stripping or photoevaporation of the outer layers of gas in the dSph, meaning that subsequent more metal enhanced star formation occurred only in the central regions.

The full abundance analysis of the FLAMES HR data (Hill et al. in prep) will provide more details of the chemical enrichment history of Scl. This will hopefully enable us to distinguish between two episodes of star formation or more continuous star formation, manifested as a gradient in velocity dispersion and metallicity from the centre of the galaxy. Fig. 3 suggests that $[\alpha/\mathrm{Fe}]$ differs for the two populations, such that the metal poor (presumably older) population has high $[\alpha/\mathrm{Fe}]$ consistent with the halo of our Galaxy, and the more metal rich population doesn't match any of the Galactic stars. The dotted line is drawn at [Fe/H]$= -1.7$ to show the proposed dividing line between the two populations, although as can be seen in Fig. 2 there is not a clear division, however the kinematics provide clear support[6].

4   Dwarf Galaxies and Galaxy Formation

It is clear from Fig. 3 that the Scl dSph does not, in the mean, have stellar abundance properties consistent with our Galaxy, and the evidence points to this also being the case for most other nearby galaxies. This suggests that nearby dwarf galaxies are not the building blocks left over from the hierarchical formation of galaxies like our own. If we wish to retain a hierarchical formalism to explain the formation and evolution of our Galaxy then a mechanism has to be found by which those objects which did merge to form our Galaxy evolved differently from the similar mass (dwarf) galaxies we see around today. It might be, for example, that the building blocks formed and evolved much closer in to the central potential and thus their star formation history and chemical evolution were affected such that these processes proceeded much more rapidly than in their more distant cousins. Although this can be understood in qualitative generalised terms there is no particular evidence to support this. You would think that a difference in the abundance patterns of stars formed close to the centre of a potential and further out would then suggest some kind of gradient in properties, or at least that today we might pick up an unabsorbed building block in the form of a dwarf galaxy. So far detailed abundances have been determined for all the dwarf galaxies in our halo (Shetrone 2004, this volume), including Sagittarius (Bonifacio 2004, this volume), which is in the process of merging with our Galaxy and they all show an astonishing uniformity of abundance ratios, for [α/Fe] as well as r- and s- process elements like [Ba/Eu], and [Y/Ba] even if [Fe/H] varies between -3 and -0.5 dex (see Venn 2004, this volume). We also see evidence for similar abundance ratios in young massive stars in more distant higher mass dwarf irregular galaxies (see Kaufer 2004, this volume) and in RGB stars in the Magellanic Clouds (Hill 2004, this volume). This uniformity in itself suggests a remarkably stable enrichment process. This is even more striking considering that all these galaxies are so different in their star formation histories, but their mean stellar abundance patterns *for stars of all ages* are very similar to each other and very different to our Galaxy.

So, although we cannot rule out the possibility that we are living at a particular time where all recent (and future) mergers will be of different types of objects than created the bulk of our Galaxy, it might be wise to start to consider some variations on the standard scenario. It is possible to argue that the most metal poor tail of stars in Scl dSph overlap the properties of the Galactic halo stars, suggesting evidence for extremely early (gas rich) merging meaning that most of the stars we see in our Galaxy were actually formed there, and only the metal poor stars in the halo may have been formed in the satellites themselves.

Acknowledgments: I am grateful for support from a fellowship of the Royal Netherlands Academy of Arts and Sciences, and the exceptional collaborators that make up DART: Vanessa Hill Mike Irwin, Pascale Jablonka, Kim Venn, Matthew Shetrone, Amina Helmi, Giuseppina Battaglia, Bruno Letarte, Andrew Cole, Francesca Primas, Patrick François, Nobuo Arimoto, Andreas Kaufer, Thomas Szeifert & Tom Abel.

References

1. Mateo M. 1998 ARA&A, 36, 435
2. Freeman K. & Bland-Hawthorn J. 2002 ARA&A, 40, 487
3. Shetrone M.D. et al. 2003 AJ, 125, 684
4. Tolstoy E. et al. 2003 AJ, 125, 707
5. Venn K. et al. 2004 AJ, 128, 1177
6. Tolstoy, E. et al. 2004 ApJL, in press (astro-ph/0411029)
7. Kunkel W.E. & Demers S. 1977 ApJ, 214, 21
8. Queloz D., Dubath P. & Pasquini L. 1995 A&A, 300, 31
9. Irwin M. & Hatzidimitriou D. 1995 MNRAS, 277, 1354
10. Bouchard A., Carignan C. & Mashchenko S. 2003 AJ, 126, 1295
11. Monkiewicz J. et al. 1999 PASP, 111, 1392
12. Hurley-Keller D., Mateo M., & Grebel E.K. 1999 ApJL, 523, 25
13. Majewski, S.R., Siegel, M.H., Patterson, R.J. & Rood, R. 1999 ApJL, 520, 33
14. Lee Y-W. et al. 2001 *Astrophysical Ages & Time Scales*, eds. T. von Hippel et al.
15. Wilkinson M. et al. 2004 ApJL, 611, 21
16. Kleyna et al. 2004 MNRAS, 354, L66
17. Mori M., Ferrara A. & Madau P. 2002 ApJ, 571, 40
18. Skillman E.D. et al. 2003 ApJ, 596, 253

Abundances in Local Group Early-Type Stars

A. Kaufer[1], K.A. Venn[2], E. Tolstoy[3], and R.P. Kudritzki[4]

[1] European Southern Observatory, Alonso de Cordova 3107, Santiago 19, Chile
[2] Macalester College, 1600 Grand Avenue, Saint Paul, MN, 55105, USA
[3] Kapteyn Institute, University of Groningen, PO Box 800, 9700AV Groningen,
   The Netherlands
[4] Institute for Astronomy, University of Hawaii at Manoa, 2680 Woodlawn Drive,
   Honolulu, Hawaii 96822, USA

Abstract. We present abundance studies of dwarf irregular galaxies in the Local Group which appear to have undergone very slow chemical evolution since they have low nebular abundances, but have had ongoing star formation over the past 15 Gyr. They are too distant for red giant abundance analyses to examine the details of their chemical evolution. However the isolated, bright blue supergiants do allow us to determine their present-day iron abundances. We compare the [α/Fe] ratios in the dwarf irregulars to those from recent analyses of red giant branch stars in dwarf spheroidal galaxies, damped Lyman α absorption systems, and the latest model predictions.

1 Dwarf Irregular Galaxies

Dwarf irregular galaxies (dIrr's) are low mass, but gas rich galaxies and are found rather isolated and spread throughout the Local Group. All dIrr's display low elemental abundances indicating that only little chemical evolution has taken place over the past 15 Gyr despite ongoing star formation. The lack of strong starburst cycles in these isolated objects may be because of little or no merger interaction. Therefore, in the context the cold dark matter scenarios of hierarchical galaxy formation by merger of smaller structures [16], dwarf galaxies and in particular the isolated dwarf irregular galaxies could be the purest remnants of the proto-galactic fragments from the early Universe. Hence, the dIrr's are also one of the possible sources for the damped Lyα absorption (DLA) systems as observed in quasar spectra over a large range of redshifts; see e.g. [13] for a recent compilation of DLA metallicities over $0.5 < z < 5$. The dIrr's being possible remnants of the early Universe allow us to study early galaxy evolution in great detail in the Local Group. In particular the early chemical evolution of galaxies is of interest here to shed light on the formation of the first generations of stars. The most powerful way to study the chemical evolution of a galaxy is via the elemental abundances and abundance ratios of their stellar and gas content which contain a record of the star formation histories (SFH) of the galaxy over the last 15 Gyr.

2 Abundance Analysis of dIrr's

The analysis of bright nebular emission lines of HII regions has been the most frequent approach to modeling chemical evolution of more distant galaxies to date [8]. So far, only a very limited number of elements can be examined and quantified when using this approach. The chemical evolution of a galaxy depends on the contributions of all its constituents, e.g., SNe type Ia and II, high mass stars, thermal pulsing in low and intermediate mass AGB stars. Thus, more elements than just those observed in nebulae need to be measured, since each have different formation sites which sample different constituents. The α elements like oxygen are created primarily in short-lived massive stars, while the iron-group elements are produced in SNe of both high and low mass stars. Therefore, it is expected that substantially different SFHs of individual galaxies shall be represented in different present-day α/Fe ratios as discussed e.g. in [5]. In reverse, the accurate measurement of the absolute stellar abundances and abundance ratios allows to put constraints on the SFH of the respective galaxy.

Unfortunately, most dIrr systems are too distant for the detailed study and abundance analyses of e.g. their red giant branch (RGB) stars, which are of great importance due to their large spread in age. Hence, to date few details on the chemical evolution of the dIrr galaxies have been available. In the case of the dIrr's studied in this work the tip of the RGB is found at visual magnitudes far beyond the capabilities of even the most efficient high-resolution spectrographs on today's 8 to 10-meter class telescopes. However, with the same class of latest telescopes and instrumentation, the visually brightest stars of the dIrr galaxies, i.e., the blue supergiants become now accessible to detailed spectroscopy, which allow us to determine the present-day abundances of α *and* iron group elements. The abundances of the α elements derived from the rich spectra of hot massive stars can be directly compared and tied to the nebular α element abundances because both nebulae and massive stars have comparable ages and the same formation sites.

3   Stellar Abundances in dIrr's

The first abundance studies based on high-resolution spectra of A-type supergiant stars were carried out by our group for the nearby Local Group dIrr's NGC 6822 and WLM [18,19]. α element and iron group abundances could be derived for two stars in both dIrr galaxies. Recently, by pushing the observational capability of UVES at the VLT to its limit, the analysis of three A-type supergiant stars with $V \approx 19.5$ in the more distant and more metal-poor dIrr galaxy Sextans A could be added to the sample [6]. Also some preliminary results from one blue supergiant star in the dIrr galaxy GR8 are available but not discussed in detail here.

For NGC 6822, WLM, Sextans A, and GR8, the [α/Fe] ratios are in good agreement with the *solar* ratio, cf. Fig. 1. This result is somewhat surprising since all four galaxies have ongoing star formation and different star formation histories as determined from their stellar populations (e.g. [4,10,2]). This α/Fe ratio is much lower than in Galactic stars of similar metallicity [3,11]. At the metallicity of the Sextans A and GR8 stars, [Fe/H] ≈ -1, the Galactic thick disk and Galactic halo stars overlap with a [α/Fe] $\approx +0.3$, cf. Fig. 1. The low α/Fe

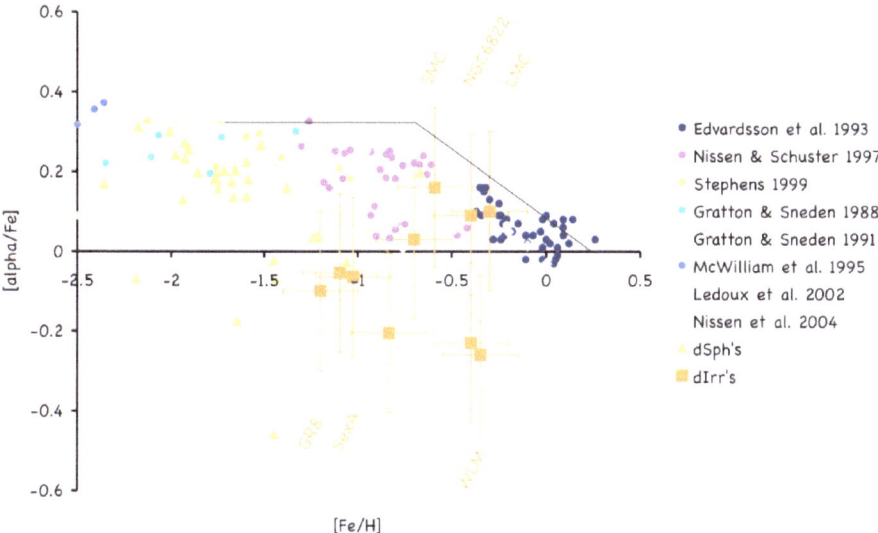

Fig. 1. [α/Fe] ratios as function of metallicity for Galactic stars, DLA's, dSph's, and dIrr's. For a discussion cf. text.

ratio in the Sextans A and GR8 stars further follows the trend found from the abundance analyses of stars in two other dIrr's, NGC 6822 and WLM [18,19], but extends the metallicity range now sampled from [Fe/H] = −0.5 in those galaxies to [Fe/H] = −1.0. At these lower metallicities Sextans A and GR8 now overlap the upper metallicities sampled by red giant stars in dwarf spheroidal galaxies, which also show much lower α/Fe ratios than Galactic metal-poor stars ([14,15,17] and Shetrone's contribution to these proceedings). In the dwarf spheroidal galaxies, the low [α/Fe] ratios near [Fe/H] = −1.0 are most likely due to the lower star formation rates, thus slower chemical evolution. The metallicity range of the dwarf irregular galaxies also overlaps with the damped Lyα absorption systems [12,7], which are also recognized to have low α/Fe ratios. This might suggest that low α/Fe ratios are a generic effect at low metallicities.

The star formation histories for Sextans A, GR8, WLM, and NGC 6822 are globally similar as derived from their CMDs. A comparison can be seen in Fig. 8 of [10]. Accordingly, all four dIrr's had significant star formation at ancient times, > 5 − 10 Gyr ago, with a hiatus at intermediate ages, 1 − 5 Gyr ago, and recent star formation events in the past 1 Gyr. [9] has recently emphasized that the SFH of a galaxy affects the *absolute* elemental abundances, but is only of minor importance for the abundance *ratios*, the latter being primarily determined by the stellar lifetimes, initial mass functions (IMF) and the stellar nucleosynthesis. Thus, the impact of different SFHs on the α/Fe ratio is related to the star formation efficiency, and manifests itself as a short plateau in the [α/Fe] ratio at low metallicities in galaxies with low star formation rates, like irregulars (either in bursts or continuous), while the plateau is maintained to higher metallicities for systems with high star formation rates like bulges and ellipticals. This scenario is sketched in Fig. 1 for dIrr/dSph's (orange) and Galactic stars (black). Thus, our finding that the [α/Fe] ratios in three dIrr's with similar SFHs is significantly lower than in the metal-poor Galactic stars can be explained in this scenario. The three dIrr's with a metallicity range from $-1.2 < $ [Fe/H] $ < -0.4$ have already left the [α/Fe] plateau and presently display the same lower [α/Fe] ratio. That this ratio is similar to *solar* suggests that the integrated stellar yields from a star formation event is consistent with the solar abundance ratios, probably due to the *universal* nature of stellar lifetimes, IMFs, and stellar nucleosynthesis. That the high-redshift DLA absorption systems also show low [α/Fe] ratios suggests that they are also systems with low star formation rates, but this could be either irregular galaxies or the outer parts of spirals. A recent study of published DLA system abundances in comparison with chemical evolution models of different galaxy morphologies by [1] has identified the irregular and spiral galaxies as the possibly ideal sites to create the DLA systems, while ellipticals can be ruled out.

References

1. F. Calura, F. Matteucci, G. Vladilo: MNRAS **340**, 59 (2003)
2. A.E. Dolphin, A. Saha, E.D. Skillman: AJ, **125**, 1261 (2003)
3. B. Edvardsson, J. Andersen, B. Gustafsson, et al.: A&A, **275**, 101 (1993)
4. C. Gallart, A. Aparicio, G. Bertelli, C. Chiosi: AJ, **112**, 1950 (1996)
5. G. Gilmore, R.F.G. Wyse: ApJ, **367**, L55 (1991)
6. A. Kaufer, K.A. Venn, E. Tolstoy, C. Pinte, R. Kudritzki: AJ, **127**, 2723 (2004)
7. C. Ledoux, J. Bergeron, P. Petitjean: A&A, **385**, 802 (2002)
8. F. Matteucci, M. Tosi: MNRAS, **217**, 391 (1985)
9. F. Matteucci: APSS, **284**, 539 (2003)
10. M. Mateo: Annual Review A&A, **36**, 435 (1998)
11. P.E. Nissen, W.J. Schuster: A&A, **326**, 751 (1997)
12. P.E. Nissen, Y.Q. Chen, M. Asplund, M. Pettini: A&A, **415**, 993 (2004)
13. J.X. Prochaska, E. Gawiser, A.M. Wolfe, et al.: ApJ, **595**, L5 (2003)
14. M.D. Shetrone, P. Côté, W.L.W. Sargent: ApJ, **548**, 592 (2001)
15. M.D. Shetrone, K.A. Venn, E. Tolstoy, et al.: AJ, **125**, 684 (2003)
16. M. Steinmetz, J.F. Navarro: ApJ, **513**, 555 (1999)
17. E. Tolstoy, K.A. Venn, M. Shetrone, et al.: AJ, **125**, 707 (2003)
18. K.A. Venn, A. Kaufer, J.M. McCarthy, et al.: ApJ, **547**, 765 (2001)
19. K.A. Venn, E. Tolstoy, A. Kaufer, et al.: AJ, **126**, 1326 (2003)

The Extreme Ends of the Metallicity Distribution in dSph Galaxies

M. Shetrone

University of Texas, McDonald Observatory, Fort Davis, Tx 79734, USA

Abstract. This paper reviews recent abundance results of local dSph giants. All dSph systems seem to show evidence of slow star formation rates, compared to the Milky Way, based on the most metal-rich stars exhibiting low [even-Z/Fe] ratios and high [s-process/r-process] ratios. The most metal-poor stars in the Draco, Ursa Minor, Sextans and Sculptor dSphs seem to show a split in their light even-Z, Mg and O, and the heavier even-Z, Ca and Ti, abundance ratios, where the light even-Z are halo like and the heavier even-Z elements exhibit sub-halo abundance ratios. This split remains a mystery. A review of the first dSph abundance results from FLAMES+GIRAFFE on the VLT and HRS on the Hobby-Eberly Telescope [1] shows that the study of chemical evolution of dSph galaxies is rapidly moving out of infancy and into an era requiring very large surveys and/or targeted studies.

1 The Most Metal-Rich dSph Stars

The most metal-rich stars in dwarf spheroidals (dSph) have been shown to have significantly lower even-Z abundance ratios than stars of similar metallicity in the Milky Way (MW). In addition, the most metal-rich dSph stars are dominated by an s-process abundance pattern in comparison to stars of similar metallicity in the MW. This has been interpreted as excessive contamination by Type Ia super-novae (SN) and asymptotic giant branch (AGB) stars (Bonifacio et al. 2000, Shetrone et al. 2001, Smecker-Hane & McWilliam 2002). By comparing these results to MW chemical evolution, Lanfranchi & Matteucci (2003) conclude that the dSph galaxies have had a slower star formation rate than the MW (Lanfranchi & Matteucci 2003). This slow star formation, when combined with an efficient galactic wind, allows the contribution of Type Ia SN and AGB stars to be incorporated into the ISM before the Type II SN can bring the metallicity up to MW thick disk metallicities.

Recent abundance ratio work in this field falls into two categories. The first category has been investigations into aspects of metal-poor AGB and Type Ia SN yields and their relationship to the chemical evolution in the dSph galaxies, e.g. McWilliam et al. (2003), Venn et al. (2004), McWilliam & Smecker-Hane (2005). In these works the abundances of specific elements are compared to

---

[1] The Hobby-Eberly Telescope (HET) is a joint project of the University of Texas at Austin, the Pennsylvania State University, Stanford University, Ludwig-Maximilians-Universität München, and Georg-August-Universität Göttingen. The HET is named in honor of its principal benefactors, William P. Hobby and Robert E. Eberly.

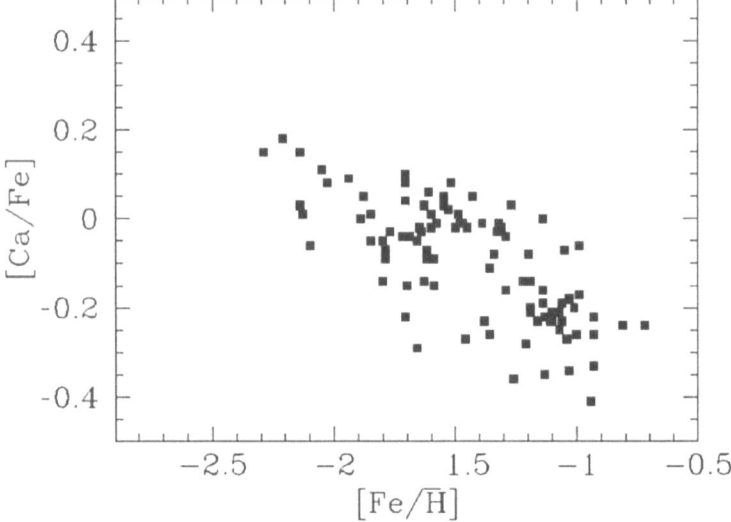

Fig. 1. An example of the new data sets coming available with the VLT+FLAMES. This data, from Hill (private communication, 2005) exhibits the steep decline of [Ca/Fe] with increasing [Fe/H] seen in all dSph. In addition, the most metal-poor stars do not exhibit the typical MW halo ratio near 0.3 dex.

predictions of yields of low metallicity SN and AGB stars. While the origins of these elements, such as Mn, Cu and Y, may seem slightly esoteric, these types of analyses will help constrain future models of SN yields.

The second category of dSph abundance investigations has been attempts to gain large enough samples to accurately model the extent of the chemical evolution, the relative contributions the Type Ia and AGB yields and to what extent galactic winds have played a roll in the chemical evolution. The new instruments that have come on-line in the last year have increased the multiplexing capabilities of these surveys. As an example, Figure 1 shows preliminary results from the Dwarf Abundance and Radial velocity Team (DART) ESO large program; using UVES FLAMES+GIRAFFE on a sample of Sculptor dSph giants, Hill (private communication, 2005) collected nearly 100 stellar spectra, a sample larger than all of the literature high resolution dSph surveys combined. This survey will be able to show subtle declines and trends that the other surveys could never detect. The decline seen in [Ca/Fe] with increasing [Fe/H] reported by Shetrone et al. (2003) and Geisler et al. (2004) are easily detected. The spread in [Ca/Fe] at a given metallicity is being investigated by DART.

2   The Most Metal-Poor dSph Stars

In a closed box or leaky-box chemical evolution model the most metal-poor stars would have formed before the majority of the Type Ia SN or the AGB stars were significant contributors to the the ISM. Thus, the abundances of these very metal-poor stars should be excellent surrogates of the Pop III and very metal-poor Pop II Type II SN products held in the dSph gravitational potential. This last point is important because if the yields of some masses of Type II SN are lost from the dSph gravitational potential then this might significantly impact abundance ratios found in the next generation of dSph stars.

The first papers in this field suggested that the overall alpha abundances found in the most metal-poor dSph stars are not similar to those found in the halo. However, a more detailed analysis of the individual elements including corrections for differences in log gf values has shown that the O and Mg abundances are consistent with those found in the MW halo, while the Ca and Ti abundances are systematically lower than those in the MW halo, see figure 1, 2 and 6 in Shetrone (2004). This can also be seen in Figure 1 which shows that the most metal-poor Sculptor stars have [Ca/Fe] ratios less than 0.2 dex while the halo median at this metallicity is roughly 0.15 dex larger.

The difficulty with studies of the most metal-poor dSph stars is actually finding the most metal-poor stars. Not only are these stars less numerous than their more metal-rich counter parts, but their spatial distribution is larger than the more metal-rich stars, e.g. Tolstoy et al. (2004), Palma et al. (2003), Harbeck et al. (2001), Majewski et al. (1999). The use of high resolution multi-object spectrographs such as FLAMES becomes less efficient in the search for the most-metal poor stars because of the large spatial extent, rarity and huge background contamination. For these types of studies targeting single star high resolution spectral follow-up to photometric or low resolution surveys can be more appropriate.

One star analyzed in the Shetrone et al. (1998) survey was found to be very metal-poor but with fairly low overall-alpha abundances. Unfortunately, due to the low S/N and low metallicity many of the elements had upper limits and large error bars. This single star, Draco 119, was re-observed by Fulbright et al. (2004) with much higher S/N. They were able to confirm that the Mg abundance was halo like while the Ca and Ti abundances were lower than those found in the halo by a few tenths of dex. Even more remarkable were the upper limits found for the neutron capture elements. Fulbright et al. found upper limits for [Ba/Fe] 1 dex lower than MW halo giants of similar metallicity, and, even more amazingly, the upper limit for [Sr/Fe] was found to be nearly 2 dex lower than similar MW giants. This begged the question: is the Draco 119 abundance pattern unique, i.e. due to some strange inhomogeneous mixing event, or is this pattern found in all very metal-poor Draco stars.

A search for equally metal-poor Draco stars, using the Hobby-Eberly telescope, did not turn up any Draco giants as metal-poor as Draco 119; but it did turn up a few stars just a few tenths more metal-rich. By integrating long enough to detect the strong red Ba lines in these stars using the HRS on the Hobby-Eberly telescope, Shetrone et al. (2005) did not find extremely neutron capture poor stars, see Figure 2. The abundance pattern of Draco 119 appears

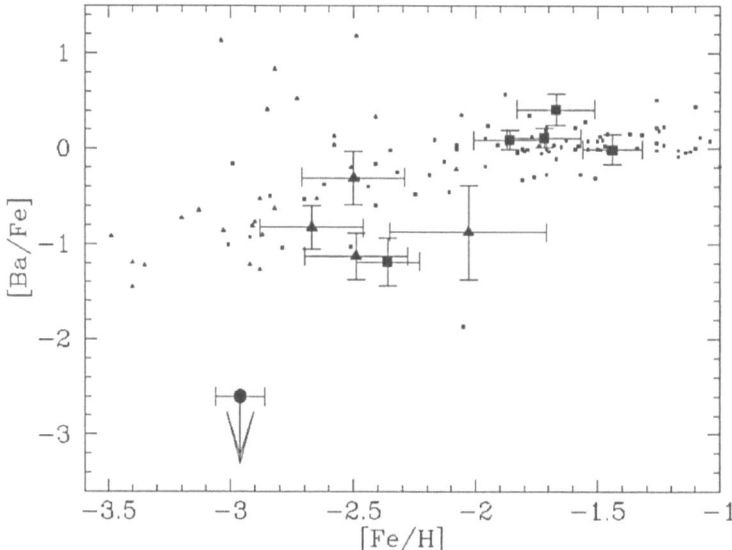

Fig. 2. The abundance ratio of Ba from the Shetrone et al. (1998), squares, Fulbright et al. (2004), circle, and Shetrone et al. (2005), triangles shown plotted against their derived metallicities. The upper limit from Fulbright et al. (2004) is more than an order of magnitude lower than that of the slightly more metal rich Draco dSph stars or comparable metallicity MW halo stars, small symbols.

to be a due to inhomogeneous mixing and is not found in all very metal-poor Draco stars.

I would like to thank Kim Venn, Andy McWilliam, Verne Smith, Jon Fulbright and the DART for preprints and invaluable discussions. I would also like to thank the NSF for support through AST-0306884; and summer REU intern John Moore and the team at the HET for their assistance in bringing these results to press.

References

1. Bonifacio, Hill, Molaro, Pasquini, Marcantonio, Santin: AA **359**, 663 (2000)
2. Fulbright, Rich, Castro: ApJ in press (2004)
3. Geisler, Smith, Wallerstein, Gonzalez, Charbonnel: in press (2004)
4. Harbeck, Grebel, Holtzman, Guhathakurta, Brandner, Geisler, Sarajedini, Dolphin, Hurley-Keller, Mateo: AJ **122**, 3092 (2001)
5. Lanfranchi, Matteucci: MNRAS **345**, 71 (2003)
6. Majewski, Siegel, Patterson, Rood: ApJL **520**, 33 (1999)
7. McWilliam, Rich, Smecker-Hane: ApJL **592**, 21 (2003)
8. McWilliam, Smecker-Hane, Cosmic Abundances as Records of Stellar Evolution and Nucleosynthesis ASP Conference Series, Ed. Bash and Barnes, (2005)
9. Palma, Majewski, Siegel, Patterson, Ostheimer, Link: AJ **125**, 1352 (2003)

10. Shetrone, Origin and Evolution of the Elements, from Carnegie Observatories Centennial Symposia. Ed. McWilliam and Rauch, p220 (2004)
11. Shetrone, Bolte, Stetson: AJ **115**, 1888 (1998)
12. Shetrone, Cote, Sargent: ApJ **548**, 592 (2001)
13. Shetrone, Moore, Abel, Tolstoy, Siegel, Winnick: in prep (2005)
14. Shetrone, Venn, Tolstoy, Primas, Hill, Kaufer: AJ **125**, 684 (2003)
15. Smecker-Hane, T., McWilliam, A . A astro/ph-0205411 (2002)
16. Tolstoy, Irwin, Helmi, Batagglia, Letarte, Jablonka, Hill, Primas, Venn, Shetrone, Cole, Arimoto, Sadakane, Kaufer, Szeifert, Abel: ApJL in press (2004)
17. Venn, Irwin, Shetrone, Tout, Hill, Tolstoy: AJ **128**, 1177 (2004)

UVES Detailed Chemical Abundances in the Sgr dSph and the CMa Overdensity

L. Sbordone[1,2]

[1] ESO – European Southern Observatory
[2] Universitá di Roma 2 "Tor Vergata"

1 Introduction

In this contribution, we present detailed chemical abundances on two stellar systems presently believed to be undergoing tidal merging with the Milky Way (MW). The first one is the Sagittarius Dwarf Spheroidal (Sgr dSph, [5]), a massive (10^8 M$_\odot$) dSph orbiting along a very short period ($< 1GYr$) almost polar orbit inside the Halo, along which is slowly dissolving in a huge stellar stream [6]. The second object is the recently discovered, and still controversial CMa dwarf galaxy [8], a heavily degraded overdensity embedded in the MW thick disk, believed to be the residual of an in-plane accretion of an object of mass comparable to the Sgr dSph. This interpretation is still controversial, since [9] claimed that the structure almost disappears, when disk warping and/or flaring is properly taken into account in the modeling of the MW contribution.

2 Data and Analysis

For the Sgr dSph we present the UVES DIC1 spectra for 12 giants. Complete analysis of two of them has already been published [2], while for the other ten only iron and α-elements abundances have been published so far (see [3]). Details on the reduction and analysis procedures, and physical parameters for the stars are provided in [3], but they can be briefly resumed here: the spectra have been analyzed by means of LTE, one dimensional atmosphere models, using ATLAS, WIDTH and SYNTHE codes (see [7] and [10]). T$_{eff}$ for the stars are in the range 4800 - 5050 K, log g between 2.3 and 2.7. We analyzed abundances of proton capture (Na, Al, Sc, V), α (Mg, Si, Ca, Ti), Iron-peak (Cr, Fe, Co, Ni, Zn) and heavy neutron-capture (Y, La, Ce, Nd) elements.

Immediately after the discovery of the CMA overdensity we obtained DDT time at VLT – FLAMES for an exploratory study of the CMa population underlying the galactic open cluster NGC 2477, identified by [1]. Here we present the results relative to the investigation of the 7 UVES spectra, the results relative to the stars observed through the MEDUSA fibers are presented in [13]. Four of the UVES stars had radial velocities or absolute magnitudes incompatible with the CMa overdensity; a paper detailing the analysis and results of the remaining three objects has been submitted (see [11]). The analysis has followed the same method above described for Sgr dSph stars. Two stars appear to be giants (log g 2.3 and 2.8, T_{eff} 4990 and 4994 K), while the third one is a subgiant (T_{eff} 5367, log g 3.5). For these stars, in addition to the elements above listed, we also analyzed Co, Cu, Ba and Eu.

3   Results

The most striking feature observed in Sgr dSph is the presence of a metal rich ([Fe/H] between -1 and 0) and young population, showing the α-elements under-abundance now recognized as typical of dSph (see [12]). Besides that, we observe a strong (up to 0.8 dex) overabundance of n-capture elements (La, Ce, Nd) with respect to iron, a significant Na and Al underabundance, and an intriguing Ni deficiency. All these chemical features appear to be typical of Sgr dSph and of its associated systems like Terzan 7 and Pal 12 (e.g. [4]). Moreover, we observed an important underabundance of Zn ([Zn/Fe] $\sim$ -0.4). This may undermine the parallelism between dSph and the DLAs, where Zn is used as a proxy for Fe.

Of the three stars in the CMa overdensity, the subgiant appears the most interesting one. First of all, as can be seen in [13], contamination of the sample is a significant issue here, and only the subgiant, with $v_r \sim 135 kms^{-1}$, has a very low likelihood to belong to the MW. This star has slightly over-solar iron ([Fe/H]=0.15), and shows underabundant alpha elements ([α/Fe]=-0.27) and strongly overabundant n-capture elements (e.g. [CeII/Fe]=0.75). All these seem to be signatures of an extragalactic formation. At the same time, it shows a significant Cu overabundance ([Cu/Fe]=0.25), highly unusual for a disk star. One of the giants appears to be almost surely a MW star (no significant depar-tures from thick-disk abundances), while the other shows, to a lesser extent, the same signatures found in the subgiant ([Fe/H]=0, [α/Fe]=-0.15, [Cu/Fe]=0.23, [LaII/Fe]=0.31, [CeII/Fe]=0.21). From these findings, we are lead to believe that a population of extragalactic origin actually exist inside the Canis Mayor overdensity, although the sample is too limited to allow us to push further the interpretation.

References

1. Bellazzini, M., et al., 2004, MNRAS, in press
2. Bonifacio P., et al., 2000, A&A , 359, 663
3. Bonifacio, P., et al., 2004, A&A, 414, 503
4. Cohen, J. G., 2004, AJ, 127, 1545
5. Ibata, R. A., Gilmore, G., & Irwin, M. J., 1995, MNRAS, 277, 781I
6. Majewski, S. R., et al. , 2003, ApJ, 599, 1082
7. Kurucz R. L., 1993, CDROM 13, 18
8. Martin, N. F., et al., 2004, MNRAS, 348, 12
9. Momany, Y., et al. , A&A, 421, L29
10. Sbordone, L., Bonifacio, P., Castelli, F., & Kurucz, R. L., 2004, MSAIS, 5, 93
11. Sbordone, L., et al., 2004, submitted to A&A letters.
12. Venn, K. A., et al., 2004, AJ, 128, 1177
13. Zaggia, S., et al., 2004, these proceedings.

The Metallicity Distribution and Its Gradients in the LMC via Calcium Triplet Spectroscopy

R. Carrera[1], C. Gallart[1], E. Pancino[2], R. Zinn[3], and E. Hardy[4]

[1] Instituto de Astrofísica de Canarias, Spain
[2] INAF - Osservatorio Astronomico di Bologna, Italy
[3] Yale University, USA
[4] N.R.A.O., Chile

Abstract. We present a new calibration of the CaII triplet as metallicity indicator based in 4 globular and 11 open clusters which cover a range of metallicity -2≤[Fe/H]≤+0.1 and age 13≤(Age/Gyr)≤0.25. We use it to derive the metallicity distribution in two fields situated at 5 and 8 degrees from the center of the LMC. We show that the mean [Fe/H] of the LMC field decreases as we move away from the bar.

1 A New Calibration of the Ca II Triplet

The infrared CaII triplet (CaT) is a economic way of obtaining global stellar metallicities for large numbers of stars. However, previous works derive relations valid only for metal-poor, old populations (e.g [5]). The very recent work by [2] has demonstrated that this calibration can be extended to more metal-rich regimes and to younger ages. The relation by [2] covers the range of metallicity -2≤[Fe/H]≤-0.2 and age 13≤(Age/Gyr)≤2. However, this calibration has to be further extended to be used in a galaxy such as the LMC, which has had star formation in the last Gyr and possibly contains stars as metal-rich as solar metallicity. For this purpose we have observed 4 globular and 11 open clusters which cover a metallicity range -2≤[Fe/H]≤+0.1 and age 13≤(Age/Gyr)≤0.25. The reduced equivalent width (W') of the CaT lines for each cluster, as defined in [2] shows a linear correlation with [Fe/H] for the whole range of ages and metallicities (fig. 1a). The main error source in the calibration is the metallicity determination for most of the open clusters.

2 The Metallicity Distribution in the LMC

We have deep color-magnitude diagrams in 4 fields situated at 3, 5, 6 and 8 degrees from the LMC bar (see [3]), in which we are performing CaT follow-up spectroscopy with HYDRA at the CTIO 4m telescope. As it is discussed in [3], the age of the population is older when we move away from the bar. At the moment we have low-resolution spectroscopy for the fields situated at 5 and 8 degrees. The histogram of the metallicities for each field is shown in fig. 1b.

We can compare our results with previous works (bar: [6]; 2°: [1]; 8^o5: [4]), also using CaT. However, since each of them are obtained with different calibrations, we only can consider this comparison as a first approximation. If we plot all these data together we find that the metallicity decreases when we move away from the bar. This gradient could have two interpretations: (i) the overall age-metallicity relation changes across the galaxy, with stars of the same age being more metal-poor as we move outwards or (ii) the age-metallicity relation is the same, but simply there are less intermediate-age, more metal-rich stars in the outer part.

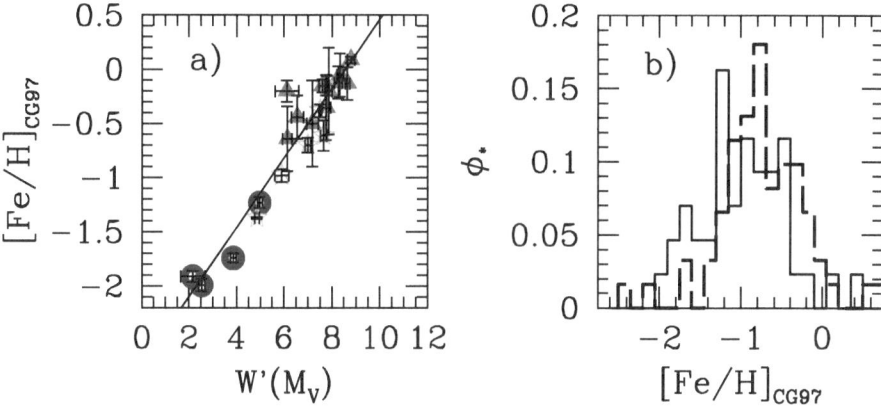

Fig. 1. a) Linear correlation between metallicity, in the Carreta & Gratton (1997) scale versus W'. Triangles are the open clusters, open circles are the globular clusters and open stars are the data by [2]. b) Metallicity distribution for the field situated at 5° (dashed line) and 8° (solid line) from the bar. The amount of stars with metallicity bellow [Fe/H]=-1 increases at larger distances from the bar.

References

1. Cole, A.A., Smecker-Hane, T.A. & Gallagher, J.S., III 2000, AJ, 120, 1808
2. Cole, A.A., et al. 2004, MNRAS, 347, 367
3. Gallart, C., Stetson, P.B., Hardy, E., Pont, F., Zinn, R. 2004, AJL, accepted
4. Olszewski, E.W. 1993, in ASP 48, ed. Graeme H. Smith, and Jean P. Brodie
5. Rutledge, G.A., Hesser, J.E., Stetson, P.B. 1997, PASP, 109, 907
6. Smith, V.V., et al. 2002, AJ, 124, 3241

Abundances in Sagittarius Stars

P. Bonifacio[1], S. Zaggia[1] L. Sbordone[2,3], P. Santin[1], L. Monaco[1], S. Monai[1],
P. Molaro[1], G. Marconi[3], L. Girardi[1], F. Ferraro[4], P. di Marcantonio[1],
E. Caffau[1], and M. Bellazzini[5]

[1] Istituto Nazionale di Astrofisica – Osservatorio Astronomico di Trieste,
    Via Tiepolo 11, I-34131 Trieste, Italy
[2] Università Tor Vergata, Roma
[3] European Southern Observatory, Casilla 19001, Santiago, Chile
[4] Universitá degli Studi di Bologna, Via Ranzani, Bologna, Italy
[5] Istituto Nazionale di Astrofisica – Osservatorio Astronomico di Bologna,
    Via Ranzani, Bologna, Italy

Abstract. The Sagittarius dwarf spheroidal is a very complex galaxy, which has undergone prolonged star formation. From the very first high resolution chemical analysis of Sgr stars, conducted using spectra obtained during the commissioning of UVES at VLT, it was clear that the star had undergone a high level of chemical processing, at variance with most of the other Local Group dwarf spheroidals. Thanks to FLAMES at VLT we now have accurate metallicities and abundances of alpha-chain elements for about 150 stars, which provide the first reliable metallicity distribution for this galaxy. Besides the already known high metallicity tail the existence of a metal-poor population has also been highlighted, although an assessment of the fraction of Sgr stars which belong to this population requires a larger sample. From our data it is also obvious that Sagittarius is a nucleated galaxy and that the centre of the nucleus coincides with M54, as already shown by Monaco et al.

1   A Closer Look at the Harassed Neighbour

Sagittarius, the nearby galaxy which is being harassed by the Milky Way has been the object, in recent years, of several investigations which have highlighted the complexity of this galaxy [1–5]. We have used a part of the guaranteed time of the ITAL-FLAMES consortium to begin a wide field spectroscopic study of this galaxy. In May 2003 we used FLAMES on two fields, one centered on M54 and the other centered on the "field 1" of [1], which is 22′ West of M54. The targets were selected from the photometry of [6] in such a way as to sample the whole width of the Red Giant Branch, in order to allow us to fully capture the metallicity spread of Sagittarius (see figure 1 of [5], to get a visual impression of the selection). We used both UVES and Giraffe, the analysis of the UVES spectra is described by Monaco in this volume[7]. We obtained spectra using two high resolution settings of Giraffe, setting HR09, centered at 525.8nm and setting HR14, centered at 651.5nm. The grating we used has been decommissioned in October 2003, it was less efficient than the one currently used, but provided a higher resolution (about 12 kms^{-1}), the price paid was a lower S/N ratio and a smaller spectral range of the two settings. Not all stars in the M54 field have been observed in both settings, some have only one of the two settings.

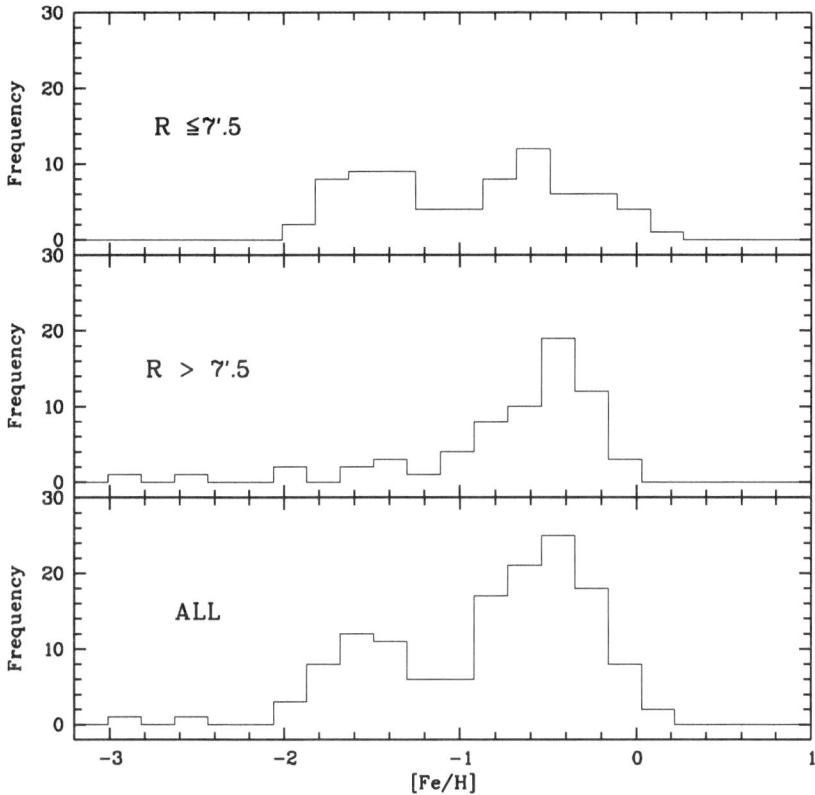

Fig. 1. The metallicity distribution of Sgr. The bottom panel shows the metallicities for all our sample of stars, the second peak at metallicity about –1.5 is due to the presence of M54, in the middle panel only stars *outside* the tidal radius of M54 are histogrammed, and in the top panel only stars *inside* the tidal radius of M54.

Radial velocities were measured by cross-correlation, using a synthetic spectrum as template. Individual spectra were shifted to rest wavelength and coadded. Effective temperatures were derived from the $(V-I)_0$ colours by means of the Alonso calibration [8]. We assumed log g = 2.0 for all stars (estimated from isochrones) and with these parameters we fed the spectra to our automatic procedure for the determination of abundances [9]. We found that the S/N ratio was too low to be able to determine reliably the microturbulent velocities, the weak Fe I lines could not be measured on many spectra. This resulted in a marked dependence of derived abundances on microturbulent velocities. It is well known that microturbulence is not a truly independent parameter but correlates with surface gravity and, more mildly also with effective temperature. By considering the large sample of stars studied by [10] one can be convinced that for all stars with $1.5 \leq \log g < 3.0$ (20 stars) there is no marked dependence from either $T_{\rm eff}$ or log g, and the mean value of the microturbulent velocity is 1.6 kms^{-1}. For this reason we fixed the microturbulent velocity at 1.6 kms^{-1}.

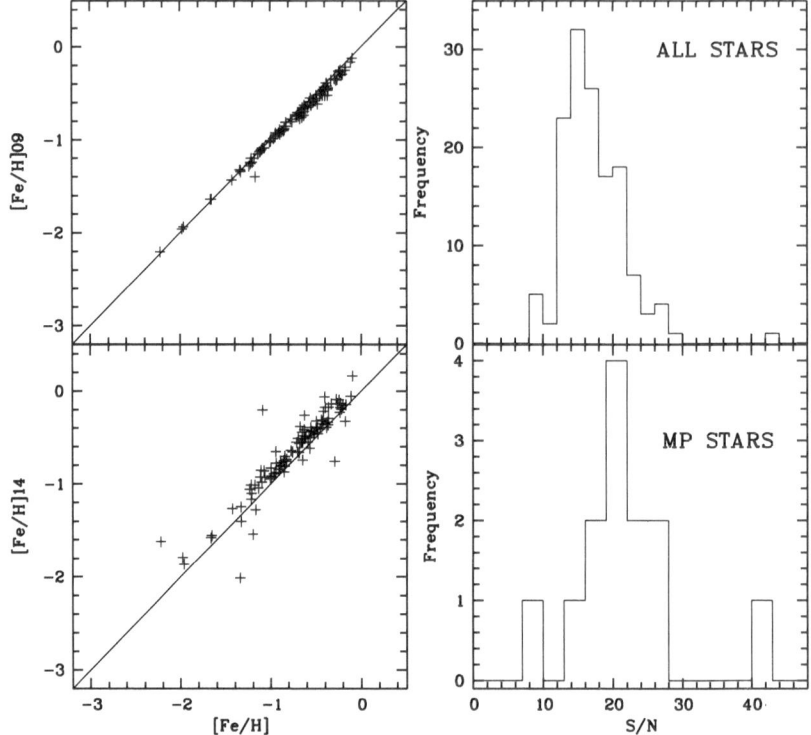

Fig. 2. Sanity checks: the left panel shows the comparison between [Fe/H] derived from both HR09 and HR14 setting versus [Fe/H] derived from setting HR14 and HR09 only; the right panel shows the S/N ratios for all Sgr stars (top) and for the Sgr stars with [Fe/H]≤ -1.72 (bottom).

2   Results and Discussion

For the present work we define as Sgr radial velocity members all stars with radial velocities in excess of 100 kms^{-1}. This is a quite crude assumption, we know, from Galactic models, that a low contamination of Bulge and Disc stars should be present even at these high radial velocities. As a matter of fact we already purged the sample from a few high velocity K dwarfs. A more sophisticated analysis which allows to "clean" the sample from contaminating populations is in progress, however from our preliminary results, we estimate the contamination to be quite low. The derived [Fe/H] values for our Sgr radial velocity members are histogrammed in Fig. 1. The presence of a metal–poor tail, extending down to [Fe/H]~ -3.0 is obvious. This feature is detected here for the first time, however the size of the sample is too low to try to estimate the percentage of Sgr stars which belong to this tail. Clearly with such small numbers the issue of contamination, even low, is crucial, however we believe it is unlikely that the metal-poor tail is due entirely to interlopers, unrelated to Sgr. By assuming, naively, that metallicity is related to age, the metal-poor tail, taken at face value, suggests that the Sgr hosts a population which is as old as the oldest Galactic populations, thus Sagittarius and the Milky Way are coeval. This is also consistent with the blue horizontal branch of Sgr.

Another feature which is outstanding in Fig. 1 is that when we look *inside* the tidal radius of M54 we still find the metal rich dominant population of Sgr. This has already been demonstrated by Monaco et al. [11] from an analysis of the colour-magnitude diagram and it is comforting that we recover the result by looking at the metallicities of the stars. It is therefore clear that Sagittarius is a nucleated galaxy with the centre of the nucleus coinciding with that of M54.

We performed a couple of sanity checks, shown in Fig. 2. In the first place we checked that the metallicity derived by using only setting HR09 or setting HR14 is essentially the same as that derived by using both settings. When only setting HR09 is used the correlation is excellent, there might be instead a very small offset, at the level of 0.1 dex when using only setting HR14, however for the present purposes this may be ignored. In the second place we checked that the metal-poor tail does not comprise only stars with very poor S/N ratios. Figure 2 shows that this is not the case. In fact the metal-poor stars have on average higher S/N ratios than the more metal-rich stars, this simply reflects the fact that they are brighter than the more metal rich stars, as predicted by theoretical isochrones. These checks convinced us that the metal-poor tail is not due to a bias in our analysis.

The extreme complexity of Sgr, which we are unveiling, calls for observations of larger samples of stars. Unfortunately our observations in period 73 were plagued by bad weather, we hope to be more lucky in period 75!

References

1. Marconi, G., Buonanno, R., Castellani, M., Iannicola, G., Molaro, P., Pasquini, L., & Pulone, L. 1998, A&A, **330**, 453
2. Bonifacio P., Hill V., Molaro P., Pasquini L., Di Marcantonio P., Santin P. 2000, A&A , **359**, 663
3. Smecker–Hane T.A., McWilliam A., 2002, ApJ submitted, astro-ph/0205411
4. Bonifacio, P., Sbordone, L., Marconi, G., Pasquini, L., & Hill, V. 2004, A&A, **414**, 503
5. Zaggia S., Bonifacio P., Bellazzini M., Caffau E., Di Marcantonio P., Ferraro F., Marconi G., Monaco L., Monai S., Santin P., Sbordone L., 2004, MSAIS, **5**, 291
6. Monaco, L., Ferraro, F. R., Bellazzini, M., & Pancino, E. 2002, ApJ, **578**, L47
7. Monaco L., 2004 in *Chemical Abundances and Mixing in Stars in the Milky Way Galaxy and its Satellites*, L. Pasquini & S. Randich eds., Springer-Verlag, Berlin, p.
8. Alonso, A., Arribas, S., & Martínez-Roger, C. 1999, A&AS, **140**, 261
9. Bonifacio, P. & Caffau, E. 2003, A&A, **399**, 1183
10. Fulbright, J. P. 2000, AJ, **120**, 1841
11. Monaco, L., Bellazzini, M.,Ferraro, F. R., & Pancino, E. 2004, MNRAS in press, astro-ph/0411107

The Composition of the Sagittarius Dwarf Spheroidal Galaxy and Implications for Nucleosynthesis and Chemical Evolution

A. McWilliam[1] and T. Smecker-Hane[2]

[1] Carnegie Observatories, 813 Santa Barbara Street, Pasadena, CA, USA
[2] Dept. of Physics & Astronomy, University of California at Irvine, Irvine, CA, USA

Abstract. We have measured the chemical composition of 14 stars in the Sagittarius dwarf spheroidal galaxy (Sgr dSph) using high S/N Keck HIRES echelle spectra. For the Sgr dSph stars with [Fe/H]≥-1 the abundances are highly unusual, showing a striking enhancement in heavy s-process elements, increasing with [Fe/H], deficiencies of the α-elements (O, Si, Ca, and Ti), deficiencies of Al and Na, and deficiencies of the odd-numbered iron-peak elements Mn and Cu. Our abundances suggest that the composition of the metal-rich Sgr dSph stars is dominated by the ejecta of an old, metal-poor population, including products of AGB stars and type Ia supernovae (SN).

The simplest scenario to explain our abundance results is that chemical enrichment of the Sgr dSph occurred over long timescales (several Gyr), during which time this galaxy experienced significant gas loss. This led to the situation where injection of newly synthesized material was dominated by low-metallicity, low-mass, AGB stars and type Ia SN. Since mass-loss from low-mass galaxies is generally expected to have occurred, on the basis of their low mean metallicity, we anticipate that our unusual abundance patterns may be quite common among low-mass galaxies. Published abundance results for the Fornax dSph and the Galactic globular cluster ω Cen are similar to Sgr dSph, which suggests similarities in the chemical enrichment histories of all three systems. Filled circles represent our Sgr dSph results; others are solar neighborhood.

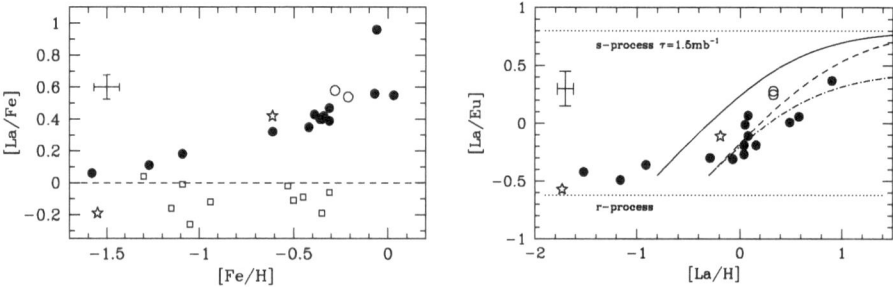

Fig. 1. *Left:* Enhancements up to 1 dex in [La/Fe]. *Right:* Solid and dashed lines show the predicted evolution with halo mix plus pure s-process material; the dot-dash line indicates 95% s-process. Our data indicate s-process dominated after [Fe/H]~-0.6.

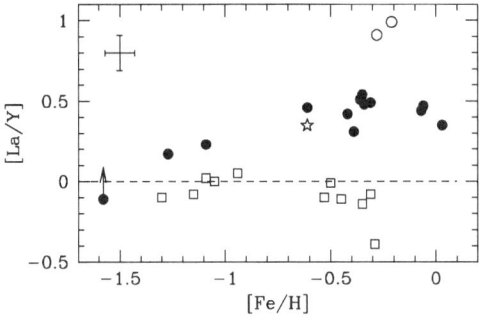

Fig. 2. Enhancements of [La/Y], by ∼0.5 dex, implicate metal-poor AGB s-process.

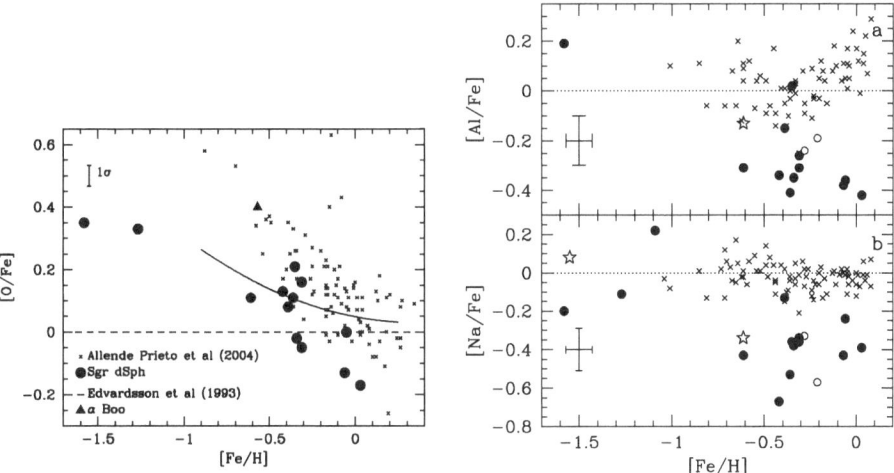

Fig. 3. Depletions of O, Al and Na suggest enhanced type Ia supernova contribution.

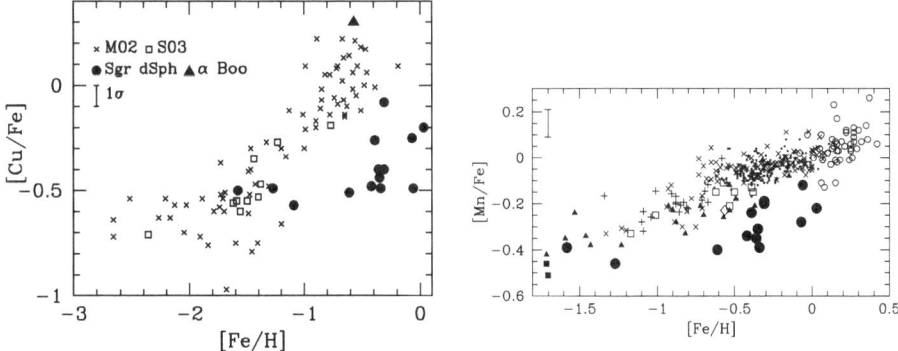

Fig. 4. Mn and Cu depletions indicate enhancement of low-Z SNIa synthesis products.

The Metallicity-Luminosity Relation for Local Group Dwarf Galaxies

E.K. Grebel[1], J.S. Gallagher[2], and D. Harbeck[3]

[1] Astronomical Institute, University of Basel, CH-4102 Binningen, Switzerland
[2] Department of Astronomy, University of Wisconsin, Madison, WI 53706, USA
[3] Space Sciences Lab, University of California, Berkeley, CA 94720, USA

1 Local Group Dwarf Galaxies

Hierarchical structure formation scenarios postulate that large galaxies form by continuous accretion of smaller subunits. Consequently, the properties of large galaxies (esp. of their old populations) should also reflect the properties of the accreted building blocks. The few surviving dwarf galaxies, in turn, hold the fossil record of the early conditions of galaxy evolution. The Local Group provides a convenient nearby laboratory to study the resulting predictions in detail.

Of the 38 currently known Local Group galaxies, 35 are satellites and/or dwarf galaxies, i.e., possibly leftover building blocks. The galaxies mainly belong to one of two basic types of dwarf galaxies: gas-rich dwarf irregular (dIrr) galaxies with ongoing star formation and masses $> 10^8 \, M_\odot$, and gas-deficient dwarf spheroidal (dSph) galaxies without recent star formation and $\sim 10^7 \, M_\odot$. Both dIrrs and dSphs contain ancient stellar populations. Where measurable, these appear to be coeval with the oldest datable Galactic Population II stars indicating a common epoch of early star formation (Grebel & Gallagher 2004).

The dSphs are the most numerous type of dwarf galaxy. Three basic formation scenarios tend to be invoked to explain their possible origin: (1) *Ab initio formed low-mass structures that exhausted their gas* (e.g., through star formation, ram pressure stripping, UV radiation from massive galaxies, or re-ionization squelching). Problems with this scenario (missing satellites, angular momentum loss, low intergalactic medium density, isolated dSphs, missing signature of re-ionization squelching) are discussed in Grebel, Gallagher, & Harbeck (2003; hereafter GGH03). (2) *Tidal fragments or tidal dwarfs* (problems: the apparently high dark matter content and lack of a large depth extent; Klessen, Grebel, & Harbeck 2003). (3) *Formerly more massive, tidally transformed dwarfs.* This scenario seems to be supported by the widely observed morphological segregation and the trend of increased H I masses with increasing distance from massive primaries (GGH03). The few fairly isolated dSphs – Cetus, Tucana, KKR25 – are then unexpected. On the other hand, we still lack reliable orbits for all dSphs.

2 The Metallicity-Luminosity Relation

It is well-known that dSphs and dIrrs differ in their metallicity-luminosity relations. This finding is based on stellar metallicities in gas-poor dwarfs and nebular

H II region abundances in gas-rich dwarfs. Richer, McCall, & Stasinska (1998) compared dIrr H II region O abundances with O abundances of planetary nebulae (PNe) in dSphs. While the offset persisted, PNe have only been detected in the two most luminous dSphs and trace primarily intermediate-age populations as opposed to the present-day abundances in H II regions.

We therefore made an effort to "compare apples with apples", i.e., to compare stars of the *same* population (and thus *similar age*) and with mean metallicities (commonly abbreviated by [Fe/H] or [Me/H]) based on *stellar metallicity* tracers. In GGH03, we compiled the most homogeneous data set of metallicities for 40 nearby dwarfs currently available using (1) *only old populations* (red giants in dSphs and in the outskirts of dIrrs), (2) *spectroscopic abundances* wherever available (from own Keck and literature data), and (3) *photometric abundances* elsewhere (from comparison with globular cluster fiducials). The resulting metallicity-luminosity relation for metallicities of old populations (below) shows that even for old populations, there is a considerable offset between dSphs and dIrrs. The error bars indicate the measured metallicity *spread* in each galaxy.

At the same luminosity dSphs are more metal-rich than dIrrs even for metallicities of similar populations. In contrast to dIrrs, dSphs must have experienced fairly rapid early enrichment (GGH03). These evolutionary differences (and other factors, see GGH03) make normal dIrrs unlikely progenitors of dSphs, whereas dIrr/dSph transition-type dwarfs seem quite plausible progenitors.

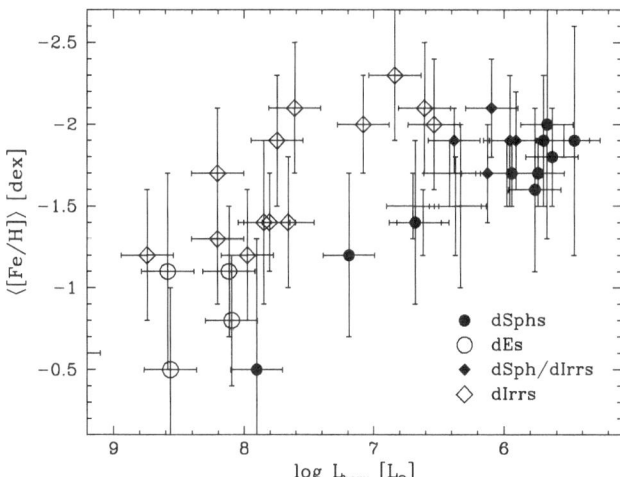

References

1. E.K. Grebel & J.S. Gallagher: ApJ, **610**, L89 (2004)
2. E.K. Grebel, J.S. Gallagher, & D. Harbeck: AJ, **125**, 1926 (2003) [GGH03]
3. R.S. Klessen, E.K. Grebel, & D. Harbeck: ApJ, **589**, 798 (2003)
4. M.G. Richer, M.L. McCall, & G. Stasinska: A&A, **340**, 67 (1998)

Chemical Abundances and Milky Way Formation

G. Gilmore[1] and R.F.G. Wyse[2]

[1] Institute of Astronomy, Madingley Road, Cambridge, UK
[2] Department of Physics and Astronomy, Johns Hopkins University,
Baltimore, MD 21218, USA

Abstract. Stellar chemical element ratios have well-defined systematic trends as a function of abundance, with an excellent correlation of these trends with stellar populations defined kinematically. This is remarkable, and has significant implications for Galactic evolution. The source function, the stellar Initial Mass Function, must be nearly invariant with time, place and metallicity. Each forming star must see a well-mixed mass-averaged IMF yield, implying low star formation rates, with most star formation in at most a few regions of similar evolutionary history. These well-established results are difficult to reconcile with standard hierarchical formation models, which assemble many stellar units: galaxy evolution seems to have been dominated by gas assembly, with subsequent star formation. Recent results, and some new ones, on the Galactic bulge, stellar halo and thick disk that justify this conclusion are presented.

1 The Context: Abundances and Galaxy Formation

The ΛCDM paradigm for structure formation in the Universe, described in many hundreds of published papers, is very effective at reproducing observed large scale structure, based on a boundary condition of a scale-free Gaussian random power spectrum. Yet ΛCDM contains no information on the physics of whatever makes up CDM, and remains deficient in its description of galaxies and small-scale structures: thus it is on galaxy scales and smaller where we can still learn the most, and hopefully attach some (astro-)physics to an *ab initio* power spectrum.

Among the very many studies which emphasise the current progress in studies of galaxy-scale predictions, we note as recent examples Glazebrook et al. (2004) and Cimatti et al. (2004), which highlight the buildup of massive spheroids at earlier times than predicted; de Blok (2004) showing CDM mass profiles on small scales are shallower than predicted; D'Onghia & Lake (2004), on the substructure problem, showing it extends from single galaxies to galaxies in groups; and Abadi et al. (2003) illustrating how detailed comparisons of simulations and data for the Solar neighbourhood are becoming feasible.

How will we identify the extra astrophysics required to reconcile the properties of CDM dark haloes with those of luminous galaxies? We can start by developing knowledge of the evolutionary history of at least one place in at least one galaxy. We would be unlucky if that place were far from the norm: alternatively, any theory that predicts such a history to be very unusual might be suspect - the galaxian Copernican principle. Kinematics and current spatial location are of course critical parameters, so that traditional stellar populations analyses are prerequisites. However, kinematical histories, as represented by velocity dispersions, are at best confused by phase-wrapping over time, and can be largely lost if a virialisation process is associated with significant mergers. Kinematics do provide valuable statistical information, but, apart from the single (approximately, sometimes) conserved quantity of angular momentum, provide direct information only about very recent mergers, a minor aspect of Galaxy formation in any model. The additional complementary information is best available from chemical abundances, especially including element ratios.

It is remarkable that chemical abundances are a valid, and relatively robust, tracer of galactic evolution: it is worth considering why this is possible.

Figure 1 (from Venn et al. 2004) is a compilation of most of the recent high-quality element ratio data for Galactic field stars (together with data for some stars in several satellite galaxies). The remarkable implication of this figure is that the overwhelming majority of Galactic field stars have element ratios relative to Fe which have a scatter which is always relatively small, over a range of some 4 dex in [Fe/H]. This small scatter is quite contrary to much simple expectation, and has profound implications.

One implication follows from the continuity in the pattern of the elemental ratios as a function of [Fe/H]: at every metallicity, all forming stars have element ratios which require that they are being formed from gas which has a common well-mixed history. At every metallicity, the star-forming gas must be well-mixed, and only mildly different in chemical element-ratio enrichment than that gas which formed the previous stellar generation. This seriously restricts the role of significant inflows of gas over an extended period of time, as inflows of gas with a very different enrichment history would induce scatter. However, inflow of metal-free gas (as often invoked to 'solve' the local disk 'G-dwarf problem') would reduce [Fe/H] while leaving [α/Fe] unchanged. A second implication follows from the dependence of element production on the initial mass of the main-sequence progenitor of the supernova. The observed small dispersion in element ratios, even under the extreme assumption of perfect ISM mixing, requires that each new-forming star sees an approximately invariant, and mass-averaged, IMF (eg Wyse & Gilmore 1992; Nissen et al. 1994). The rate of star formation must be low enough to allow time for element creation followed by large-scale mixing.

That is, the straightforward interpretation of abundance data for Galactic field stars in terms of stellar populations is feasible only because the Galaxy apparently acquired its gas early, or at a rate which was well-matched to the star formation rate across the whole volume now sampled by local halo stars, and kept this gas well-mixed; and because the stellar IMF is (close to) invariant over time and metallicity. Neither deduction was obvious, nor is the underlying physics understood. However, these two deductions apply so well they have become assumed: authors use any violation to rule out some possible Galaxy merger histories, as in the Venn et al. analysis from which Figure 1 is taken.

We now briefly consider in turn the abundance constraints on the major Galactic stellar populations, highlighting the poorly known aspects.

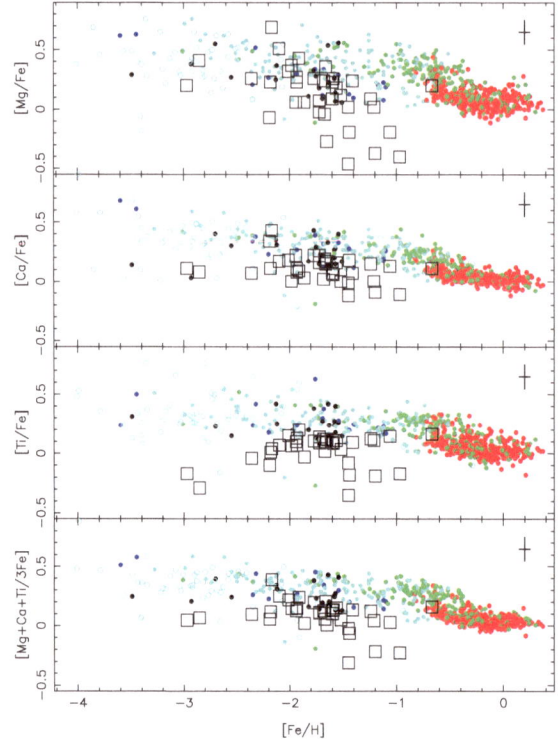

Fig. 1. Element ratio data for Galactic field stars (dots) compiled by Venn et al. (2004). [The open squares are stars in dSph galaxies.] The important point to notice here is the very small scatter in element abundances for Galactic field stars at any given [Fe/H] value, over 4dex in [Fe/H].

2   The Galactic Bulge

It is quite astonishing how little observational information is available on the chemistry and kinematics of the Galactic Bulge. A massive and exciting opportunity for the current multi-object spectrographs is being overlooked here.

The central regions of the Galaxy, including the Bulge, should accumulate any tightly-bound high phase-space density material associated with mergers (since the most-bound regions of substructure end up in the most-bound regions; Zurek, Quinn & Salmon 1988); low-angular momentum material in general; material driven in by bar asymmetries in the gravitational potential; and any accreted gas which can cool efficiently and lose angular momentum. The earliest stars formed will be there, even in the merger scenario for build-up of bulges (e.g. Kauffman 1996), while the whole will be affected in some as yet undefined way by feedback from the central Super-Massive Black Hole. It should be a very complex place, well worth detailed investigation. The luminosity scale height of the Bulge is of order 300pc, so that even the best studied 'inner' field, Baade's

Window at coordinates ($\ell = 1$, $b = -4$) or a projected Galactocentric distance of $\gtrsim 500$ pc, is itself rather far from the heart of the matter.

In so far as these properties are determined, the outer bulge, beyond 2 scale lengths, is apparently predominately mildly metal-rich (perhaps 0.5 Solar in the mean) with a very broad abundance distribution function, is older than ~ 10Gyr, ie is indistinguishable from the old metal-rich globular clusters, which may be related, and is alpha-element enhanced (e.g. Ibata & Gilmore 1995a, 1995b; Sadler, Rich & Terndrup 1996; Zoccali et al. 2003; Fulbright, Rich and McWilliam 2004): this does not favour merger origin models, but does suggest predominant in-situ star formation. A star formation rate of $10 M_\odot yr^{-1}$ is implied. The angular momentum distribution function is dominated by very low angular momentum, and is strongly dissimilar to the corresponding distribution for the Galactic disk: this argues against a formation process through disk-bar evaporation (Figure 2, and cf. Ibata & Gilmore 1995b), and suggests a close connection between bulge and halo in formation.

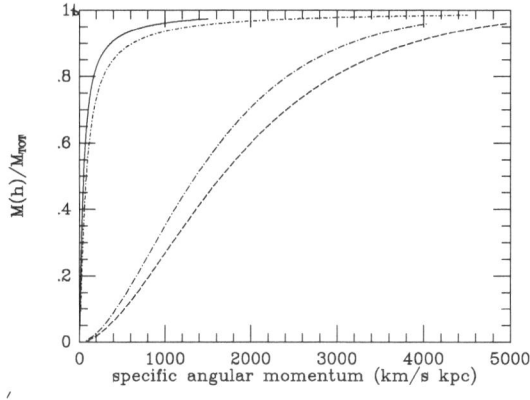

Fig. 2. The distribution of specific angular momentum in the dominant Galactic stellar populations. This indicates the similarity in this fundamental parameter, and presumably in origin, between bulge and halo (the two curves near the top left of the figure), and the quite distinct bahaviour of the Galactic thin and thick disks (the two curves through the centre of the figure). This figure is from Wyse & Gilmore 1992.

The structure of the inner Bulge, where most of the mass is, is inevitably confused by the inner disk, and by the extremely dense stellar super-cluster with small scale length concentrated at the very centre with its associated black hole (Schodel, Ott, Genzel et al. 2003). Recent studies in the IR have hinted at the considerable complexity of the central stellar populations (van Loon, Gilmore, Omont et al. 2003), with young high-mass star and cluster formation at solar metallicity (Najarro et al. 2004), a vast reservoir of molecular gas to feed continuing star formation (Pierce-Price et al. 2000), and complex bar and disk structure (Bissantz, Debattista & Gerhard 2004). Extant analyses have usually assumed the bar is associated with the bulge, rather than with the inner old disk, or, more commonly, do not make any distinction between these various central populations, even though they may well have very different kinematic, age and abundance distribution functions, and have formed in very different ways. Unravelling their inter-relationships is not yet possible due to the limited data. At present one needs to be careful to define terms, such as 'main bulge' (cf. Wyse, Gilmore & Franx 1997)

Considerable progress is currently being made in mapping the very inner Galaxy. In a recent study, Babusiaux & Gilmore (2004) have used near-IR photometry to map the inner galaxy in the Galactic Plane (figure 3), using red clump stars as tracers. Their photometric work suggests that the bulge stars 1-deg from the Plane (ie, 4 times closer in than Baade's Window) are metal-rich and alpha-enhanced, while the stars which define the bar itself, measured within 0.2deg of the Plane, best match intermediate-age isochrones without alpha-enhancement, The stars in the Bar itself seem more disk-like than those stars seen in Baade's window and other fields at similar high latitudes. Direct spectroscopic confirmation, to ensure age-metallicity degeneracies are not confusing the photometric analyses, should be possible, now that the individual bar clump giants have been identified.

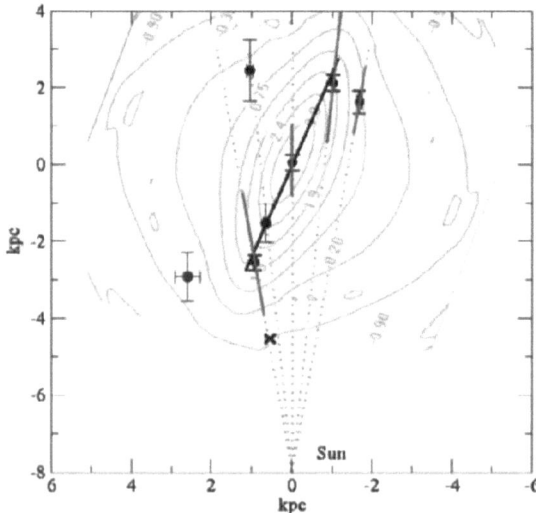

Fig. 3. The 2-D structure of the inner Galactic bar, based on direct photometric analysis of red clump stars as distance indicators. Now these individual bar stars are identified, detailed study of their kinematics and chemical abundances will be possible. Preliminary photometric indications suggest the bar is more closely disk than bulge in its populations. This figure is from Babusiaux & Gilmore 2004.

3   The Stellar Halo

Figure 2 above illustrates a similarity in at least one fundamental parameter between the Bulge and the stellar halo. The stellar halo traced by stars passing near the Sun is some 30% of its total mass of $\sim 2 \times 10^9 M_\odot$. These stars are predominantly old and metal poor. They show remarkable uniformity in their relationship between the element ratios and the total metal abundance (Figure 1), with implications noted above.

It has been known for many years (eg Unavane et al 1996; Gilmore & Wyse 1998), and has been re-emphasised recently (this meeting) based on much better data, that the stars in the stellar halo are unlike those in surviving Galactic satellites. Considerable progress is being made in determining the properties of the current Galactic satellites, especially the fundamental properties, mass density and dark matter profile, as well as in analysing the tracers of star formation histories, CMDs and chemical abundances. All the dSph satellites will have good kinematic maps complete within the near future, almost all based on FLAMES data, and all providing useful chemical abundance maps (UMi: Kleyna et al. 2003; Draco: Kleyna et al. 2001; Sextans: Kleyna et al. 2004; LeoI, LeoII, Carina: Gilmore et al. in prepn; Sgr: Ibata, Wyse, Gilmore et al. 1997; Scl: Tolstoy et al. this meeting; Fornax: Tolstoy et al. in prepn). All have similarities and complexities, yet none provides a source for a 'typical' field halo star. The different patterns of elemental abundances in the dwarf galaxies, which apparently rules them out as 'building blocks' for the Milky Way, can be understood in terms of their different star formation histories (cf. Gilmore & Wyse 1991).

This suggests strongly that the stellar halo formed in place early in Galactic history - a conclusion apparently required by the data, yet strongly in disagreement with most current galaxy formation simulations. The simulations imply significant late accretion of galaxies into the Milky Way: do we see this? A discovery which encouraged much current enthusiasm was that of the Sgr dwarf (Ibata, Gilmore & Irwin 1994, 1995): it is salutary to recall that Sgr was recognised as peculiar in real time at the AAT (during the Ibata/Gilmore bulge study) because both its velocity and its stellar populations were quite unlike any expected (or observed) field star. Sgr does however single-handedly make a substantial perturbation to the stellar mass (and age mix) of the stellar halo. More recently a second, and possibly a third, structure has been identified in the outer halo from SDSS photometry (Newberg et al. 2002); their relationship to the Sgr dwarf is unclear, but most confirmed structure may be ascribed to this galaxy (cf. Majewski et al. 2003). A firm conclusion is that in another Hubble time, the stellar halo will be relatively younger than it is today (Wyse 1996).

The halo globular cluster system also provides valuable information, since accurate distances, and hence reliable ages, can be derived. Mackey & Gilmore (2004) recently acquired and compiled a new, nearly complete, internally consistent set of photometric studies of the globular cluster population in both the Milky Way and its satellite galaxies: they deduce, from analysis of HB morphology, age, abundance and structural information, somewhat more relaxed limits on accretion of the halo than earlier analyses, but still preclude late accretion as a dominant factor.

Study of the relative importances of assembly of stellar systems and in situ star formation remains extremely active, and will presumably soon converge as the large area CCD photometric surveys cover the whole sky, and accurate spectroscopic surveys begin to acquire large samples.

3.1   M31 and Its Halo/Bulge/Thick Disk

An interesting case here is M31, which has long been suspected of having tidally stripped M32 (Faber 1973; Choi, Guhathakurta & Johnson 2002), and now we have kinematic evidence that it is also currently accreting from NGC 205 (McConnachie et al. 2004). However, the equivalent of the Milky Way's halo population II, a metal-poor component to the 'halo', traced by RR Lyrae stars and by metal-poor giants, is not dominant even at very many photometric scale lengths from the centre of M31 (e.g. Durrell, Harris & Pritchet 2004). The relation to accreted satellites is clearly complex (e.g. Ferguson et al. 2002). The dominant non-thin-disk population in M31 is rather metal-rich, with a mean of ~ -0.6 dex (Mould & Kristian 1986; Durrell et al. and references therein). Wyse & Gilmore (1988) discuss the apparent equivalence between this extended intermediate population in M31 and the Galactic thick disk, an equivalence which has become stronger as studies of M31 have developed (Sarajedini & van Duyne 2001; Brown et al. 2003). A fuller interpretation in terms of the mass assembly history of M31 awaits the current wide-area coverage with a combination of deep CMDs for age constraints, together with spectroscopic kinematics and metallicities.

4   The Old Disks

The thick disk probably was caused by, and/or is the remnant of, one or more (early?) merger events. The (local) mean thick disk metallicity of -0.6 dex and its old age suggest a very massive satellite – comparable at least to the LMC? – was destroyed if the (local) thick disk is its remnant. Simulations (eg Abadi et al. 2003) suggest at least the old thick disk will be the remnant of several merged sub-galaxies, with the relative mix changing with radius. This may be consistent with recent AAT 2dF survey results (Gilmore, Wyse & Norris 2002) which suggest that the distribution function of angular momentum, and perhaps metallicity, a few kpc from the Plane is not consistent with simple local extrapolations.

One of the more challenging things to understand in these multiple-fragment models is the evolution of the present-day surviving thin old disk, particularly since a recent accretion of the thick disk could destroy the thin disk. Further, in these models stars should not form in a thin disk at all until after most of the merging is complete, since otherwise disks have too small a scale-length (Navarro & Steinmetz 1997). A solution to both these issues may be that the local old thin disk is also accreted in these models. Here the recent results of many authors (cf

Figure 1 above) have shown there is a very small scatter in element ratios at any [Fe/H] value, particularly within a stellar population, yet there is an extremely large scatter in the age-[Fe/H] relation at every age (Nordstrom et al. 2004). Linking the local and global remains a challenge.

References

1. Abadi, M., Navarro, J., Steinmetz, M. & Eke, V. 2003 ApJ 597 21
2. Babusiaux, C., & Gilmore, G. 2004 MNRAS submitted
3. Bissantz, N., Debattista, V. & Gerhard, O. 2004 ApJ 601 L155
4. Brown, T. et al. 2003 ApJL 592 L17
5. Choi, P., Guhathakurta, P. & Johnson, K. 2002 AJ 124, 310
6. Cimatti. A., Daddi, E., Renzini, A., Vanzella, E., et al. 2004 Nature 430 184
7. de Blok, W.J.K. 2004, IAU Symp 220 eds Ryder et al. (ASP, San Francisco) p69
8. D'Onghia, E., & Lake, G., 2004 ApJ 612 628
9. Durrell, P., Harris, W. & Pritchet, C. 2004 AJ 128 260
10. Faber, S.M. 1973 ApJ 179 423
11. Ferguson, A. et al. 2002 AJ 124 1452
12. Fulbright, J., Rich, R.M. & McWilliam, A., 2004 astroph-0411041
13. Gilmore, G., Wyse, R.F.G. 1998 AJ 116 748
14. Gilmore, G., Wyse, R.F.G. & Norris, J.E. 2002 ApJL 574 L39
15. Glazebrook, K., Abraham, R., McCarthy, P. et al. 2004 Nature 430 181
16. Ibata, R., Gilmore, G. & Irwin, M. 1994 Nature 370 194
17. Ibata, R. & Gilmore, G. 1995a MNRAS 275, 591
18. Ibata, R. & Gilmore, G. 1995b MNRAS 275, 605
19. Ibata, R., Gilmore, G. & Irwin, M. 1995 MNRAS 277 781
20. Ibata, R., Wyse, R.F.G., Gilmore, G., Suntzeff, N. & Irwin, M. 1997 AJ 113 634
21. Ibata, R. et al. 2004 MNRAS 351 117
22. Kleyna, J., Wilkinson, M., Evans, N.W. & Gilmore, G. 2001 ApJ 563 L115
23. Kleyna, J., Wilkinson, M., Evans, N.W. & Gilmore, G. 2004 astroph-0409066
24. Kleyna, J., Wilkinson, M., Gilmore, G. & Evans, N.W. 2003 ApJ 588 L21
25. Kauffmann, G. 1996 MNRAS 281 487
26. van Loon, J., Gilmore, G., Omont, A. et al. 2003 MNRAS 338 857
27. Mackey, A.D. & Gilmore, G. 2004 MNRAS 354 470
28. Majewski, S., Strutskie, M., Weinberg, M. & Ostheimer, J. 2003 ApJ 599 1082
29. McConnachie, A. et al. 2004 MNRAS 351 L94
30. Mould, J. & Kristian, J. 1986 ApJ 305 591
31. Najarro, F., Figer, D., Hillier, D. & Kudritzki, R. 2004 ApJ 611 L105
32. Navarro, J. & Steinmetz, M. 1997 ApJ 478 13
33. Nissen, P. E., Gustafsson, B., Edvardsson, B., Gilmore, G. 1994 A&A 285 440
34. Newberg, H., Yanny, B., Rockosi, C., et al. 2002 ApJ 569 245
35. Nordstrom, B., Mayor, M., Andersen, J. et al. 2004 A&A 418 989
36. Pierce-Price, D., et al. 2000 ApJ 545 L121
37. Reitzel, D., Guhathakurta, P. & Rich, R.M. 2004 AJ 127 2133
38. Sadler, E., Rich, R.M. & Terndrup, D. 1996, AJ 112 171
39. Sarajedini, A. & van Duyne, J. 2001, AJ 122 2444
40. Schodel, R., Ott, T., Genzel, R. et al. 2003 ApJ 596 1015
41. Unavane, M., Wyse, R.F.G. & Gilmore, G. 1996 MNRAS 278 727
42. Wyse, R.F.G. 1996 ASP Conf series 88, eds Trimble & Reisenegger p128

43. Wyse, R.F.G. & Gilmore, G. 1988 AJ 95 1404
44. Wyse, R.F.G. & Gilmore, G. 1992 AJ 104 144
45. Wyse, R.F.G., Gilmore, G. & Franx, M. 1997 ARAA 35 637
46. Zoccali, M. et al. 2003 A&A 399 931
47. Zurek, W., Quinn, P. & Salmon, J. 1988 ApJ 330 519

Chemical Evolution in the Carina Dwarf Spheroidal

A. Koch[1], E.K. Grebel[1], D. Harbeck[2], M.I. Wilkinson[3], J.T. Kleyna[3,4], G.F. Gilmore[3], R.F.G. Wyse[5], and N.W. Evans[3]

[1] Astronomical Institute of the University of Basel, CH-4102 Binningen, Switzerland
[2] Department of Astronomy, University of Wisconsin, Madison, WI 53706, USA
[3] Institute of Astronomy, Cambridge University, Cambridge CB3 0HA, UK
[4] Institute for Astronomy, University of Hawaii, Honolulu, HI 96822, USA
[5] The Johns Hopkins University, Baltimore, MD 21218, USA

Abstract. We present metallicities for 487 red giants in the Carina dwarf spheroidal (dSph) galaxy that were obtained from FLAMES low-resolution Ca triplet (CaT) spectroscopy. We find a mean [Fe/H] of -1.91 dex with an intrinsic dispersion of 0.25 dex, whereas the full spread in metallicities is at least one dex. The analysis of the radial distribution of metallicities reveals that an excess of metal poor stars resides in a region of larger axis distances. These results can constrain evolutionary models and are discussed in the context of chemical evolution in the Carina dSph.

1 Introduction

Analyses of the faint Carina dSph have revealed that it contains a variety of stellar populations (e.g., [4]), exhibiting prominent old (>11 Gyr) and intermediate-age (5–6 and 3 Gyr) populations. This implies that Carina must have undergone several star forming (SF) episodes with at least three significant pulses. Despite this wide spread in ages, its colour-magnitude diagram features a remarkably narrow RGB. The reason for that can be an age counteracting spread in metallicities, where metal rich, young stars have colours comparable to the older, more metal poor ones. Such a possible age-metallicity degeneracy can be overcome if accurate and independent [Fe/H] measurements are obtained so that the remaining parameter of age can be estimated from isochrones. Moreover, the overall shape and spatial variations of the metallicity distribution function (MDF) itself contain valuable implications for analysing Carina's unusual SF history.

2 Observations, Reduction and Calibration

In the course of an ESO Large Programme, we observed 1257 red giants covering five fields in Carina out to the tidal radius. Our observations were performed during 23 nights with the multi-object spectrograph FLAMES at the VLT in low-resolution mode, centered at the near infrared CaT. The data were reduced using the standard FLAMES reduction pipeline. Since sky contamination from bright emission lines is strong in the CaT region, we calculated an average sky-spectrum from 20 dedicated skyfibres, which then was carefully subtracted from the science spectra. Typical signal-to-noise (S/N) ratios for our spectra lie between 25 and 150 pixel^{-1}. After rejection of foreground stars and stars with too low a S/N, 487 radial velocity members were analysed. Finally, metallicities were derived from the reduced equivalent width W' ([1],[5]).

3   The Observed and Predicted Metallicity Distributions

The resulting MDF for our 487 red giants is shown in the left panel of Fig. 1. It is peaked at an average [Fe/H] of -1.91 ± 0.01 dex, which is in excellent agreement with the spectroscopic study of [6], who found a value of -1.99 ± 0.08 dex. We find a dispersion of $\sigma=0.25$ dex and as Fig. 1 implies, the full spread of [Fe/H] is at least one dex. This suggests that the wide range in age is in fact counteracted by a wide spread in metallicity, which can explain Carina's narrow RGB and confirms that a complex mixture of stellar populations as in Carina does not necessarily contradict the RGB's narrowness. The different stellar populations in Carina have also different spatial distributions, where, the intermediate age red clump stars are clearly concentrated towards the center compared to the old HB (see [2]). Thus we plot in Fig. 1 (right panel) the error-weighted MDFs for three different regions along the galaxy's axes. While the curves for the central region and larger minor axis distances do not differ largely and also resemble those along the major axis, there is a shift of the MDF's peak towards the metal poor end by 0.2 dex when concentrating on the outer, northwestern region (lower axis distances). Fig. 1 also overplots a model curve from [3], showing the predicted MDF for Carina. This model uses a SF efficiency of $0.1 \, \mathrm{Gyr}^{-1}$ and a galactic wind efficiency at the lower end of wind rates in dSphs. Such a lower wind rate nicely explains the smooth decline of the observed MDF towards the metal rich end, as a higher wind rate would expel enriched gas more efficiently thus preventing the formation of this metal rich tail. Secondly, the low SF rate is able to explain

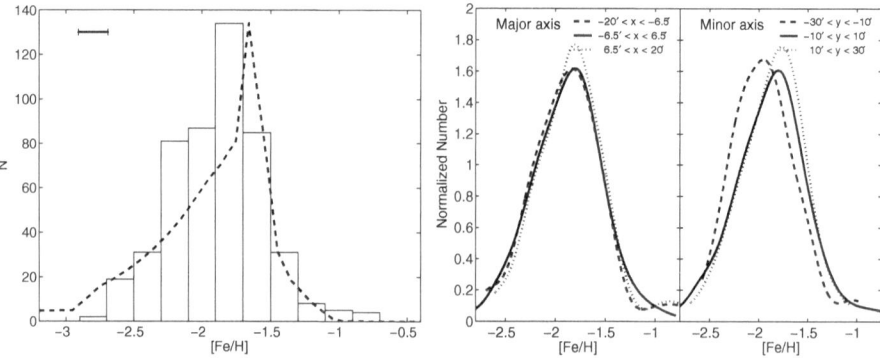

Fig. 1. Left panel: Metallicity distribution for our 487 red giants in Carina. The dashed line is the "best-fit model" prediction from [3]. The right plot shows spatially separated MDFs in three different regions of Carina, measured along the major and minor axes.

the metal poor peak of the MDF in terms of a moderate enrichment history. However, remaining caveats are the higher fraction of predicted extremely metal poor stars and the lower number of metal poor stars around [Fe/H]≈ -2, which can be understood in the context of the low infall timescale of gas assumed in dSphs ([3]). With our accurate [Fe/H] measurements at hand we will in the next steps pursue the underlying age distribution and use the abundance ratios from our multi-object high-resolution spectroscopy to further unravel the mechanisms governing Carina's evolutionary history.

References

1. Cole, A.A., et al., MNRAS, 347, 367
2. Harbeck, D., et. al. 2001, AJ, 122, 3092
3. Lanfrachi, G., & Matteucci, F. 2004, MNRAS, 351, 1338
4. Monelli, M., et al. 2003, AJ, 126, 218
5. Rutledge, G.A., Hesser, J.E., Stetson, P.B. 1997, PASP, 109, 907
6. Smecker-Hane, T.A., et al. 1999, ASP Conf. Ser., 192, 159

Chemical Connections Between Stars in the Galaxy and its Satellites

K.A. Venn[1,2], M. Irwin[3], and M. Shetrone[4]

[1] Macalester College, Department of Physics & Astronomy,
    Saint Paul, MN 55105, USA
[2] University of Victoria, Department of Physics & Astronomy,
    Victoria, BC, V8P 1A1, Canada
[3] Institute of Astronomy, University of Cambridge, Cambridge, CB3 0HA, UK
[4] McDonald Observatory, University of Texas at Austin, Austin, TX, 78712, USA

1 Introduction

Detailed elemental abundances are now available for several individual stars in the Galaxy's dwarf satellites (Shetrone *et al.* 2001, 2003; Geisler *et al.* 2005; also see the reviews in this proceedings). A comparison of these abundance ratios to those of stars in the Galaxy can be used to address several questions related to galaxy formation and evolution, as well as stellar nucleosynthesis.

2 Metallicity Comparisons

To compare the chemistries of stars in dSph (and dIrr) galaxies to those in the Galaxy, we collected and examined the elemental abundances for over 700 stars with both chemical and kinematic information from the literature. The kinematic data was used to assign stars to a Galactic population (thin disk, thick disk, halo, extreme retrograde halo) based on a standard Bayesian classification scheme using Galactic Gaussian velocity ellipsoid components (see Venn *et al.* 2004). Fig. 1

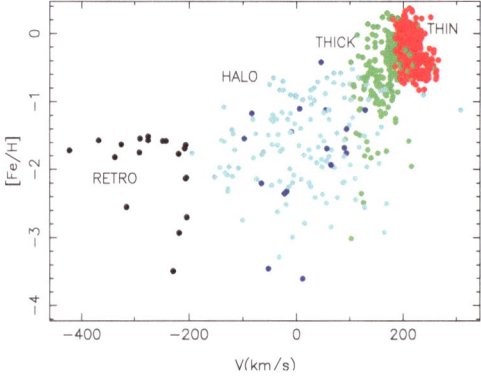

Fig. 1. Variation in [Fe/H] versus Galactic rotational velocity for stars assigned to Galactic stellar populations based purely on their kinematics; thin disk (red), thick disk (green), halo (cyan), plunging orbits (blue), extreme retrograde orbits (black).

demonstrates the expected general trend of metallicity for the different Galactic components, but also the large scatter and overlap in metallicity between the different Galactic components, most notably between the metal-rich thick and thin disks and the metal-poor thick disk and halo (as discussed previously by Unavane *et al.* 1996, and Gilmore & Wyse 1998). Intriguingly, Fig. 1 also shows a clump in metallicity of the majority of extreme retrograde stars suggesting an unexpected uniformity in their chemical properties (also reflected in their [α/Fe] and other abundance ratios), however these stars show no relationship in phase space as would be expected for a stellar stream.

3   Other Chemical Comparisons

The evolution of the chemical elements in a galaxy is intimately linked to its star formation history, and one ratio of particular interest is [α/Fe] due to the relative contributions to these elements from SN II and SN Ia events (see Pagel 1998). In Fig. 2, the [α/Fe] ratios (for an average of Mg/Fe, Ca/Fe, and Ti/Fe) are plotted with respect to metallicity. As expected, the metal-poor halo stars show high [α/Fe] ratios, while thin disk stars show an [α/Fe] ratio that approaches solar with increasing metallicity. The dSph stars are well separated from most the majority of Galactic disk and halo stars, though there is possible overlap with the extreme retrograde stars which also seem to have slightly lower [α/Fe] ratios and fall in the same metallicity range as the dSph stars. Of course, the extreme retrograde component is the most likely to come from satellite accretion events, based on kinematics alone.

Variations in star formation history should be imprinted on the s- and r-process ratios as well, however their interpretation can be more complicated because of uncertainties in their exact sources (and thus yields). Y and Ba trace the first and second peak in neutron magic number, respectively, and can be used to examine r-process yields in very metal-poor stars. However, they also have a significant contribution from the s-process in AGB stars, which dominates their production with increasing metallicity. Since AGB s-process yields are thought

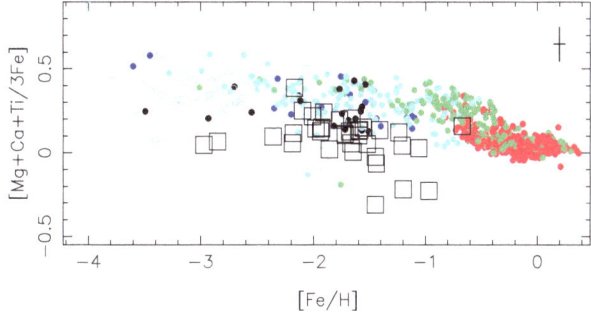

Fig. 2. Variation in [α/Fe] in Galactic stars (color coded by stellar population as in Fig. 1) and dSph stars (black squares).

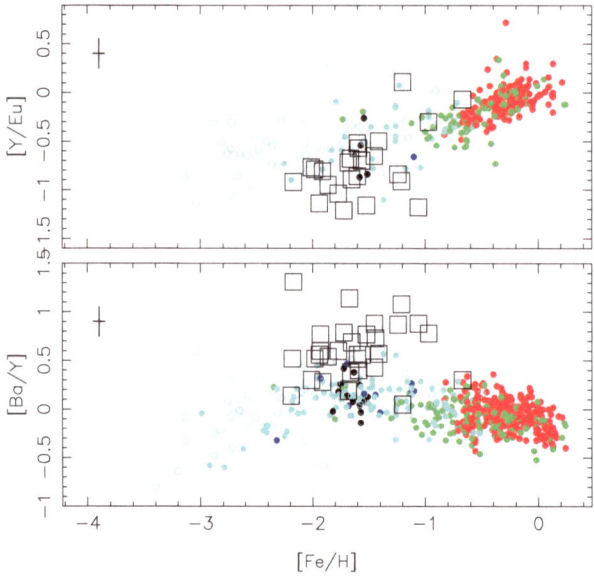

Fig. 3. Variation in [Y/Eu] and [Ba/Y] in Galactic stars (color coded by stellar population as in Fig. 1) and dSph stars (black squares).

to be metallicity dependent, including the first-to-second peak yields (Travaglio *et al.* 2004), then the [Ba/Y] ratio allows us to study differences in AGB yields between Galactic and dSph stars (presumably due to differences in star formation histories). Meanwhile, Eu is a heavy r-process dominated element, thus the [Y/Eu] ratio allows us to study the differences in s- and r-process contributions.

The ratios of [Y/Eu] and [Ba/Y] for Galactic and dSph stars appear to differ for all Galactic stellar populations, as shown in Fig. 3, including the extreme retrograde halo stars. This is most likely due to the abundances in the dSph stars coming primarily from metal-poor AGB stars. Metal-poor AGB stars are expected to expel more second (and third) peak s-process elements (like Ba), than first-peak elements (like Y), since there are fewer iron seed nuclei to absorb the wealth of available neutrons. Thus, the yield of Y is lower than Ba relative to more metal-rich AGB stars. Since the lower mass dSph galaxies appear to require more time to build up their iron abundances, and have mostly old stellar populations, then we can expect their AGB stars to typically have lower metallicities when they pollute their interstellar medium for the next generation of star formation. In the Galactic stars, it appears that these elements come from AGB stars with a wider range in metallicities, e.g., higher metallicity AGB stars are expected to have higher Y yields.

The α-process? Could the low [α/Fe] and low [Y/Eu] ratios in dSph stars be related by the α-process? The α-process (or α-rich freeze out) occurs when a neutron-rich, α-rich gas is out of nuclear statistical equilibrium and is thought to be important in the formation of ^{44}Ca (Woosley & Weaver 1995), ^{48}Ti (Nakamura *et al.* 2001), and possibly the light r-process elements like Y (Woosley & Hoffman 1992). Nakamura *et al.* (2001) also suggested the α-process occurs in hypernovae (with energies $\geq 10^{52}$ erg). Shetrone (2003) showed that [Ca/Fe] and [Ti/Fe] appear flatter and lower than [Mg/Fe] and [O/Fe] in dSph stars, particularly when compared to the Galaxy. This chemical signature would be expected if a significant amount of Ca and Ti are formed in the α-process, and the source of this process were missing in dSph galaxies. Possibly this chemical signature suggests a lack of hypernova in the low mass dSph galaxies, which would be consistent with the suggestion by Tolstoy *et al.* (2003) that dSphs may have an effectively or statistically truncated IMF.

4   Conclusions

Regardless of the nucleosynthetic explanations for the low [α/Fe], low [Y/Eu], and high [Ba/Y] ratios in most stars in dSph galaxies, clearly the chemical signatures of these stars are not the same as similar metallicity stars in the Galaxy. Also, the chemical signatures of the stars in the dSphs are broadly similar to one another, suggesting that the differences in their chemical evolution are small relative to the differences in the chemical evolution of the Galaxy. Together, these observations strongly imply that no significant component of the Galaxy formed primarily though the merger of galaxies similar to these low mass dSphs, at least not after they began to form stars. Of course, these conclusions are currently based on a very small number of stars per dSph galaxy, and from a variety of dSphs each with its own star formation history. With the new FLAMES spectrograph on UT2 at the VLT, we will soon be able to examine the detailed abundances of ≥ 100 stars in each of the southern Galactic dSph satellites to examine our assumptions about nucleosynthesis as well as hierarchical galaxy formation.

Acknowledgements: Many thanks to Luca Pasquini, Sofia Randich, and Claudia Travaglio for organizing such a pleasant and stimulating meeting.

References

1. D. Geisler, G. Wallerstein, V. Smith, *et al.* 2005, AJ, in press
2. G. Gilmore, R.F.G. Wyse, 1998, AJ, 116, 748
3. T. Nakamura, H. Umeda, K. Iwamoto, *et al.*, 2001, ApJ, 555, 880
4. M.D. Shetrone, P. Cote, W.L.W. Sargent, 2001, ApJ, 548, 592
5. M.D. Shetrone, K.A. Venn, E. Tolstoy *et al.*, 2003, AJ, 125, 684
6. M.D. Shetrone, 2003, in *Origin and Evolution of the Elements* from Carnegie Observatories Centennial Symposia, Cambridge University Press, Eds. A. McWilliam and M. Rauch, p. 220
7. E. Tolstoy, K.A. Venn, M.D. Shetrone, *et al.*, 2003, AJ, 125, 707
8. C. Travaglio, R. Gallino, E. Arnone, *et al.*, 2004, ApJ, 601, 864
9. M. Unavane, R.F.G. Wyse, G. Gilmore, *et al.*, 1996, MNRAS, 278, 727
10. K.A. Venn, M. Irwin, M.D. Shetrone, *et al.*, 2004, AJ, 128, 1177
11. S.E. Woosley, Weaver T.A., 1995, ApJS, 101, 181
12. S.E. Woosley, Hoffman R.D., 1992, ApJ, 395, 202

Abundances in Damped Lyα Galaxies

P. Molaro

Osservatorio Astronomico di Trieste-INAF, Via G.B. Tiepolo 11,
I-33100 Trieste, Italy

Abstract. Damped Lyα galaxies provide a sample of young galaxies where chemical abundances can be derived throughout the whole universe with an accuracy comparable to that for the local universe. Despite a large spread in redshift, HI column density and metallicity, DLA galaxies show a remarkable uniformity in the elemental ratios rather suggestive of similar chemical evolution if not of an unique population. These galaxies are characterized by a moderate, if any, enhancement of α-elements over Fe-peak elemental abundance with [S/Zn]≈ 0 and [O/Zn]≈ 0.2, rather similarly to the dwarfs galaxies in the Local Group. Nitrogen shows a peculiar behaviour with a bimodal distribution and possibly two plateaux. In particular, the plateau at low N abundances ([N/H] < -3), is not observed in other astrophysical sites and might be evidence for primary N production by massive stars.

1 Introduction

Any slab of intervening material along the line of sight of a background source with hydrogen column density high enough to produce damping wings in Lyα, conventionally $\log N(HI) > 20.3 \ cm^{-2}$, is producing a Damped Lyα galaxy (DLA). Damped systems hold a large fraction of neutral gas at high redshift and are considered the progenitors of present day galaxies. Holding neutral gas and being free from ionization effects, DLAs provide column densities at very high precision (≈ 10 %), which together with the fact that they are observed up to redshift ≈ 4.5 or equivalently at a look-back time of ≈ 12 Gyr, expand to almost the entire universe the possibility of a detailed chemical investigation. Observed metallicities are generally low, with [Fe/H] varying between -2.5 and -1.0, but never below -2.5 and with a mild evolution with redshift. The elemental abundances, which we discuss here in more detail, resemble very closely those observed in the dwarf and irregular galaxies of the Local Group which are the subject of this conference.

2 Dust

Dust is probably present in the DLA and significantly affects the observed abundances. I counted 55 systems for which both Fe and Zn are measured, which are plotted in Fig. 1. The abundance of Fe is always found below that of Zn. By analogy with the interstellar medium, this behaviour is interpreted as the effect of some Fe being locked into dust grains. Other indicators for the presence of dust are the correlation of [Fe/Zn] with H_2 and the reddening excess of QSOs behind Damped galaxies. The figure shows a clear trend of Fe depletion with metallicity. There is also evidence for a sort of threshold line [Fe/Zn] ≈ -(2 +[Zn/H]) with no Fe depletion below it, which points out to the presence of a forbidden region for dust formation. This behaviour may be relevant for the understanding of dust formation in low metallicity environments.

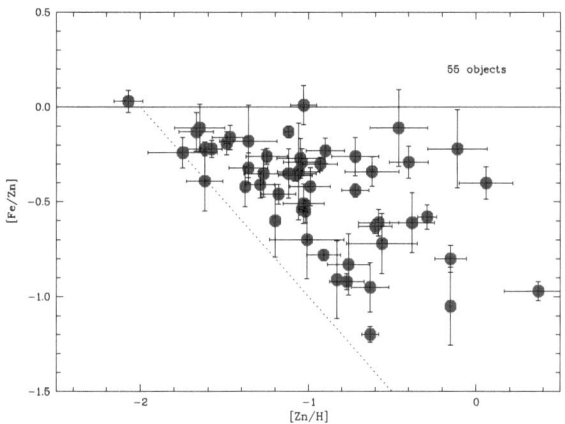

Fig. 1. Dust evidence in the DLAs

3   The [α/Zn] Ratio

The presence of dust affects abundances of elements participating into dust grains, and it needs either to be corrected or to be avoided. Considering the uncertainties involved in dust formation, the second approach looks safer. S and O are α elements showing little affinity with dust, while Zn is, rather uncomfortably, the only one available among the iron-peak group. In Fig 2 the [S/Zn] elements for 24 Damped galaxies are shown. The average value is consistent with zero: $< [S/Zn] > = 0.01 \pm 0.14$. There is also a hint of decreasing [S/Zn] with the increasing of metallicity. O has been measured in a limited number of objects and those in which also Zn is available are plotted in Fig 3, together with a couple of particularly low [O/Fe], which are a significant upper bound to the [O/Zn] value. QSO 0347-3819 is generally considered the best case for evidence of α enhancement. However, the figure shows a new [O/Zn] measure from a revised measurement of ZnII by Levshakov et al (log $N(ZnII) = 12.26$ cm $^{-2}$, 2004, private com.). The revised ratio is [O/Zn]=0.1±0.1 and there is no more strong α enhancement in this DLA. The case of Q 0812+32, with [O/Zn]=0.35, should be probably considered separately since this DLA is rather peculiar and is considered the progenitor of an elliptical. Thus, with the exception of the QSO

0812+32 case, all the DLA galaxies show mild or absent α enhancement, reminiscent of what found in the stars of the Local Group galaxies. To increase the statistical significance of the result one can use Si that, despite being refractory, tracks S rather well on average. In 40 DLA galaxies [Si/Zn] $\approx$ 0, with only few exceptions of particularly dusty objects, showing that the absence of α enhancement is not due to poor statistics but is a rather general property as was first suggested by Molaro et al (1996)

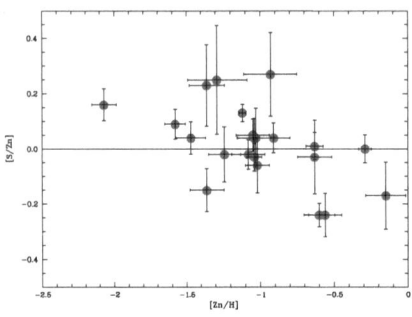

Fig. 2. [S/Zn] in DLAs

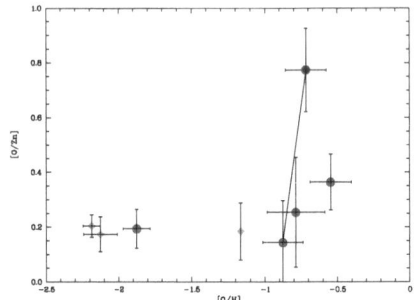

Fig. 3. [O/Zn] in DLAs, with some [O/Fe]

3.1   Nitrogen

N shows little affinity with dust and is thought to be produced by intermediate mass stars in the 4-8 M_0 range which undergo HBB in the AGB phase. N has a dual nature since it is primary at low metallicities and secondary at high metallicities, as observed in the extragalactic HII regions. Being produced by relatively small masses it is produced with a time delay compared to O or other α a fact that can explain the rather scattered [N/Si] at low abundances. It has been suggested by Prochaska et al (2002) that the DLAs show a bimodal distribution rather than a pure scatter behaviour. From the 2 systems initially considered by them there are now 5 systems and 2 upper limits that support the proposal for a bimodal distribution. The existence of two different regimes can be better appreciated when [N/Si] is plotted versus [N/H] as shown in Fig. 4. The plot clearly shows that the in DLA low and high [N/Si] occupy two separate regions. A possible explanation is that the low values are young systems where N comes from massive stars and the AGB have not yet started to produce the bulk of N. A primary N production by massive stars is not foreseen by standard models but it is present in the zero metallicity models of Chieffi & Limongi (2002) and the predicted [N/Si] ratios after integration over an appropriate IMF show ratios close to the observed ones. It would be rather appealing to find this N as *the smoking gun* of zero metal stars. In Fig 4 the [N/Si] of DLAs are shown together with the recent N determinations provided in halo stars. Stellar values

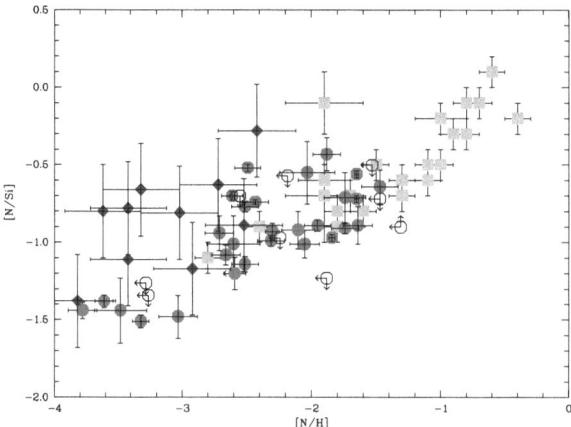

Fig. 4. Nitrogen in DLAs, circles, and halo stars from Spite et al (2004), lozenge, and from Israelian et al (2004), squares.

are in rough agreement with this picture, showing no values below [N/Si]=-1.5, although they populate also the portion of the diagram with [N/H] < -3.0.

3.2   A Single Population?

Determination of relative abundances of elements produced differently by Type I, Type II SNae and IMS probes the early chemical build-up. Chemical evolutionary models characterized by a relatively low star formation rate have been proposed to explain the mild $[\alpha/Zn]$ and the [N/Si] ratios in the DLA. A low SFR is commonly found in dwarfs, irregulars, LSB galaxies or in the external part of rotational disks. Observationally the DLAs show a remarkable constancy in the observed elemental ratios. So far there are 15 DLAs for which we have the whole triplet of elements N, Si and Zn. Si in this case is used as a fair proxy for O or S. The observed elemental ratios are: $< [S/Zn] > = 0.01 \pm 0.18$ and $< [N/Si] > - -0.79 \pm 0.16$, i.e. remarkably uniform considering that the metallicity in this sample is varying from -2.0 to -0.5, redshift from 1.6 to 3.4 and N(HI) from 20.0 to 21.5. Such a homogeneity in the ratios is strongly suggestive of very similar evolutionary properties of the DLAs if not of a specific population of galaxies, which might dominate the DLA class at least at high redshift.

References

1. A. Chieffi, M. Limongi Ap.J. **577**, 281 (2002)
2. G. Israelian, A. Ecuvillon et al: A&A **421**, 649 (2004)
3. P. Molaro, S. D'Odorico et al: A&A **308**, 1 (1996)
4. J. Prochaska, R.B.C. Henry et al PASP **114**, 1197 (2002)
5. M. Spite, R. Cayrel et al: astroph-0409536 (2004)

Metallicities and Kinematics
for Dwarf Spheroidals in the Local Group

G. Battaglia[1], E. Tolstoy[1], A. Helmi[1], M. Irwin[2], and the DART Team

[1] Kapteyn Institute, University of Groningen, 9700AV Groningen, NL
[2] Institute of Astronomy, Cambridge University, Cambridge CB3 0HA,UK

Abstract. We present the first results of CaII triplet observations from VLT/FLAMES for Sculptor, Fornax and Sextans dSphs. For each galaxy, we obtained accurate velocity and metallicity measurements for hundreds of stars out to and beyond the tidal radius. In each case, we find clear evidence for the presence of two distinct stellar components with different spatial distribution and kinematics: the metal rich component is more centrally concentrated and kinematically colder than the metal poor one.

1 Introduction

Dwarf Spheroidal galaxies are the smallest and faintest galaxies known. They are typically dominated by old stellar populations (e.g. Sculptor and Sextans), but some of them (e.g. Fornax) exhibit more recent star formation episodes (2-8 Gyr ago). Analysis of the horizontal branch morphology shows that Red HB stars are more centrally concentrated than Blue HB stars which could be interpreted either as an age or a metallicity gradient or both ([1]). Only spectroscopic observations can unambiguously separate metallicity gradients and make a link with the kinematics.

2 Results

As part of the DART large program at ESO, we took FLAMES CaT spectra of hundreds of red giant branch stars out to and beyond the tidal radius for Sculptor (Scl), Fornax (Fnx) and Sextans (Sxt) dSphs. We obtained accurate velocities ($\sim \pm 2\,\mathrm{km\,s^{-1}}$) and [Fe/H] ($\sim \pm 0.1\,\mathrm{dex}$) measurements for about 300, 600 and 230 radial velocity members for Scl, Fnx and Sxt respectively. We found a metallicity gradient in all the 3 dSphs, such that the average metallicity is higher in the central region and decreases at about 2 core radii (for Scl see [2]). In Fig.1 we show the velocity distribution at different spatial bins for Scl and Fnx where the [Fe/H] distribution is divided between "metal-poor" and "metal-rich". The metallicity cut, [Fe/H]= -1.7 for Scl and [Fe/H]= -1.4 for Fnx, corresponds to the peak of the [Fe/H] distribution. The two metallicity components have different spatial and kinematical distributions. The inner regions are dominated by the metal rich component; at larger radii metal rich stars disappear almost completely. The metal poor population typically exhibits a larger velocity dispersion than the metal rich one. The analysis of our Sxt data leads to the same

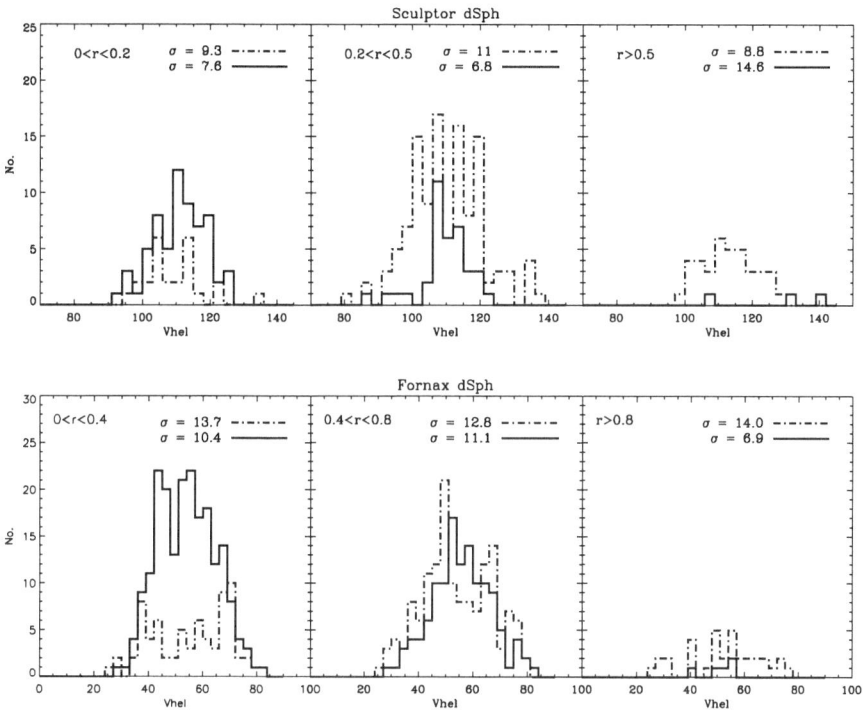

Fig. 1. Velocity histogram for different spatial samples for Scl (*top*) and Fnx (*bottom*) dSphs. The solid line shows the higher metallicity component and the dot-dashed line the lower metallicity one; r is the elliptical radius in degrees. For Scl the metallicity cut is at $[Fe/H] = -1.7$; for Fnx at $[Fe/H] = -1.4$

findings. We are therefore observing the same feature in 3 galaxies with different star formation histories, metallicity and mass.

Our results suggest that dSphs contain multiple stellar components with different spatial, kinematic and metallicity distributions, and rule out formation scenarios involving single bursts of star formation.

References

1. D. Harbeck et al.: AJ **122**, 3092 (2001)
2. E.Tolstoy:'Abundances as Tracers of the Formation and Evolution of (Dwarf) Galaxies'.In:Chemical Abundances and Mixing in Stars, 13-17 Sep. 2004, Castiglione della Pescaia, Italy, L. Pasquini, S. Randich (eds.)

Chemical Analysis of Extragalactic Carbon Stars

P. de Laverny[1], C. Abia[2], I. Domínguez[2], B. Plez[3], and O. Straniero[4]

[1] Observatoire de la Côte d'Azur, Dpt. Cassiopée, UMR/CNRS 6202, BP4229, 06304 Nice cedex 4, France (laverny@obs-nice.fr)
[2] Universidad da Granada, Spain
[3] GRAAL, Université Montpellier II, France
[4] Osservatorio di Collurania, Teramo, Italy

Abstract. We have performed the chemical analysis of extragalactic carbon stars from VLT/UVES spectra. The derived individual abundances of metals and s-elements as well as the well known distance of the selected stars in the Small Magellanic Cloud and the Sagittarius dwarf galaxies permit us to test current models of stellar evolution and nucleosynthesis during the Asymptotic Giant Branch phase in low metallicity environments.

1 Introduction

AGB stars constitute excellent laboratories to test the theory of stellar evolution and nucleosynthesis. Their particular internal structure allows two important processes to occur in them. First is the so-called 3th dredge-up (3DUP), a mixing mechanism in which the convective envelope penetrates the interior of the star after each thermal instability in the He-shell (thermal pulse, TP). The other is the activation of the s-process synthesis from alpha captures on ^{13}C or/and ^{22}Ne nuclei that generate the necessary neutrons which are subsequently captured by iron-peak nuclei. The repeated operation of TPs and the 3DUP episodes enriches the stellar envelope in newly synthesized elements and transforms the star into a carbon star, if the quantity of carbon added into the envelope is sufficient to increase the C/O ratio above unity. In that way, the atmosphere becomes enriched with the ashes of the above nucleosynthesis processes which can then be detected spectroscopically.

In order to test the current evolution and nucleosynthesis models predicting the formation and the yields of such carbon stars, we have collected high-resolution spectra of stars located in the SMC and the Sagittarius dwarf galaxy, extragalactic systems with low average metallicity and well known distances.

2 Observations and Chemical Analysis

The selected extragalactic carbon stars have been observed with the VLT/UVES instrument in service mode. The spectral resolution was around 40 000 over the domains 420-500 nm and 670-900 nm. We used carbon-rich MARCS model atmospheres and specific linelists in order to derive the abundances of metals,

Table 1. Main characteristics of the observed carbon stars

Galaxy	Star	M_{bol}	T_{eff}(K)	[M/H]	C/O	$^{12}C/^{13}C$
SMC	B30	-5.5	3100	-1.1	1.1-1.2	>200
Sgr dSph	C1	-3.1	3750	-0.8	1.05-1.1	45
Sgr dSph	C3	-4.9	3500	-0.3	1.02-1.05	30

the C/O and $^{12}C/^{13}C$ ratios and the abundances of low (Sr, Y, Zr, Nb, Ru) and high mass (La, Ce, Pr, Nd, Sm, Hf) s-elements. The main characteristics of the observed stars are summarized in Table 1.

3 Results

The chemical analysis has revealed that rather low C/O ratios are found in metal-poor extragalactic carbon stars, as found for galactic carbon stars of the solar vicinity. Furthermore, the three analyzed stars show similar s-elements enhancements: [ls/Fe]=0.8-1.3 and [hs/Fe]=1.1-1.7. This leads to new constraints for evolutionary models. For instance, the derived C/O and $^{13}C/^{12}C$ ratios are lower than model predictions at low metallicity. On the contrary, theoretical predictions of neutrons exposures for the production of the s-elements are compatible with observations (see Fig. 1). Finally, from their known distances, we have estimated the luminosities and masses of the three stars. It results that SMC-B30 and Sgr-C3 are most probably intrinsic carbon stars while Sgr-C1 could be extrinsic.

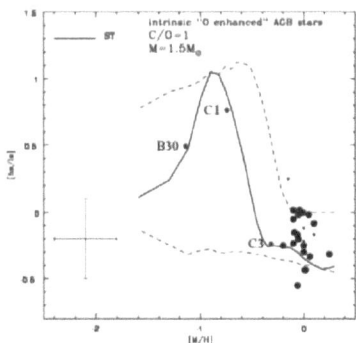

Fig. 1. Heavy over light s-elements ratio versus metallicity. Lines are theoretical predictions for a 1.5M$_\odot$ AGB stars with C/O=1.1 and three different choices for the neutron exposure rate (see Busso et al., 2001, ApJ, 557, 802). Black dots are for galactic carbon stars analyzed in Abia et al. (2002, ApJ 578, 817).

Should Chemical Abundance Distributions of Satellites Look Like Their Stellar Halos?

A.S. Font[1], J.S. Bullock[2,3], and K.V. Johnston[1]

[1] Wesleyan University, Middletown CT 06457
[2] Harvard Smithsonian CfA, Cambridge, MA 02138
[3] Hubble Fellow

Abstract. Observations suggest systematic differences between chemical abundances of stars in satellite galaxies and those in the Milky Way halo. These results are difficult to understand at present in the context of hierarchical structure formation, in which dwarf galaxies are believed to be the building blocks of galaxy formation. In this study, we model the accretion and disruption of dwarf galaxies in a ΛCDM Universe using, in combination, a semi-analytical code and numerical simulations. We conclude that differences between the accretion times of surviving satellites versus stars in the local halo, as well as the effect of feedback processes, may account for the observed differences in the chemical abundance distributions.

1 Introduction. The Method

The accretion history of a parent galaxy is constructed using a semi-analytical code. The full phase-space evolution during each accretion event is then followed separately with numerical simulations [2]. Star-formation and chemical evolution models are implemented within each satellite. The star formation prescription matches the number and luminosity of present-day galaxies in the Local Group, whereas the chemical evolution model takes into account the metal enrichment of successive stellar populations as well as feedback processes. Below we present results of a sample of four such simulated galaxy halos, denoted as Halos H1, H2, H3 and H4.

2 Results

Satellite galaxies in the Local Group exhibit a well defined stellar mass-metallicity relation ([5], [6]). Figure 1a shows the average metallicity of surviving satellites in the simulated halos versus their corresponding stellar mass. We find that, in order to match the observed relation in [6] (shown with dashed line in Fig.1a), low mass galaxies need to lose a large fraction (up to 90%) of their metals through galactic winds, whereas the large mass satellites can retain most of their metals. Similar results have also been obtained in theoretical studies which model in detail the supernova blow-out in dwarf galaxies [9].

The metallicity distribution function (MDF) in galaxy halos contains other important clues about the process of galaxy formation. In recent years, a wealth of MDF data have been collected for the Milky Way (eg. [3], [4]), M31 (eg. [1]),

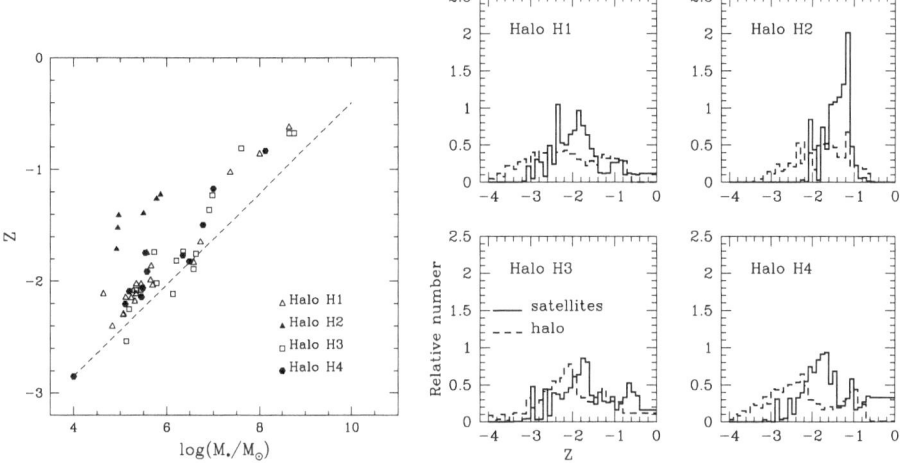

Fig. 1. a) The relation between the stellar mass and the average metallicity of surviving satellites in the simulated halos. Dashed line represents the best fit of [6] to the observations. b) MDF of surviving satellites versus that of the halo in the simulations.

as well as for dwarf galaxies (both satellites and in the field) in the Local Group [8]. The observations seem to suggest that the MDF of satellite stars is narrower than that of the Milky Way halo. Figure 1b shows a comparison between the MDF of the surviving satellites and that of the halo in our set of four simulations. In general, we do see differences between satellite and halo distributions, and the satellite MDF tends to be less extended.

3   Future Work

Future work [7] will include a modeling of phase-space variations of α elements in galaxy halos and their satellites, using the chemical enrichment code of [10].

References

1. Bellazzini, M. et al. 2003, MNRAS, 405, 867
2. Bullock, J. S. & Johnston, K. V., in prep.
3. Carney, B. et al. 1996, AJ, 112, 668
4. Chiba, M & Beers, T. C. 2000, AJ, 119, 2843
5. Dekel, A. & Silk, J. 1986, ApJ, 303, 39
6. Dekel, A. & Woo, J. 2003, MNRAS, 344, 1131
7. Font, A. S. et al., in prep.
8. Grebel, E. K., Gallagher, J. S. & Harbeck, D. 2003, AJ, 125, 1926
9. Mac Low, M. -M. & Ferrara, A. 1999, MNRAS, 303, 301
10. Robertson, B. et al, in prep.

Germanium in Damped Lyman-α Systems and Copper Connection

R. Gallino[1], M. Pignatari[1], S. Bisterzo[1], J.X. Prochaska[2], Y. Fenner[3], A. Heger[4], and S.E. Woosley[2]

[1] Dipartimento di Fisica Generale, Universitá di Torino, Via P. Giuria 1, 10125 Torino, Italy
[2] UCO/Lick Observatory, University of California, Santa Cruz, CA, USA
[3] Centre of Astrophysics & Supercomputing, Swinburne University of Tecnology, Melbourne, Australia
[4] Los Alamos National Laboratories, Los Alamos, NM 87545, USA

From the analysis of recent high-resolution spectroscopy data of Cu in unevolved galactic stars at various metallicities, as compared with nucleosynthesis predictions in massive stars exploding as SNII, ([9], [6], [10]), we deduced that most of galactic Cu is built up by neutron capture in massive stars ([3], [1]). Its production depends linearly on the metallicity (see also Fig. 3 by [2]). Neutrons are mainly released by the ^{22}Ne$(\alpha,n)^{25}$Mg reaction during pre-explosive hydrostatic core He burning followed by convective shell C burning. The production zone is essentially the oxygen-rich region. A large abundance of ^{22}Ne results from the conversion of the original CNO nuclei into ^{14}N during previous H burning, followed by ^{14}N$(\alpha,\gamma)^{18}$F$(\beta^+)^{18}$O$(\alpha,\gamma)^{22}$Ne during the early phase of core He burning. Very similar predictions may be inferred for the elements Ga, Ge, As, Se, Br, Kr, and Rb using the SNII yields of two models of initial 15 and 25 $M_\odot$ and solar composition, calculated with a full nuclear network extending up to Bi ([10], http://www.ucolick.org/ alex/nucleosynthesis/data.shtml). This result is

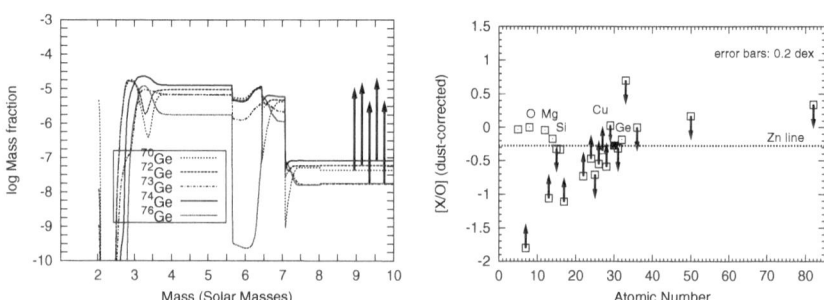

Fig. 1. *Left panel.* Post-explosive yields versus mass of the Ge isotopes for a 25 $M_\odot$ of solar composition by [10]. Arrows represent a production factor of 200 over the initial mass fraction of each isotope. *Right panel.* Logaritmic abundances relative to O and to solar ratio observed in the DLA-B/FJ0812+32 System (dust corrected) [5]. The observed [Zn/O] value is represented by a full square.

characteristic of the weak sr(p) process taking place in massive stars (Pignatari et al., these Proceedings). In the 25 $M_\odot$ post-explosive yields shown in Fig. 1, the Ge isotope yields in the mass region $\sim$2.7 to 3.5 $M_\odot$ are the result of explosive nucleosynthesis, by photodisintegration processes on previous ashes. The region from 3.5 to 5.7 $M_\odot$ contains the ashes of hydrostatic convective C-burning shell. The region from 5.7 to 6.5 $M_\odot$ contains the ashes of core He burning. The region from 6.5 to 7 $M_\odot$ is the convective He shell, where interesting explosive nucleosynthesis effects toke place. Beyond 7 $M_\odot$ is the H-rich envelope. The arrows represent a factor 200 over the initial mass fraction of each isotope. Note that ^{16}O is produced (as a primary isotope) in the same zone by a factor 100 of its initial abundance. Germanium is composed of 5 isotopes. According to the phenomenological analysis of the s process, ^{70}Ge is an s-only isotope (20.5% of solar Ge), shielded against the r-process by ^{70}Zn, ^{72,73,74}Ge are of mixed s-r origin, ^{76}Ge (7.8% of solar Ge) is an r-only isotope. Instead, as is evident from Fig. 1, all Ge isotopes are produced in massive stars in almost equal solar proportion. The main s-process in AGB stars accounts for a minor contribution to these elements: $\sim$5% of solar Cu, $\sim$10% of solar Ga, Ge, As, $\sim$15% of solar Se, Br, and $\sim$30% of solar Kr. Minor contributions are expected by the classical r-process. A negligible production is expected from SNIa ([4], [8]). In the Damped Lyman-α system FJ0812406+320808 ([5], [2]), at high redshift z=2.626, massive stars were predominantly at work with a high star formation rate in a short time scale of less that 2.5 Gyr, such as to produce a metal-rich galaxy of about 1/3 solar metallicity showing α-element enhancements (Fig. 1, right panel). The observed upper limits of Cu, Ga, As and Kr are not incompatible with the above predictions. As to zinc, the most abundant isotope, ^{64}Zn (48.9% of solar Zn), is produced in a primary-like way by massive stars, all other other Zn isotopes are produced by the secondary-like weak sr(p) process ([1], Bisterzo et al., these Proceedings). AGBs contribute only 3% of solar Zn and SNIa produce a negligible amount. The high upper limit of the Pb abundance may reflect the strong s-process production of lead by AGBs at low metallicity (Delaude et al., these Proceedings). Work partly supported by the Italian MIUR-Cofin2002 Project Pregalactic Chemical Evolution for the Heavy Elements.

References

1. S. Bisterzo, R. Gallino et al. Mem. Soc. Astron. It., in press
2. Y. Fenner, J.X. Prochaska, B.K. Gibson: ApJ **606**, 116 (2004)
3. R. Gallino, S. Bisterzo, et al., in XII Workshop on Nuclear Astrophysics, ed. E. Mueller, Ringberg, MPA/P14, 58 (2004)
4. K. Iwamoto, F. Brachwitz, et al.: ApJS **125**, 439 (1999)
5. J.X. Prochaska, J.C. Howk, A.M. Wolfe: Nature **423**, 57 (2003)
6. F.X. Timmes, S.E. Woosley, T.A. Weaver: ApJS **98**, 617 (1995)
7. C. Travaglio, R. Gallino, et al.: ApJ **601**, 864 (2004)
8. C. Travaglio, W. Hillebrandt, M. Reinecke, F.-K. Thielemann: A&A **425**, 1029 (2004)
9. S.E. Woosley, T.A. Weaver: ApJ **101**, 181 (1995)
10. S.E. Woosley, A. Heger, T.A. Weaver: Rev. Mod. Phys. **74**, 1015 (2002)

VLT/UVES Abundances of Individual Stars in the Fornax Dwarf Spheroidal Globular Clusters

B. Letarte[1], V. Hill[2], P. Jablonka[2], and E. Tolstoy[1]

[1] Kapteyn Instituut, Postbus 800, 9700 AV Groningen, Nederlands
[2] Observatoire de Paris, 5 place Jules Janssen, 92190 Meudon, France

Abstract. We present high resolution abundance analysis of nine stars belonging to three of the five globular clusters (GCs) of the Fornax dwarf galaxy. The spectra were taken with UVES at a resolution of 43 000. We find them to be slightly more metal-poor than what was previously calculated with other methods [1]. Fornax cluster 1 is now the most metal-poor globular cluster ever observed, with a [Fe/H] = −2.6. We find evidence of deep mixing in two stars belonging to cluster 1 and cluster 3. Also, Fornax globular clusters seem to have similar α/Fe ratios relative to Galactic globular clusters, though Ca and Ti may show slightly lower α-enhancement.

Alpha elements (like Ca and Mg, the top two panels of Fig. 1) are typically the products of supernovae type II (SNe II) explosions. The α/Fe ratio is usually overabundant in GC and halo stars relative to disk stars, which flatten at low-metallicity. These plots also show that the three Fornax field stars have quite a different signature from the Fornax GC stars, confirming that the GCs probe a different population than the bulk of the (more metal-rich and more chemically evolved) field population in Fornax. On the other hand, Fornax GC follow the same patterns as Galactic GC stars, suggesting that the old GC formation (and maybe all the oldest populations ?) was very similar in these two very different galaxies. The overabundance of the α-elements seen in GCs stars compared to Fornax field stars may be interpreted as simple SNe II yields which are later diluted by SNe Ia Fe production in the field stars.

In evolved red giant stars, internal mixing will bring CNO processed material to the surface of the star. Normal giants stars exhibiting only "shallow" mixing will therefore show abundance anomalies of C and N, but none of heavier elements. A good tracer of deep mixing is the anti-correlation in O-Na, as well as that in Mg-Al. In our galaxy, the O-Na and Mg-Al anti-correlations are commonly observed in globular cluster giants of various metallicities, but never in similar giant field stars. Almost certainly these anti-correlations are the result of proton-capture chains that convert O, N, Ne to Na, and Mg to Al in the H fusion layer of evolved cluster stars. It is believed that environmental effects within the cluster are responsible for the differences. In the bottom panel of Fig. 1 is the Na abundance. The two Fornax GC stars (from different clusters) with enhanced Na also exhibit depleted Mg (middle panel) and O (not shown), supporting deep mixing in these stars, as in their in Galactic counterpart.

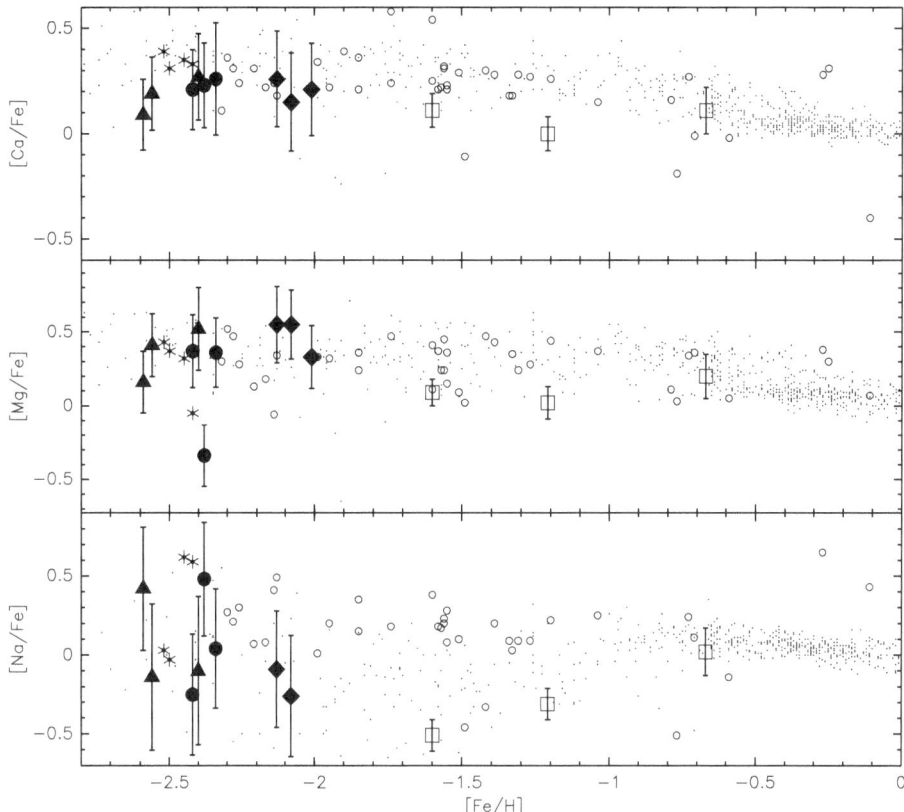

Fig. 1. Plot of Ca, Mg and Na ratios over Fe in Fornax GC (filled symbols) and Fornax field (empty squares) [2] stars. We show data for individual stars in cluster 1 (triangles), cluster 2 (circles) and cluster 3 (diamonds). The other points comes from 2 review papers, the small dots are galactic stars from [3] and the small open circles are Galactic GCs, from [4]. The asterisk are M15 stars, our reference cluster.

References

1. Strader, J., Brodie, J. P., Forbes, D. A., Beasley, M. A., Huchra, J. P., 2003, AJ, 125, 1291S.
2. Shetrone, M. D., Venn, K. A., Tolstoy, E., Primas, F., Hill, V., Kaufer, A., 2003, AJ, 125, 684S.
3. Venn, K. A., Irwin, M., Shetrone, M. D., Tout, C. A., Hill, V., Tolstoy, E., 2004, ApJ, Accepted.
4. Pritzl, B. J., Venn, K. A., Irwin, M., 2004, AJ, submitted.

Chemical Abundances of RGB-Tip Stars in the Sagittarius Dwarf Spheroidal Galaxy

L. Monaco

INAF - Osservatorio Astronomico di Trieste, via Tiepolo 11, 34131 Trieste, Italy

Abstract. We present preliminary iron abundances and α element (Ca, Mg) abundance ratios for a sample of 22 Red Giant Branch (RGB) Stars in the Sagittarius galaxy (Sgr), selected near the RGB-Tip. The sample is representative of the Sgr dominant population. The mean iron abundance is [Fe/H]=-0.49. The α element abundance ratios are slightly subsolar, in agreement with the results recently presented by [2].

1 Target Selection and Abundance Analysis

The Sagittarius dwarf Spheroidal galaxy (Sgr dSph) is currently disrupting under the strain of the Milky Way (MW) tidal field. The study of the Sgr chemical composition allows us to study at the same time the star formation history of a dwarf galaxy and the relevance of the hierarchical merging process for the formation of large galaxies such as the MW.

In May 2003, we obtained spectra for 24 Sgr stars using the high resolution spectrograph FLAMES-UVES@VLT. The target selection has been performed using the 2Mass infrared photometry (see, e.g., [6]) where the Sgr Red Giant Branch stands out very clearly from the contaminating MW field. In this way we selected stars belonging to the Sgr dominant population near the RGB-Tip. Such a selection turned out to be very efficient: 23 out of 24 stars are Sgr radial velocity members.

Up to date we derived the iron abundance for 22 stars and alpha element abundance ratios (Mg, Ca) for 20 stars of the sample. Temperatures have been derived from the (V-I) color using the calibration of [1] and E(B-V)=0.14 [4]. Gravities were determined superposing theoretical isochrones [3] to the observed optical color magnitude diagram while microturbulent velocities have been derived minimizing the dependence of the derived abundance from the measured equivalent width. A model atmosphere has been computed for each star using the ATLAS 9 code. Equivalent widths have been measured by using the standard IRAF task SPLOT and, finally, abundances were determined using the WIDTH code.

As can be seen from the metallicity distribution (see Fig. 1, upper panel), the mean metallicity of the Sgr dominant population is <[Fe/H]>=-0.49±0.19, in excellent agreement with the results derived in [5] from the magnitude of the RGB-bump and the shape of the Red Giant Branch. A component extending towards lower metallicities is also present. The α element abundance ratio (see Fig. 1, bottom panel) is slightly subsolar, <[α/Fe]>=-0.20±0.10, in agreement with [2] and, apart from a small offset, also with [7].

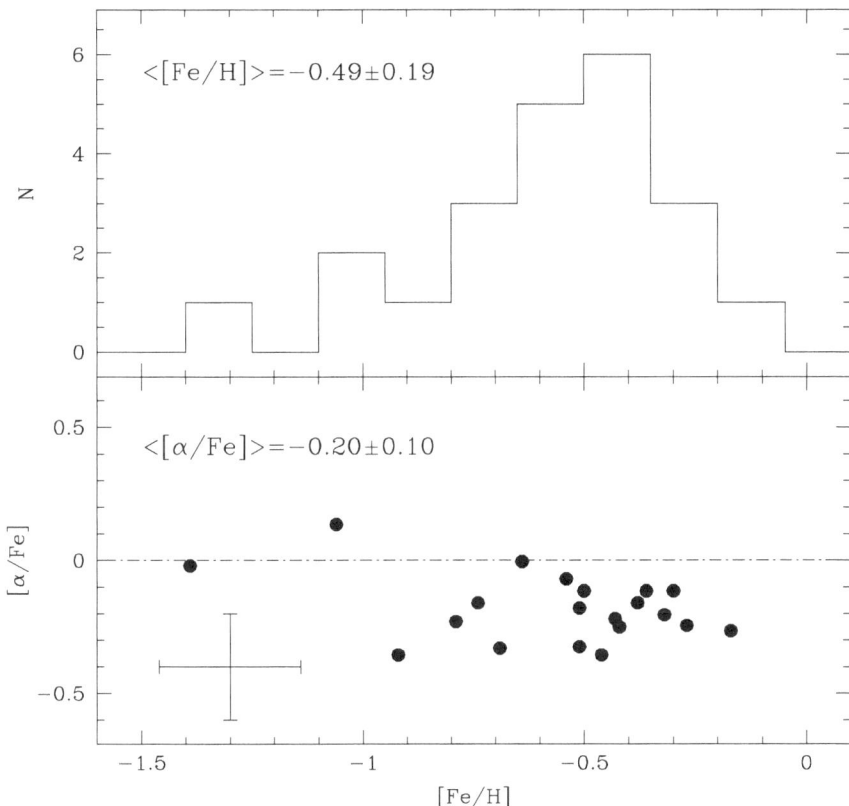

Fig. 1. Upper panel: metallicity distribution for the 22 stars analyzed so far. Lower panel: α element abundance ratio *vs* iron abundance for 20 stars of the sample. The $[\alpha/\text{Fe}]$ abundance ratio is the average of $[\text{Mg}/\text{Fe}]$ and $[\text{Ca}/\text{Fe}]$. A typical errorbar is also plotted.

References

1. Alonso A., Arribas S., Martínez-Roger C., 1999, A&AS, 140, 261
2. Bonifacio P., Sbordone L., Marconi G., Pasquini L., Hill V., 2004, A&A, 414, 503
3. Girardi L., Bertelli G., Bressan A., Chiosi C., Groenewegen M. A. T., Marigo P., Salasnich B., Weiss A., 2002, A&A, 391, 195
4. Layden, A.C., & Sarajedini, A., 2000, AJ, 119, 1760 (LS00)
5. Monaco L., Ferraro, F.R., Bellazzini, M., Pancino, E., 2002, ApJ, 578, L47
6. Monaco L., Bellazzini, M., Ferraro, F.R., Pancino, E., 2004, MNRAS, 353, 874
7. Smecker-Hane & McWilliam, 2002, ApJ submitted, astro-ph/0205411

Photometry and Spectroscopy of Bright Stars in the Carina dSph Galaxy

M. Monelli[1], G. Bono[1], M. Nonino[2], C.R. James[3], P. François[4], S. Moehler[5], R. Buonanno[1], G. Nevils[3], A. Recio-Blanco[6], H.A. Smith[7], P.B. Stetson[8], F. Thévenin[6], and A.R. Walker[9]

[1] INAF - Osservatorio Astronomico di Roma, via Frascati 33, 00040 Monteporzio Catone, Roma, Italy
[2] INAF - Osservatorio Astronomico di Trieste
[3] Sam Houston State University, Department of Physics
[4] Paris Observatory
[5] Institut für Theoretische Physik und Astrophysik der Universität, Kiel
[6] Observatoire de la Côte d'Azur
[7] Dept. of Physics, Michigan State University
[8] Dominion Astrophysical Observatory, Herzberg Institute of Astrophysics
[9] Cerro Tololo Inter-American Observatory

Abstract. We present the results of a spectroscopic analysis of bright stars in the Carina dSph galaxy. We collected low-resolution FORS2@VLT spectra of $\approx$ 200 stars. Our spectroscopic targets have been selected among the evolved Carina stars, in particular we selected low-mass, old Red Giant, Asymptotic, and Horizontal Branch stars, as well as intermediate-age stars of the Red Clump. We present preliminary estimates concerning the radial velocities of old and intermediate-age populations.

1 Introduction

The Carina dSph galaxy is an intriguing satellite of the Galaxy. Deep wide field photometry ([3], [1]) has disclosed a complex Star Formation History (SFH) characterized by at least three well separated bursts, at $\approx$11, $\approx$5, $\leq$1 Gyr ago. Nevertheless, the chemical enrichment in this galaxy is still under debate. Even though the Red Giant Branches (RGB) of old and intermediate-age populations are mixed, RGB stars cover a narrow range in color, at fixed metallicity. This seems to support a very small spread in metallicity ([1]). However, low-resolution spectroscopy ([4]) of RGB stars support a spread of $\approx$0.25 dex. Moreover, there is evidence that the radial distributions of the old and intermediate-age populations are different. Intermediate-age stars appear more centrally concentrated, while the old population is distributed in a sort of broader "halo" surrounding Carina. We collected low-resolution spectra of $\approx$200 stars belonging to both old and intermediate-age populations, to investigate their kinematical and chemical properties.

2   Data Reduction and Discussion

The targets were selected on the basis of deep, wide field optical photometry, and four fields around the Carina centre were observed with FORS2@VLT, with a spectral resolution R≈2000. The GRIS_1400V+18 grism was adopted.

Standard data reduction, i.e. bias and flat field correction, has been performed with Iraf. The Iraf task APEXTRACT/APALL was used to extract the spectra, with interactively selected background sampling, in order to avoid contamination for the star spectrum. The wavelength calibration has been done using daily He, Ne, HgCd arcs, and, in order to improve the calibration, wavelengths values for the transitions used were taken from http://physics.nist.gov/.

The radial velocities were estimated by identifying a few lines (3 to 6, depending on the quality of the spectra) and fitting a Gaussian profile to each line to find the line core. We estimated the average radial velocity, and we present here only the stars with standard deviation of the mean less than 7 km s^{-1}.

Fig. 1 (left panel) shows the radial velocity distribution of selected stars. We estimate a main peak of V_r ≈220±7 km s^{-1}, in good agreement with previous measurements based on RGB stars [2]. A secondary peak appears at V_r ≈180±7 km s^{-1}. This dichotomy is shown in the right panel of Fig. 1, where the radial velocity is plotted as s function of the distance from the centre.

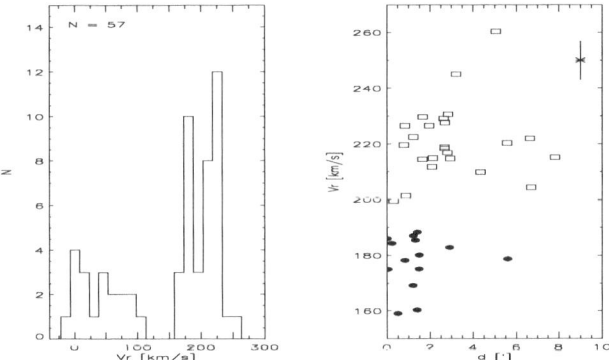

Fig. 1. Left : distribution of the radial velocities. Right : radial velocity plotted as a function of the distance from the Carina center

These preliminary results will be further investigated with the complete sample of stars.

References

1. Monelli, M. et al., 2003, AJ, 126, 218
2. Mateo et al., 1993, AJ, 106, 510
3. Smecker-Hane, T. et al., 1996, ASP Conf. Ser. 98, 328
4. Smecker-Hane, T. et al., 1999, ASP Conf. Ser. 192, 159

Part V

Mixing Processes – Models

Rotation and Internal Gravity Waves in Low-Mass Stars

C. Charbonnel[1,2]

[1] Geneva Observatory, 51, chemin des Maillettes, CH-1290 Sauverny, Switzerland
[2] LATT, CNRS UMR 5572, 12, av.E.Belin, 31400 Toulouse, France

Abstract. We discuss the successes and failures of self-consistent rotating models of main sequence and slightly evolved low-mass stars. We focus in particular on the strongest observational constraints which are the lithium surface abundances and the solar rotation profile deduced from helioseismology. We recall that the hot side of the so-called lithium dip is well explained by hydrodynamical models where the transport of angular momentum is carried out by meridional circulation and shear turbulence. For cooler stars however the transport of angular momentum is dominated by another process. We show that internal gravity waves are the best candidate and we explain how the mass dependence of this mechanism is expected to resolve the enigma of the lithium dip in terms of rotational mixing, forming a coherent picture in main sequence stars of all masses.

1 Rotating Models and the Blue Side of the Li Dip

In many locations in the Hertzsprung-Russell diagram, stars exhibit signatures of processes that require challenging modeling beyond the standard stellar theory. In this context, rotation has become a major ingredient of modern stellar models, especially when abundance anomalies have to be accounted for. In the most sophisticated theoretical developments (i.e., Zahn 1992, Maeder 1995, Talon & Zahn 1997, Maeder & Zahn 1998, Palacios et al. 2003), the internal rotation law evolves as a result of contraction, expansion, mass loss, meridional circulation (hereafter MC) and shear turbulence, and the mixing of the chemicals is directly linked to the rotation profile. Such a self consistent treatment has been successfully applied in various parts of the HR diagram over a large range of stellar masses and evolutionary stages (see the review by Maeder & Meynet 2000, and the papers by Maeder, Meynet and Palacios in this volume).

One of the most striking signatures of transport processes in low-mass stars is the so-called Li dip (Fig. 1). This drop-off in the Li content of MS F-stars in a range of 300K in Teff centered around 6700K was first discovered in the Hyades by Wallerstein et al. (1965) and latter confirmed by Boesgaard & Tripicco (1986). This feature appears in all galactic clusters older than $\sim$ 200 Myrs as well as in field stars (Balachandran 1995). Boesgaard (1987) noticed that at the Teff of Li dip in the Hyades also occurs a sharp drop in rotational velocities (Fig. 1). Rotation was then suggested to play a dominant role in the build up of the Li dip.

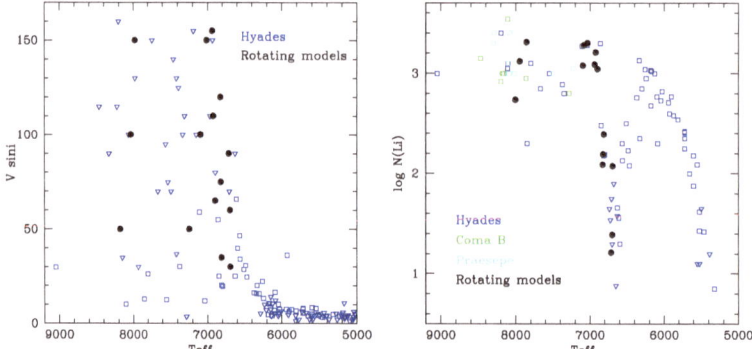

Fig. 1. Projected rotational velocity in the Hyades and lithium versus effective temperature in 3 open clusters of ~ the same age (Hyades, ComaB, Praesepe). The open symbols are the observed data (Kraft 1965, Stauffer et al. 1987, Mermilliod 1992, Burkhart & Coupry 1998, 2000, Boesgaard 1987). The black symbols are the predictions of the rotating models at the age of the Hyades (compilation of Talon & Charbonnel 1998 and Charbonnel & Talon 1999)

Different investigations of the possible connection between rotation and the Li dip have appeared in the literature. Most relied on highly simplified descriptions of the rotation-induced mixing processes. In the MC model of Tassoul & Tassoul (1982) used by Charbonneau & Michaud (1988), the feed-back effect due to angular momentum (hereafter AM) transport as well as the induced turbulence were ignored. Following Zahn (1992), Charbonnel et al. (1992, 1994) considered the interaction between MC and turbulence induced by rotation, but the transport of AM was not treated self-consistently.

In Talon & Charbonnel (1998, TC98), Charbonnel & Talon (1999, CT99) and Palacios et al. (2003) we went one step further : We included in the models the most complete description currently available for rotation-induced mixing, and we computed self-consistently the transport of the chemicals and that of AM due to wind-driven MC. We used the same input physics than that used with success by the Geneva group to explain several observational patterns of more massive stars (e.g. Maeder & Meynet 2000 and Talon & Charbonnel 2003 and references therein).

In this framework one obtains a very good agreement between the models and the observations on the blue side of the Li dip as can be seen in Fig. 1. According to the observations of the velocity distribution in open clusters of various ages, the stars hotter than ~ 6900K are not slowed down by a magnetic torque. These stars soon reach a regime with no net AM flux in which MC and turbulence counterbalance each other. The weak mixing resulting is just sufficient to counterbalance microscopic diffusion, except in the slowest rotators where its signature is visible. For Teff between 6600 and 6900K, a weak magnetic torque slows down the outer stellar layers. As the magnetic torque increases with decreasing Teff, MC has much more AM to transport to the surface, leading to a larger destruction of Li. Consequently at the age of the Hyades rotational mixing perfectly explains the shape of the blue side of the Li dip, as well as the observed dispersion. These models also account very nicely for the CNO abundance patterns in the Hyades (Takeda et al. 1998, Varenne & Monier 1999) which are additional clues of the transport processes which counteract atomic diffusion in these stars (see Palacios et al. 2003). Last but not least, these rotating models explain the Li patterns in stars originating from the hot side of the Li dip including the subgiant and the giant phases in older open clusters like IC 4651 (Pasquini et al. 2004; see Fig. 2) and in the field.

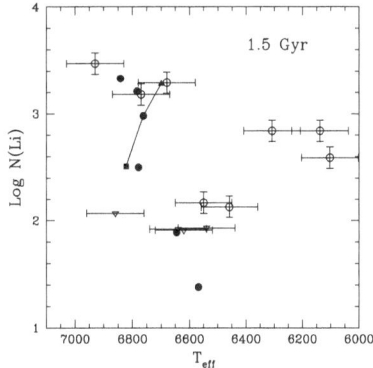

Fig. 2. Li abundance in main-sequence and turnoff stars in the open cluster IC 4651 aged of ∼1.5-1.7 Gyr vs Teff from photometry (open points and triangles for actual determination and upper limits respectively). Black points show rotating model predictions (CT99 and Palacios et al. 2003) at 1.5 Gyr for an initial rotation velocity of 110 km.sec^{-1} and for various stellar masses. For 1.5M$_\odot$ models for an initial rotation of 50 and 150 km.sec^{-1} are also shown (black triangle and square). This provides an estimate of the expected dispersion for stars inside the Li dip. See Pasquini et al. (2004) for more details

2 Rotating Models and the Red Side of the Li Dip

What happens for cooler (i.e. less massive) stars on the red side of the Li dip? As we shall see now, the stellar mass or the effective temperature of the dip is a transition point for stellar structure and evolution. First of all it is a transition as far as the rotation history of the stars is concerned. Indeed the physical processes responsible for surface velocity are different, or at least operate with different timescales on each side of the dip. At the age of the Hyades, the stars hotter than the dip still have their initial velocity while cooler stars have been efficiently spun down (Fig. 1). This behavior is linked to the variation of the thickness of the superficial H-He convection zone which gets rapidly deeper as Teff decreases from 7500 to 6000K (e.g. TC98). Below ∼ 6600 K, the stars have a sufficiently deep surface convection zone to sustain a strong magnetic field which spins down the outer layers efficiently. On the red side of the lithium dip, the magnetic torque strengthens as the convective envelope grows. If we assume there that all the momentum transport is assured by the wind-driven circulation, and if we keep the same parameters that explain the Li abundances on the blue side of the Li dip as well as the chemical anomalies in more massive stars (§ 1), then we obtain too much Li burning compared to the observations. In other words the rise of Li abundances on the right side of the dip is not explained within this framework.

3   Internal Gravity Waves and Their Mass Dependence

TC98 proposed that the Li dip corresponds to a transition region where other(s) physical mechanism(s) for AM transport start to become more efficient than MC and turbulence. This suggestion was linked to another observation which fails to be reproduced by the hydrodynamical models relying on MC and turbulence, namely the solar rotation profile given by the helioseismic observations (e.g. Brown et al. 1989, Basu this volume) : At the solar age, those models still have large Ω gradients (i.e. a rapidly rotating core) which are not present in the Sun (Chaboyer et al. 1995, Matias & Zahn 1997). That conclusion was reached independently by two different groups, using different descriptions for the transport processes : On one hand, the Yale group who computed the evolution of AM in low mass stars with a simplified version of the action of MC which was considered as a diffusive rather than an advective process (Pinsonneault et al. 1989, see Pinsonneault this volume). On the other hand Matias & Zahn (1997) who performed a complete study for the evolution of the Sun's AM, where they took into account the advective nature of MC. They concluded that MC and turbulence are not efficient enough to enforce the flat rotation profile measured by helioseismology.

These complementary observational constraints indicate that another process participates to the transport of AM in solar-type stars, while MC and turbulence are successful in more massive stars. The two most likely candidates are the large-scale magnetic field which could be present in the radiative zone and the internal gravity waves (hereafter IGW) which are generated by the external convective zone. As we just explained, the observations suggest that the efficiency of this process is linked to the growth of the convective envelope. This is a characteristics of IGW.

Gravity waves are generated by turbulent motions in the convective regions. Excitation can be related to internal stresses that correlate with the mode's eigenfunction (e.g., Goldreich et al. 1994, Kumar et al. 1999) or to penetration below a convection zone (Press 1981, García López & Spruit 1991, Zahn et al. 1997). In real stars both sources would contribute to wave generation and are thus additive. The exact properties of the wave spectra remain somewhat uncertain and it is relatively hard to precisely determine their absolute magnitude. Their differential properties as a function of the stellar mass (or effective temperature) is however reliable and crucial in the present context.

Talon & Charbonnel (2003) showed indeed that both the generation and the efficiency of IGW in extracting AM depend on the structure of the convective envelope which varies strongly with Teff. In fact the total momentum luminosity in waves rises with effective temperature up to a maximum around 6000 K and then decreases when one moves towards the region of the dip. As a result the net momentum extraction associated with IGW has the proper effective temperature dependence to explain the Li dip in terms of rotational mixing (Fig. 3) : On the hot side of the dip and in more massive stars the transport of AM and of chemicals by MC and shear turbulence explains the Li pattern as well as the abundance anomalies of He and CNO in more massive stars. In lower mass (i.e., cooler) main sequence stars, IGW dominate the transport of AM, thereby reducing the magnitude of MC and shear and shaping the Li pattern on the red side of the dip. These cooler objects should thus reach quasi-solid body rotation on relatively short timescales. Talon et al. (2002) showed indeed that, through differential filtering, IGW are able to shape the solar rotation profile in $\sim 10^7$ yrs.

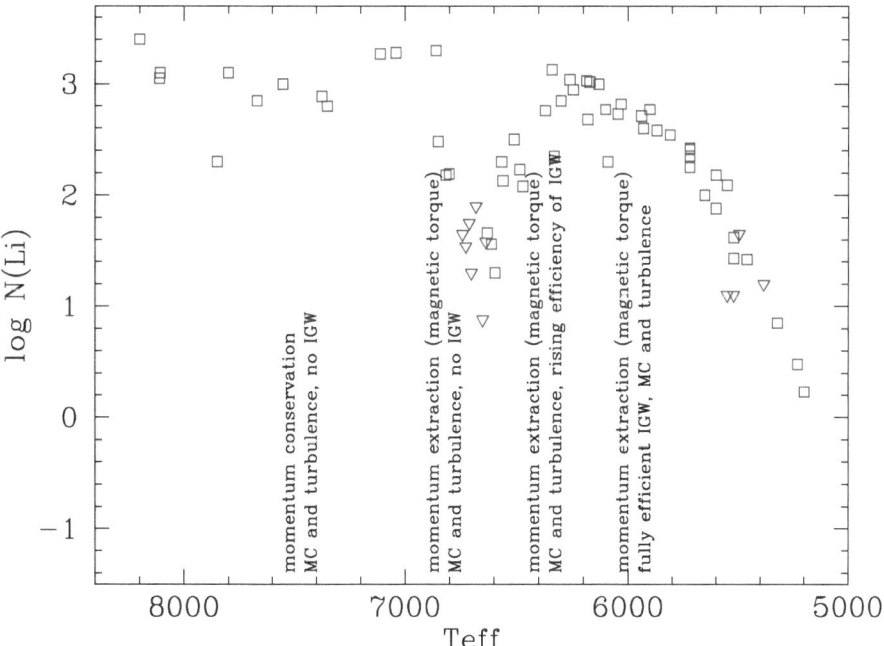

Fig. 3. Measured Li abundances in the Hyades. Superposed is the pattern of AM transport required to produce the Li dip in terms of rotational mixing

4   What Remains To Be Done?

The examination of the differential properties of the generation and filtering of internal gravity waves in low-mass main-sequence stars indicates that we have in hand a very interesting process to resolve the enigma of the Li dip in terms of rotational mixing, forming a coherent picture of mixing in various parts of the HR diagram (Fig. 3).

We have now to go one step further and to build stellar evolution models where the transport of angular momentum will be followed self-consistently under the action of meridional circulation, shear turbulence, and internal gravity waves. In this path some important aspects still need to be clarified : Can we better describe the excitation mechanisms and evaluate in a more reliable way the quantitative properties of the wave spectra? What is the direct contribution of IGW to the transport of chemicals, especially in the dynamical shear layer produced just below the convective envelope by the wave-mean flow interaction? What is the influence of the Coriolis force on IGW? How do IGW interact with a magnetic field? Work is in progress in this direction.

References

1. Balachandran C.: ApJ **446**, 203 (1995)
2. Boesgaard A.M.: PASP **99**, 1067 (1987)
3. Boesgaard A.M., Tripicco M.J.: ApJ **302**, L49 (1986)
4. Brown T.M., Christensen-Dalsgaard J., Dziembowski W.A., Goode P., Gough D.O.: ApJ **343**, 526 (1989)
5. Burkhart C., Coupry M.F.: A&A **220**, 197 (1998)
6. Burkhart C., Coupry M.F.: A&A **354**, 216 (2000)
7. Chaboyer B., Demarque P., Pinsonneault M.H.: ApJ **441**, 865 (1995)
8. Charbonneau P., Michaud G.: (1988)
9. Charbonnel C., Talon S.: A&A **351**, 635 (1999) CT99
10. Charbonnel C., Vauclair S., Maeder A., Meynet G., Schaller G.: A&A **283**, 155 (1994)
11. Charbonnel C., Vauclair S., Zahn J.P.: A&A **255**, 191 (1992)
12. García López R.J., Spruit H.C.: ApJ **377**, 268 (1991)
13. Goldreich P., Murray N., Kumar, P.: ApJ **424**, 466 (1994)
14. Kraft (1965)
15. Kumar, P., Talon, S., Zahn, J.-P.: ApJ **520**, 859 (1999)
16. Matias J., Zahn J.-P.: "Sounding solar and stellar interiors", IAU Symposium 181 (1997)
17. Maeder A.: A&A **299**, 84 (1995)
18. Maeder A., Meynet G.: ARA&A **38**, 143 (2000)
19. Maeder A., Zahn J.P.: A&A **334**, 1000 (1998)
20. Mermilliod J.C.: private communication (1982)
21. Palacios A., Talon S., Charbonnel C., Forestini M.: A&A **399**, 603 (2003)
22. Pasquini L., Randich S., Zocalli M., Hill C., Charbonnel C., Nordtröm B.: A&A **424**, 951 (2004)
23. Pinsonneault M.H., Kawaler S.D., Sofia S., Demarque P.: ApJ 338, 424 (1989)
24. Press W.H.: ApJ **245**, 286 (1981)

25. Stauffer J.R., Hartmann L.W., Latham D.W.: ApJ **320**, L51 (1987)
26. Takeda Y., Kawanomoto S., Takada-Hidai M., Sadakane K.: PASJ **50**, 509 (1998)
27. Talon S., Charbonnel C.: A&A **335**, 959 (1998) TC98
28. Talon S., Charbonnel C.: A&A **405**, 1025 (2003)
29. Talon S., Kumar, P., Zahn J.-P.: ApJ **574**, L175 (2002)
30. Talon S., Zahn J.-P.: A&A **317**, 749 (1997)
31. Tassoul J.L., Tassoul M.: ApJS **49**, 317 (1982)
32. Varenne O., Monier R.: A&A **351**, 247 (1999)
33. Wallerstein G., Herbig G.H., Conti P.S.: ApJ **141**, 610 (1965)
34. Zahn J.P.: A&A **265**, 115 (1992)
35. Zahn J.P., Talon S., Matias J.: A&A **322**, 320 (1997)

Helioseismic Evidence for Mixing in the Sun

S. Basu

Astronomy Department, Yale University, PO Box 208101,
New Haven CT 06520-8101, USA

Abstract. We discuss the evidence gathered from helioseismology about considerable mixing below the solar convection zone. The evidence is obtained directly through inversions and also through more subtle, somewhat indirect signatures if mixing.

1 Introduction

Helioseismology is the study of the Sun using solar oscillation frequencies. The Sun oscillates in millions of modes. The modes are linear and adiabatic. Since the Sun is a spherical body (the departures from sphericity have been measured to be very small), it is most natural to describe the angular dependence of the normal-modes in terms of spherical harmonics. Each mode is described by its radial order n, which is the number of nodes in the radial direction, the degree ℓ, where $L = \sqrt{\ell(\ell+1)}$ is roughly speaking the number of wavelengths along the solar circumference, and the azimuthal order m that measures the number of nodes along the equator. The numbers n, ℓ, and m describe the mode completely and determine its frequency ν (or $\omega \equiv 2\pi\nu$). In the absence of rotation or any other agent such as magnetic field to break spherical symmetry, all modes with the same n, and ℓ have the same frequency. Thus each (n, ℓ) multiplet is $2\ell + 1$ fold degenerate. In the case of the Sun, rotation, magnetic field and other large scale flows and asymmetries are small, hence these can be assumed to be small perturbations. Thus for the Sun, the mean frequency of an (n, ℓ) multiplet is a function of solar structure alone. These can be used to probe the structure of the Sun.

2 The Evidence

The equations describing linear, adiabatic stellar oscillations are known to be Hermitian (Chandrasekhar 1964). This property of the equations is used to relate the differences between the structure of the Sun and a known reference solar model to the differences in the frequencies of the Sun and the model by known kernels. Thus by determining the differences between solar models and the Sun by inverting the frequency differences between the models and the Sun we can determine whether or not mixing took place in the Sun.

Fig 1(a) shows the relative difference in the squared sound-speed between the standard solar model (SSM) of Basu et al. (2002) and the Sun. The solar data

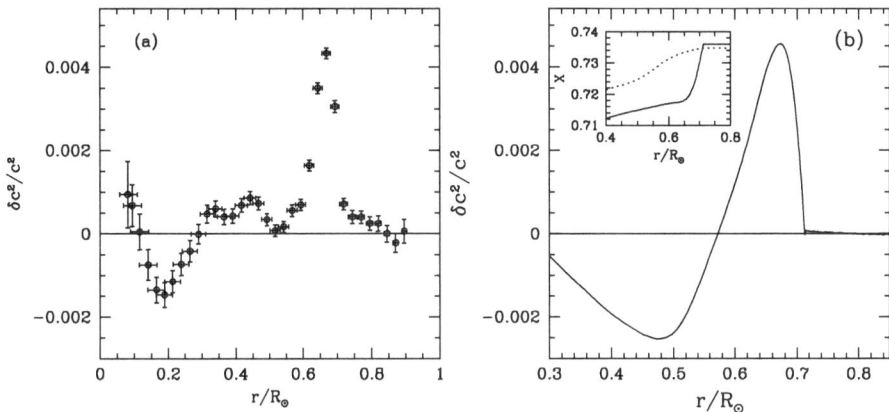

Fig. 1. Panel (a): The relative squared sound-speed difference between the Sun and the SSM of Basu et al. (2000). The differences are in the sense (Sun-Model)/Sun. Panel (b): same as Panel (a), but for two solar models, one with mixing below the CZ base and the other without (an SSM). The differences are in the sense (Mixed-SSM)/Mixed. The inset shows the the hydrogen profiles for the two models.

used are a combination of low degree data from the Birmingham Solar Oscillation Network (BiSON; cf. Chaplin et al. 1996) and intermediate degree data from the Low-ℓ (LOWL) instrument (cf. Tomczyk et al. 1995). Other data sets give similar results (see Basu et al. 2000). Note that there is a very large difference in a very localised regions just below convection zone (henceforth CZ) base (i.e., 0.713$R_\odot$; Basu 1997). This sound-speed difference is most easily interpreted as being due to lack of mixing in the SSM. The SSM incorporates the gravitational settling of helium below the CZ. In the absence of mixing, as is the case in SSMs, the helium abundance increases sharply below the CZ base, causing the sound-speed to decrease (since $c^2 \propto T/\mu$, c being the sound-speed, T the temperature and μ the mean molecular weight; μ increases with increase in helium). The fact that in this region the sound-speed of the Sun is much higher indicates that there is mixing in the Sun. This can be checked using models. Fig. 1(b) shows the sound-speed difference between a model with mixing below the CZ base and an SSM. Again we see the sharp feature in the sound-speed difference. In fact the sound-speed differences between the Sun and models that have mixing do not have the sharp feature seen in Fig. 1 (see e.g. Basu et al. 2000).

Apart from inversions, there is another way to determine whether or not there is mixing in the Sun. Any spherically symmetric, localized sharp feature or discontinuity in the Sun's internal structure leaves a definite signature on the solar p-mode frequencies. Gough (1990) showed that changes of this type contribute a characteristic oscillatory component to the frequencies $\nu_{n,\ell}$ of those modes which penetrate below the localized perturbation. The amplitude of the oscillations increases with increasing "severity" of the discontinuity, and the wavelength of the oscillation is essentially the acoustic depth of the sharp-feature. Solar modes encounter two such features, the base of the convection zone (henceforth CZ) and the He II ionization zone. such features, the base of the convection zone (henceforth CZ) and the He II ionization zone. The transition of the temperature gradient from the adiabatic to radiative values at the CZ base gives rise to the oscillatory signal in frequencies of all modes which penetrate below the CZ base. This signal increases in the presence of gravitational settling of helium when mixing is absent, again because the sound-speed changes abruptly because of the sudden increase in helium abundance. The signal is small, but can be amplified and measured by taking the 4th differences of the frequencies (see Basu 1997, Basu & Antia 1994). Models without mixing are expected to have a higher amplitude than models with mixing (see Basu 1997). Fig. 3 shows the oscillatory signal in the 4th differences of solar frequencies.

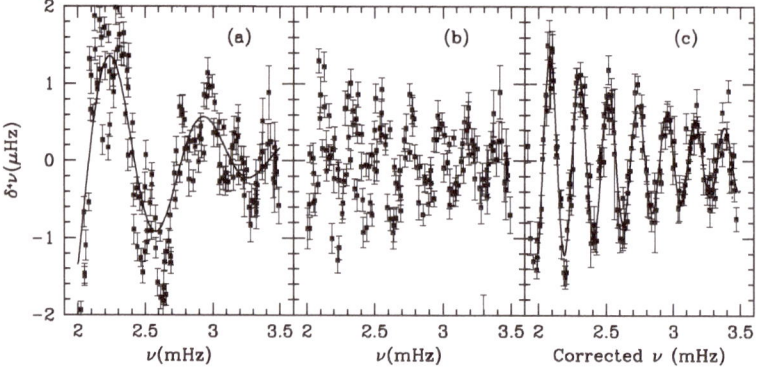

Fig. 2. The oscillatory signals in the fourth differences of solar frequencies. Panel (a): the 4th differences as points with the fit to the signal from the HeII ionization zone. Panel (b): the signal from the CZ base once the HeII signal is removed. Panel (c): the fit to the CZ-base signature with the ℓ-dependence of the signal removed.

Fig. 4 shows the average amplitude of the oscillatory function for different models and for the Sun. Both GONG and MDI data are used for the Sun. Model STD and S are SSMs that incorporate diffusion, and we can see that they have have greater amplitudes than the Sun. STD is the model shown in Fig. 1(a). ND is an old SSM which did not incorporate diffusion, so while the model does not agree well with the Sun, the abundance profile is in better agreement. Model ROT and RCVD are rotationally mixed models, and TD has ad hoc turbulent mixing below the CZ base and we can see that these models do much better. INV is a model that was constructed to have the solar abundance profile as obtained from inversions. These results re-reinforce the conclusions drawn from inversions that mixing does indeed take place in the Sun. The fact that τ, the acoustic depth of the models and the Sun do not always agree is a less serious matter. It is known that the surface properties of solar models are usually wrong, mainly because of the approximations used to describe convection. This means that the

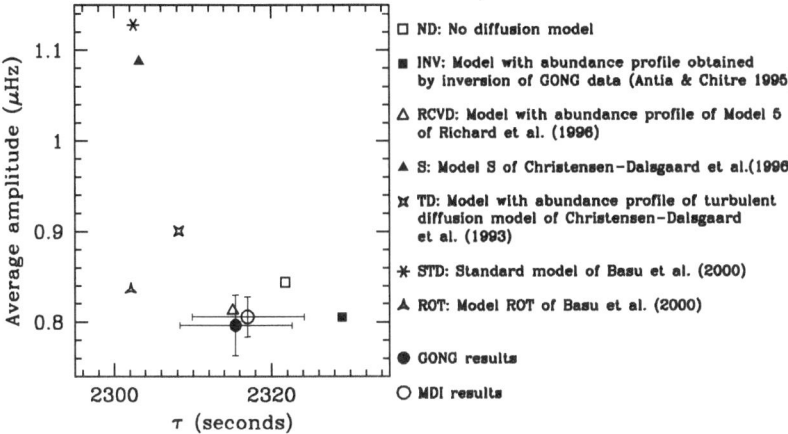

Fig. 3. The amplitude of the CZ-base signal plotted as a function of the acoustic depth for solar data and several solar models. Note that only models which incorporate mixing have amplitudes similar to that of the Sun.

sound-speed of the outer layers of the models is very different from the Sun, resulting in a different acoustic depth of the CZ base even when the position of the CZ base is correct, as in all model except ND.

Many groups have now started collecting data on oscillations of other stars. For these star, the unresolved observations mean that only low degree modes will be available. We can only obtain the structure of the cores by inverting these modes (Basu 2003). However, it should be possible to determine the presence of mixing by using the oscillatory signal. This however requires that the errors in the frequencies are of the order of 0.1μHz or less.

References

1. Antia, H.M., Chitre, S.M.: Astrophys. J., **442**, 434 (1995)
2. Basu, S.: Mon. Not. R. Astron. Soc **288**, 572 (1997)
3. Basu, S.: Astrophys. Sp. Sci **284**, 153 (2003)
4. Basu, S., Antia. H.M.: Mon. Not. R. Astron. Soc **269**, 1137 (1994)
5. Basu, S., Pinsonneault, M.H., Bahcall, J.N.: Astrophys J. **529**, 1084 (2000)
6. Chandrasekhar, S.: Astrophys. J. **139**, 664 (1964)
7. Chaplin, W. J. et al.: Solar Phys. **168**, 1 (1996)
8. Christensen-Dalsgaard, J., Proffitt, C.R., Thompson, M.J.: Astrophys J. **403**, L75 (1993)
9. Christensen-Dalsgaard, J., et al.: Science **272**, 1286 (1996)
10. Gough, D.O. in: Lecture Notes in Physics, ed. by Y. Osaki, H. Shibahashi, **367** (Springer, Berlin 1990) p283
11. Richard, O., Vauclair, S., Charbonnel, C., Dziembowski, W.A.: Astron. Astrophys. **312**, 1000 (1996)
12. Tomczyk, S. et al.: Solar Phys. **159**, 1 (1995)

Clarifying Problems in the Description of Pre Main Sequence Evolution

F. D'Antona[1] and J. Montalbán[2]

[1] INAF, Osservatorio di Roma, I-00040 Monteporzio
[2] Université de Liège, Institut d'Astrophysique et Géophysique, Allée du 6 Août, B-4000 Liège, Belgium

Abstract. We comment on a few points about modeling of pre main sequence (PMS) evolution, and try to derive some constraints on the stellar evolution during this phase, in spite of the parametric description of convection. We discuss: *i*: the role of deuterium burning and the choice of the initial starting model, *ii*: the results of models in which convection is calibrated using the 2D RHD simulation; *iii*: lithium burning and HR diagram location in the models calibrated –or not– on the solar radius. We stress that PMS binaries and lithium depletion in young clusters both point toward very inefficient convection in PMS, and that this result does not seem to be due to the use of local convection models.

1 Basic Questions

What observers would like to understand when they are in front of the numerous recent sets of pre main sequence tracks is –or should be– the following:

1. which are the reasons behind the different results? Can we discriminate model from model?
2. what is "important" and what is only marginal? That is: are there any important constraints which we can derive from observations, even in the absence of a full understanding of modeling?

More than ten years ago, when distributing the popular set of models published in D'Antona & Mazzitelli (1994) the authors added a 'readme' file asking observers to use the tracks as 'tests', more than as a benchmark for observations, in the hope that observations could throw some light on our models, which, especially in regard to convection, were –and are– parametric. After that, many other sets of tracks have been computed. In this paper we try to summarize which are the key ingredients of the published models, and whether we can discriminate between numerical uncertainties and physical uncertainties.

2 The Starting Model

How much the ages of young PMS object depend on the starting stellar evolution model? Baraffe et al. (2002) show that the first million year(s) are very uncertain for low mass stars (the statement refers to masses $\sim 0.1 M_\odot$ or smaller), as they depend on the choice of the starting model at the phase of deuterium burning. We clarify here that this choice is not independent from our understanding of how star formation proceeds. We show in Fig.1 the luminosity evolution in 0.5M$_\odot$ tracks starting from different values of central temperature, from $\log T_c = 5.7$ to 6.1. In the figure we also show the coefficient of energy generation coming from the proton plus deuterium nuclear reaction. If we take an initial central temperature of 10^6K or more, the model starts 'in the middle' of D-burning, and its luminosity evolution is certainly different from the evolution obtained if the star arrives at the ignition temperatures by hydrostatic contraction, so that D-burning is gently ignited as soon as the physical conditions allow it. But the numerical choice of the initial model does not imply that the object can really start its life at $T \simeq 10^6$K, in the middle of D-burning, so this may be a questionable assumption.

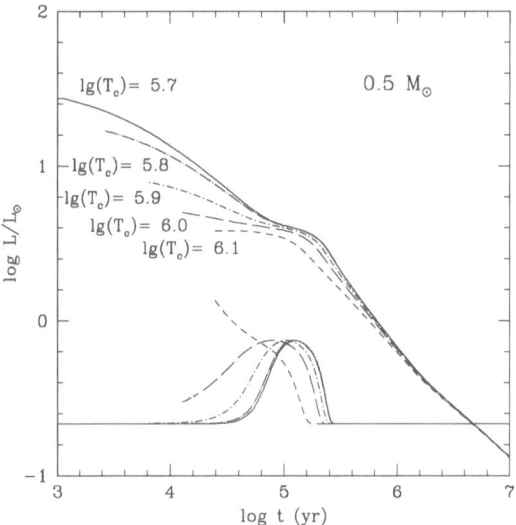

Fig. 1. Luminosity evolution of tracks of 0.5 M$_\odot$ starting at different central temperature, labelled in the figure. At the bottom we sketch the energy liberated per gram due to the deuterium fusion with protons. The tracks of $\log T \geq 6.0$ start in the middle of D-burning.

The uncertainty in the age of pre main sequence stars is therefore of the order of the thermal timescale at the luminosity of D-burning: smaller than a few times 10^5 yr for normal T Tauri, and larger than 10^6 yr for very low mass stars and brown dwarfs (BD). In fact, comparing observations spanning a wide range of masses we could even constrain the models, for example we can ascertain whether the Stahler et al. (1986) picture of collapse is valid also in the BD regime, or the very low masses evolve through hydrostatic contraction at luminosity larger than those of D-burning.

3   Model Atmospheres and Convection in Atmosphere

How much the models depend on the model atmosphere taken as boundary condition? Which is the relative role of opacity and convection model? It is difficult to disentangle the role of non gray boundary conditions from that of the convection model, which is still the dominant uncertainty in the evolutionary tracks. In Montalbán et al (2004) we have shown that it is necessary to consider both the atmospheric and envelope treatment of convection: if these treatments are different, the resulting structure depends also on the optical depth τ_{ph} at which the match between atmosphere and interior is done. Therefore, to identify a PMS track, it is not sufficient to specify the set of model atmospheres adopted and the value of α used for internal structure computation (α_{in}): if the model atmospheres have $\alpha_{atm} \neq \alpha_{in}$, also τ_{ph} must be specified. This (simple) knowledge gives a key of reading for some results in the literature, some of which, especially in the domain of normal T Tauri, cannot be attributed to 'better opacities' but to the differences in convection modeling between the atmosphere and the interior. For instance, it is not the improved opacity in NextGen atmospheres (Allard & Hauschidt, 1997; hereinafter AH97), which push the location of Baraffe et al. (1998) PMS tracks (in models fitting the Sun, with $\alpha_{in}=1.9$) to lower T_{eff} than other computations, but the fact that the match with the NextGen models, having $\alpha_{atm}=1$, is made at optical depth $\tau_{ph} = 100$: these models then adopt very low efficiency for convection in the most superadiabatic part of the envelope, and result to be placed at cooler T_{eff} than other models (Montalbán et al. 2004; D'Antona & Montalbán 2003).

The main problem is that our treatment of convection is still local and fully parametric. A way of overcoming this problem would be to have grids of 3D non local models, and, at each (T_{eff}-gravity), calibrate the average α which provides the same specific entropy jump between the atmosphere and the adiabatic region. This procedure does not give information on the structure of the superadiabatic envelope, but allows a proper computation of the interior. Ludwig, Freytag & Steffen (1999), using their 2D atmosphere models, have provided a calibration of the parameter α (as a function of $\log T_{eff}$ and $\log g$ in the domain $T_{eff} = 4300 - 7100$ K, $\log g = 2.54 - 4.74$) to be used in the computation of gray stellar models. These 2D models indicate that convection is on average 'efficient' in the atmosphere and envelope, corresponding to a large α. The idea of calibrating the average α using numerical simulations has been extended by now to a few 3D computations: Ludwig, Allard and Hauschildt (2002), for an M dwarf at $T_{eff} = 2800$ K and $\log g = 5$, find $\alpha \simeq 2.1$; and Trampedach et al. (1999), for the range of main sequence gravities and $\log T_{eff}=3.68$–3.83, find $\alpha \simeq 1.6$–1.8 in the whole range. On the other hand, Asplund et al. (2000) have compared 2D and 3D atmosphere models for Sun, and find that the 2D solar model has marginally larger gradients than the 3D one. Although an extrapolation to regions not explicitly computed is not allowed, also these few 3D models indicate efficient convection in the overadiabatic envelope.

Fig. 2 (left panel) shows that by adopting the α calibration by Ludwig et al. (1999), we obtain evolutionary tracks very similar to those obtained by means of the Full Spectrum Turbulence (FST) model (Canuto, Goldman & Mazzitelli 1996), which is also known to be a high efficiency model.

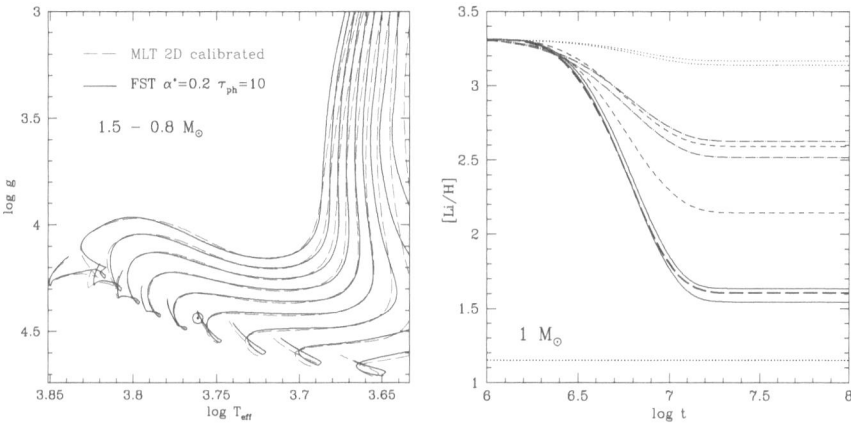

Fig. 2. (Left panel): evolutionary tracks using FST in the $\log T_{\rm eff}$ vs. log g plane (solid line; non gray models with $\tau_{\rm ph} = 10$ by Montalbán et al.,2004) and 2D calibrated MLT (dashed line).(Right panel): Lithium evolution for the solar mass with different assumptions about convection and model atmospheres. The dotted line at bottom represents today's solar lithium abundance. MLT models with AH97 model atmospheres down to $\tau_{\rm ph} = 10$ and 100 are shown dotted for $\alpha_{\rm in} = 1$ and dash–dotted for $\alpha_{\rm in} = 1.9$. The Montalbán et al. (2004) MLT models with Heiter et al. (2002) atmospheres down to $\tau_{\rm ph} = 10$ (lower) and 100 (upper) are dashed; The continuous lines show the non gray FST models for $\tau_{\rm ph} = 10$ and 100, and, in between, the long dashed model employing the 2D calibrated MLT.

4   Observational Tests of Models

Generally convection is calibrated by requiring that its free parameter(s) are chosen to reproduce the solar radius at the solar age. However, it is also possible to use models which do not fit the solar location, and in fact these seem to reproduce much better two observational constraints of the pre main sequence, namely the PMS Lithium depletion and the HR diagram location of some binary stars for which masses are known.

As we have shown in D'Antona & Montalbán (2003), only *models adopting very low efficiency convection in PMS are compatible with the lithium depletion patterns shown by young open clusters*, that is, only the models which do not fit the solar location. This conclusion is reinforced in Fig. 2 (right panel), where we plot also the strong solar lithium depletion resulting from the models computed by employing the MLT calibrated on the 2D simulations. In addition, we show in Fig.3 the location in the HR diagram of three PMS binaries for which the masses are well determined.

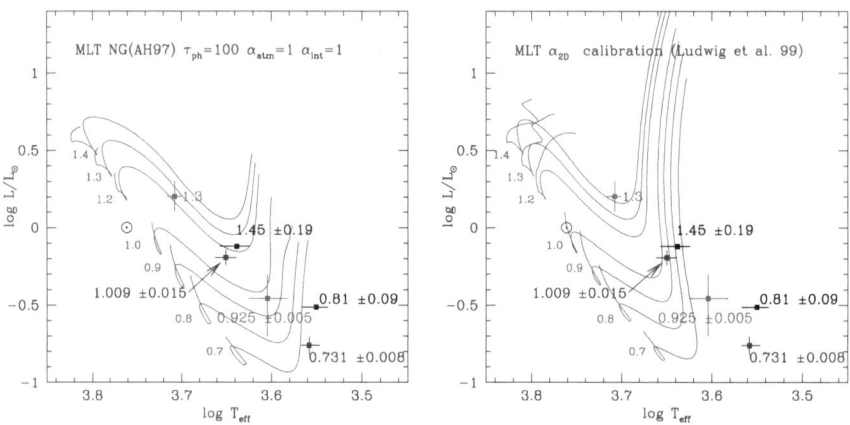

Fig. 3. Location in the HR diagram of three binaries with well determined masses, labelled in the figure: RXJ 0529.4+0041 (Covino et al. 2000), V1174 Ori (Stassun et al. 2004) and NTT 045251+3016 (Steffen et al. 2001). The PMS tracks superimposed are, on the left, the AH97 $\alpha_{in} = \alpha_{atm} = 1$ tracks, and on the right the tracks in which the MLT is calibrated on the 2D models. Notice that the left tracks, which are more consistent with the observed data, fail to fit the solar location by ~ 400 K.

We see that the models which best reproduce the location of all the six data points are the tracks which do not fit the solar location. The models whose convection is calibrated on the 2D simulation make a poor job, as the FST models and other models with 'efficient' convection do: therefore this result can not be inputed to the fact that we employ local convection models. A possibility is that we are in front of an "opacity" problem, more that in front of a convection problem. Actually we would be inclined to say that opacities are not a problem (we have shown this in Montalbán et al. (2004), by comparing models computed with Heiter et al (2002) or with AH97 model atmospheres), but something can still be badly wrong, as implied by the recent redetermination of solar metallicity (Asplund et al., 2004). A further possibility is that the inefficient convection in PMS requires the introduction of a second parameter –linked to the stellar rotation and magnetic field, as we have suggested in the past (Ventura et al., 1998; D'Antona et al., 2000), but this remains to be worked out.

References

1. Allard F., Hauschildt P., 1997 (AH97) the NextGen model grids, web location: http://dilbert.physast.uga.edu/~yeti/mdwarfs.html
2. Asplund, M., Ludwig, H.-G., Nordlund, Å., & Stein, R. F. 2000, A&A, 359, 669
3. Asplund, M., Grevesse, N., Sauval, A. J., Allende Prieto, C., & Kiselman, D. 2004, A&A, 417, 751
4. Baraffe I., Chabrier G., Allard F., Hauschildt P., 1998, A&A 337, 403
5. Baraffe, I., Chabrier, G., Allard, F., & Hauschildt, P. H. 2002, A&A, 382, 563
6. Canuto V.M., Goldman I., Mazzitelli I., 1996, ApJ 473, 550
7. Covino, E., et al. 2000, A&A, 361, L49
8. D'Antona F., Mazzitelli I., 1994, ApJS 90, 467 (DM94)
9. D'Antona, F., Ventura, P., & Mazzitelli, I. 2000, ApJL, 543, L77
10. Heiter U., Kupka F., van't Veer-Menneret C., Barban C., Goupil M.J., Garrido, R., 2002a, A&A, 392, 619
11. D'Antona, F. & Montalbán, J. 2003, A&A, 412, 213
12. Ludwig, H., Freytag, B., & Steffen, M. 1999, A&A, 346, 111
13. Ludwig, H.-G., Allard, F., & Hauschildt, P. H. 2002, A&A, 395, 99
14. Montalbán, J., D'Antona, F., Kupka, F., & Heiter, U. 2004, A&A, 416, 1081
15. Stahler, S. W., Palla, F., & Salpeter, E. E. 1986, ApJ, 308, 697
16. Stassun, K. G., Mathieu, R. D., Vaz, L. P. R., Stroud, N., & Vrba, F. J. 2004, ApJS, 151, 357
17. Steffen, A. T., et al. 2001, AJ, 122, 997
18. Trampedach, R., Stein, R.F., Christensen–Dalsgaard, J., Nordlund, Å. 1999, in *Theory and tests of convection in stellar structure*, ASP Conf. Ser. 173, 233
19. Ventura P., Zeppieri A., Mazzitelli I., D'Antona F., 1998, A&A 331, 1011

Pinning Down Gravitational Settling

A.J. Korn[1], N. Piskunov[1], F. Grundahl[2], P. Barklem[1], and B. Gustafsson[1]

[1] Uppsala Astronomical Observatory, University of Uppsala, Uppsala, Sweden
[2] Department of Astronomy, University of Århus, Århus, Denmark

Abstract. We analyse high-resolution archival UVES data of turnoff and subgiant stars in the nearby globular cluster NGC 6397 ([Fe/H] ≈ −2). Balmer-profile analyses are performed to derive reddening-free effective temperatures. Due to the limited S/N and uncertainties related to blaze removal, we find the data quality insufficient to exclude the existence of gravitational settling. If the newly derived effective temperatures are taken as a basis for an abundance analysis, the photospheric iron (Fe II) abundance in the turnoff stars is 0.11 dex lower than in the (well-mixed) subgiants.

Practically all sophisticated stellar evolution models predict the existence of processes altering photospheric abundances on long timescales (see e.g. Pinsonneault, these proceedings). For example, Richard et al. [6] predict iron abundances in turnoff stars of NGC 6397 to be lower by 0.2 dex than in red giants.

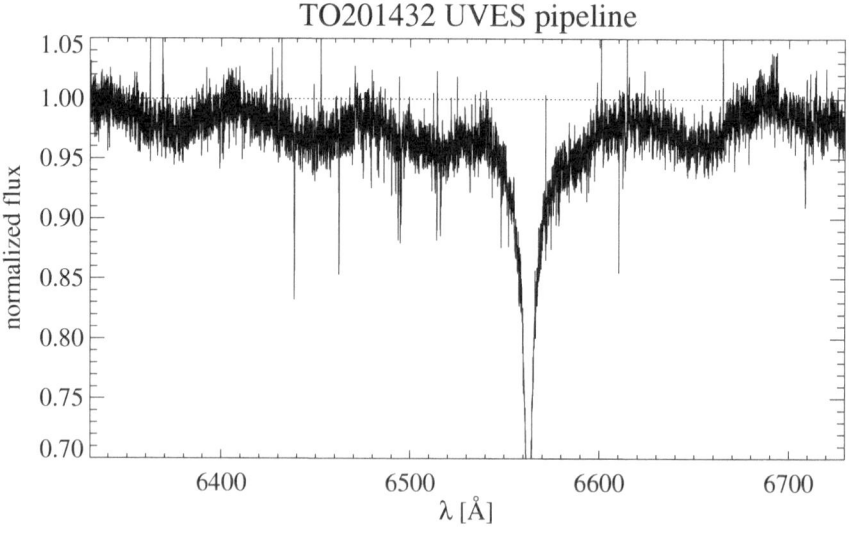

Fig. 1. Co-addition of four UVES pipeline spectra of NGC 6397/TO201432 (observing dates 2000-06-18 and 22, two spectra per night). The resulting spectrum was arbitrarily normalized at 6410 and 6690 Å. As blaze residuals are not properly accounted for in the pipeline order merging, the échelle order pattern is clearly visible in the merged spectrum. With an amplitude of 2 %, these instrumental artifacts do not allow to derive Balmer-profile temperatures to better than 200–300 K.

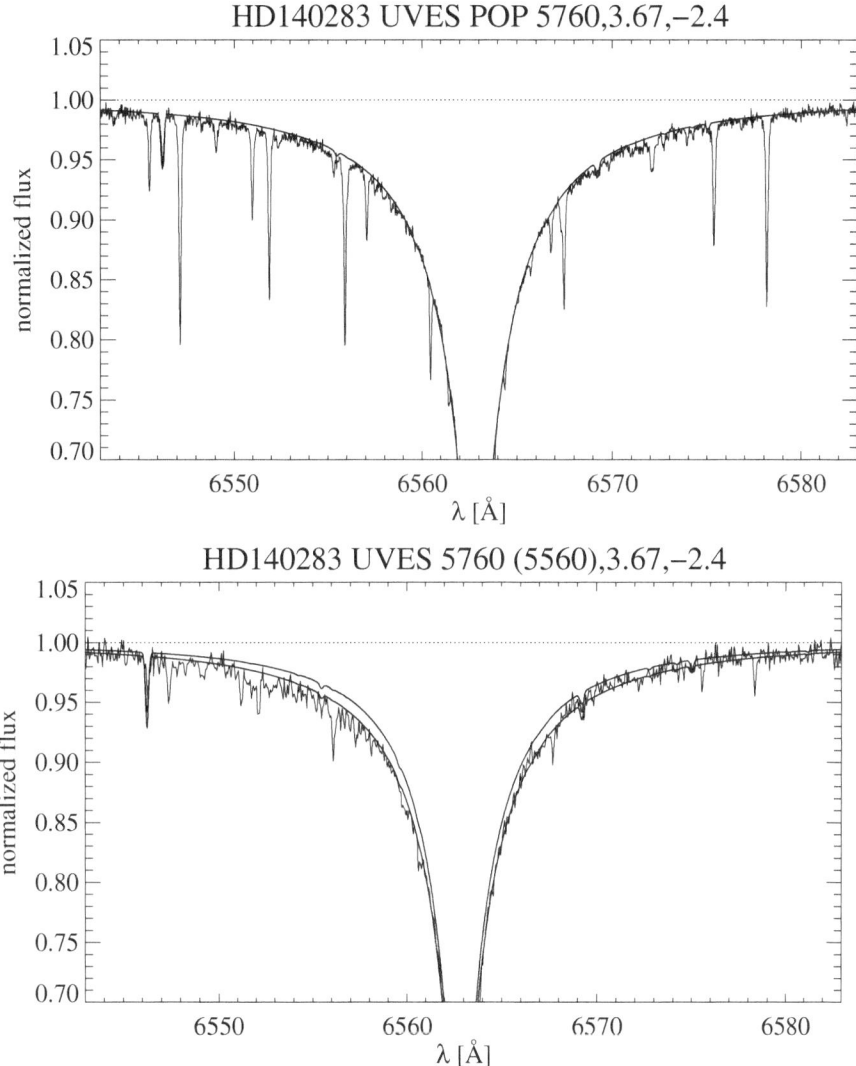

Fig. 2. Top panel: Observed (UVES POP [1]) and synthetic Hα profiles of the field star HD 140283. The unmerged data (2001-07-09, frame 570) was retrieved from [1] and rectified using a parabola interpolated from the continua of adjacent orders. A best-fit T_{eff} of 5760 K (at $\log g = 3.67$ and [Fe/H] $= -2.4$, neither critical) is indicated.

Bottom panel: Observed and synthetic Hα profiles for HD 140283 (black: 5760 K, grey: 5560 K). The data was retrieved from the VLT archive (observing date 2000-06-15), reduced using REDUCE and rectified using parabolic fits to the continua in adjacent orders. Notice the variable telluric features which somewhat suppress the blue wing. An effective temperature of 5560 K (best estimate of [3]) is clearly too low, 5760 K (derived from the UVES POP spectrum) too high. At a S/N of 150, observational systematic errors are thus of the order of at least 100 K.

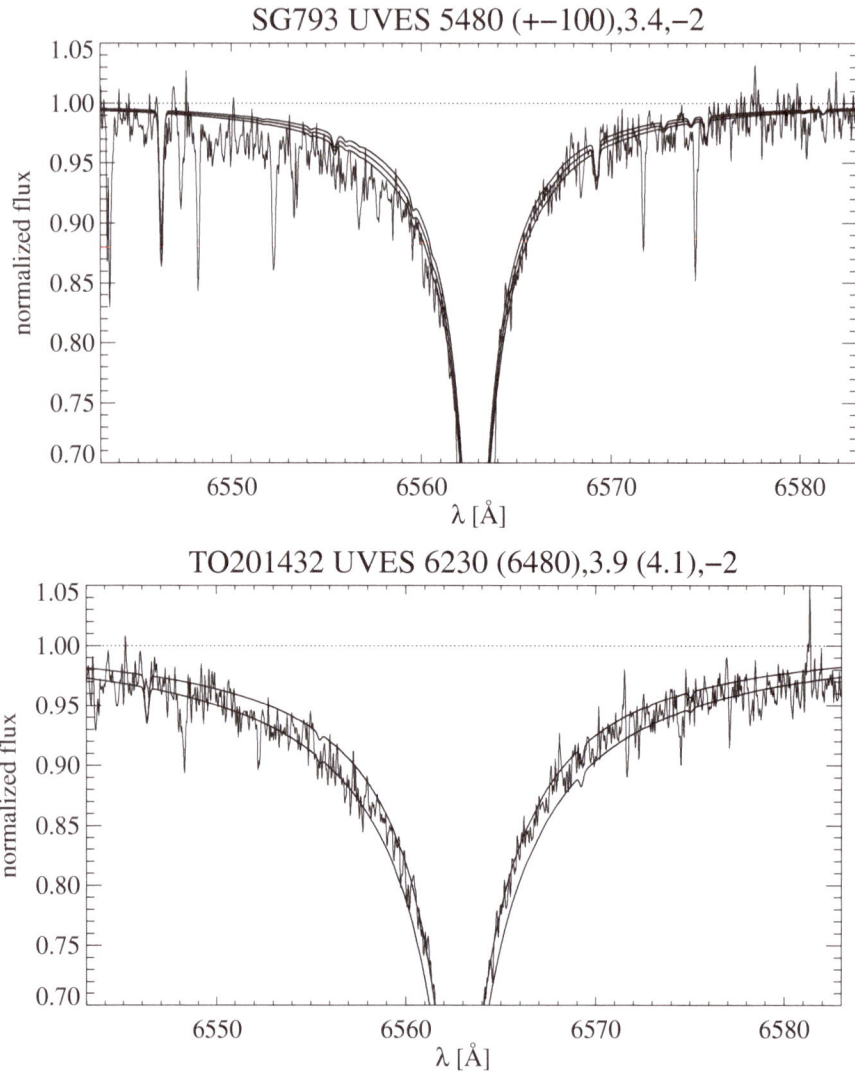

Fig. 3. Top panel: Observed and synthetic Hα profiles of NGC 6397/SG793 (black: 5480 K, grey: ±100 K). The data was retrieved from the VLT archive (observing dates 2000-06-18 and 20, one spectrum per night), reduced using REDUCE, co-added and rectified using parabolic fits to the continua in adjacent orders. An effective temperature of 5480 K is indicated, in good agreement with Gratton et al. [3].

Bottom panel: Observed and synthetic Hα profiles of NGC 6397/TO201432 (black: 6230 K, grey: 6480 K). The data was retrieved from the VLT archive (observing dates 2000-06-18 and 22, two spectra per night), reduced using REDUCE, co-added and rectified using parabola. An effective temperature of 6480 K (best estimate of [3]) is clearly too high, 6230 K is indicated instead. This lower effective temperature reduces $\Delta T_{\mathrm{eff}}(\mathrm{TO} - \mathrm{SG})$ to 750 K (to be compared with 1000 K found by [3]).

Gratton et al.[3] (VLT Large Program 165.L-0.263) pioneered high-resolution, high-S/N observations of turnoff (TO) stars in nearby GCs. In comparing iron abundances in TO and cool subgiant (SG) stars in NGC 6397 (and NGC 6752), they found excellent agreement, leaving little room for gravitational settling. Here, we re-examine this claim based on a re-analysis of the archival data using the sophisticated reduction package REDUCE of Piskunov & Valenti [5].

Observing TO stars in globular clusters is still a challenging task, even when one has access to 10m-class telescopes and efficient spectrographs like UVES. Gratton et al. [3] use Balmer-profile temperatures as a reddening-free temperature indicator. However, Balmer-profile analyses are challenging, both observationally (Korn [4]) and theoretically (Barklem et al. [2]). Fig. 1 shows some of the data used by Gratton et al. (kindly made available to us by Eugenio Carretta). It is perilous to derive Balmer-profile temperatures from such spectra.

Below, we present an Hα re-analysis of the highest S/N targets: the SG 793 (S/N $\approx$ 105) and the TO star 201432 (S/N $\approx$ 97, both Fig. 3). We include the halo subgiant HD 140283 (S/N $\approx$ 145, Fig. 2), observed in the same LP. With $\log g$ taken from a 13 Gyr isochrone, it is the *relative* T_{eff} difference (ΔT_{eff}(TO – SG)) which determines the extent of gravitational settling inferred.

It is particularly interesting to see that we derive a *higher* effective temperature for the reference field star HD 140283 (5650 K, while [3] give 5560 K), but a markedly *lower* effective temperature for the cluster turnoff star (6230 K vs. 6480 K). Seemingly, the subgiant is sufficiently cool (5480 K) such that the (weak) intrinsic profile of Hα could even be recovered from the UVES pipeline spectrum. Observational problems dominate and drastically limit the temperature determination and its reliability.

These results question the effective temperatures assigned to the cluster turnoff stars by Gratton et al. [3]. Thus, their conclusions concerning gravitational settling might also be subject to systematic effects. From this work, gravitational settling of iron of up to 0.1 dex at [Fe/H] ≈ -2 seems possible.

Clearly, better observations (preferably with a fibre-fed spectrograph) are needed to actually constrain the extent of gravitational settling. Independent temperature indicators (photometry, excitation equilibrium of e.g. Fe I) should be checked for consistency as well. It is crucial to identify and deal with systematic effects in *all* aspects of the analysis.

References

1. S. Bagnulo, E. Jehin, C. Ledoux et al.: The Messenger **114**, 10
   (ESO DDT Program ID 266.D-5655, download via http://www.eso.org/uvespop)
2. P.S. Barklem, N. Piskunov, B.J. O'Mara: A&A **363**, 1091 (2000)
3. R.G. Gratton, P. Bonifacio, A. Bragaglia et al.: A&A **369**, 87 (2001)
4. A.J. Korn: 'Rectifying Échelle Spectra – A Comparison between UVES, FEROS and FOCES'. In: *Scientific Drivers for ESO Future VLT/VLTI Instrumentation*, ed. by J. Bergeron, G. Monnet (Springer, Heidelberg 2002), pp. 199–204
5. N.E. Piskunov, J.A. Valenti: A&A **385**, 1095 (2002)
6. O. Richard, G. Michaud, J. Richer: ApJ **580**, 1100 (2002)

Extra-Mixing During the Red Giant Branch Evolution of Low-Mass Stars

A. Weiss

Max-Planck-Institut für Astrophysik, Karl-Schwarzschild-Str. 1, Garching, Germany

Abstract. It is thought that rotation leads to extra-mixing between the hydrogen-burning shell and the convective envelope in low-mass red giants. The atmospheric abundances of Li, C, and N are therefore expected to change with luminosity on the upper RGB. This theoretical prediction is in excellent agreement with the observed RGB abundance variations in both field and globular-cluster red giants. The latest observational data on the evolutionary variations of the surface chemical composition in low-mass metal-poor stars are used to constrain the basic properties of extra-mixing in the majority of upper RGB stars. A possible mechanism of this "canonical" extra-mixing is turbulent diffusion driven by rotation. The progress in modeling this process is reviewed. In addition, observational evidence for primordial composition modifications and ideas about their origin in the context of extra-mixing are discussed.

1 Canonical Low-Mass Star Evolution

The evolution of low-mass stars ($M \lesssim 2 - 2.5\,M_\odot$) starts with central hydrogen burning on the main-sequence. Due to the rather low core temperatures, proton-nucleosynthesis affects only the nuclei involved in the pp-chains and CNO-cycles. Even for the lower mass range of, say $0.7 - 0.9\,M_\odot$, the initial C $\to$ N conversion takes place, although the whole cycle (involving changes in O) is completed only rarely. Nuclei of higher cycles, such as those of the NeNa and MgAl cycles, are not affected. At the end of the main-sequence, the hydrogen-burning shell establishes itself within the former burning core, such that above the shell we find a region of composition modified by H-burning: a ^{12}C/^{13}C value close to the equilibrium value of ≈ 3, almost vanishing carbon, and strongly enhanced nitrogen abundance. In addition, ^{3}He is increased by huge factors in those regions of the former core where temperature was low enough for a high ^{3}He equilibrium abundance within the ppI-chain, but high enough for nuclear reactions establishing the equilibrium quickly [8]. Finally, ^{7}Li is strongly depleted over a large fraction of the stellar interior, due to its low p-capture temperature of $\approx 2.5 \cdot 10^6$ K.

As the star becomes cooler on the subgiant and early red giant branch, the convective envelope deepens until it touches these regions and mixes the modified material to the surface. This is the *first dredge-up* (1st du, [18]), which is the only mixing event predicted by canonical stellar evolution theory for such stars. The resulting changes in element and isotope abundances are shown in Fig. 1. After the 1st du the convective envelope retreats ahead of the advancing hydrogen-shell and no further changes in the surface abundances are expected.

When the shell encounters the point of deepest convective penetration, the evolution is slowed down, resulting in the so-called bump, and, more importantly, no molecular weight barrier between shell and convective envelope remains. Theory and observation nicely agree about the size and location (around $\log L/L_\odot = 2$) of the bump [28].

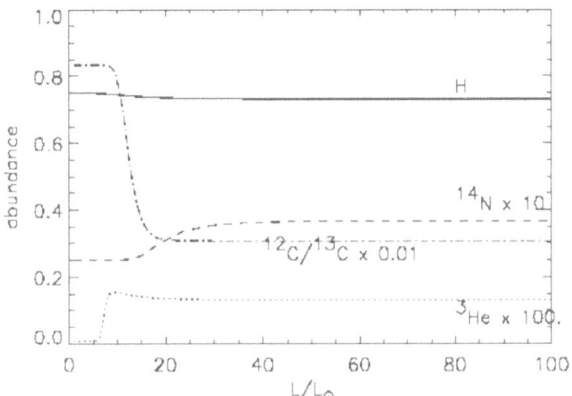

Fig. 1. Changing surface composition of a metal-poor $0.9\,M_\odot$ star during its RGB evolution (courtesy C. Charbonnel)

This picture is supported by observations, one important set of data coming from [17] for field stars with accurate *Hipparcos* parallaxes (see Fig. 2). The strength of the 1st du increases with decreasing metallicity, with mass up to $\approx 1 M_\odot$, decreasing again for higher masses, and with decreasing initial helium content. It also leads to a slight enrichment of the convective envelope with helium. For more massive stars, the energy generation of which is dominated by the CNO-cycle, a slight decrease in oxygen would be expected, too.

2   Observational Evidence for Extra-Mixing

[17] not only demonstrated the reality of the 1st du as predicted consistently by all stellar evolution calculations, but – more importantly – showed that in field stars a further decrease of ^{7}Li, C, ^{12}C/^{13}C, and an accompanying increase in N takes place along the upper RGB. This additional evolutionary effect can not be explained by canonical evolution, but can be fully accounted for by requiring an additional *extra-mixing* between the bottom of the convective envelope and the outermost layers of the advancing hydrogen-shell as demonstrated in Fig. 2, taken from [12]. The shell is always hot enough for CNO-burning being the main energy source, and thus the CN-conversion takes place in regions outside the major energy producing layers. This explanation of the abundance

changes by a postulated mixing of CN-subcycle material from the outermost shell region, had been proposed already by [24]. The observations of the field stars also demonstrate that the extra-mixing has a well-defined starting point along the RGB: the RGB-bump [7].

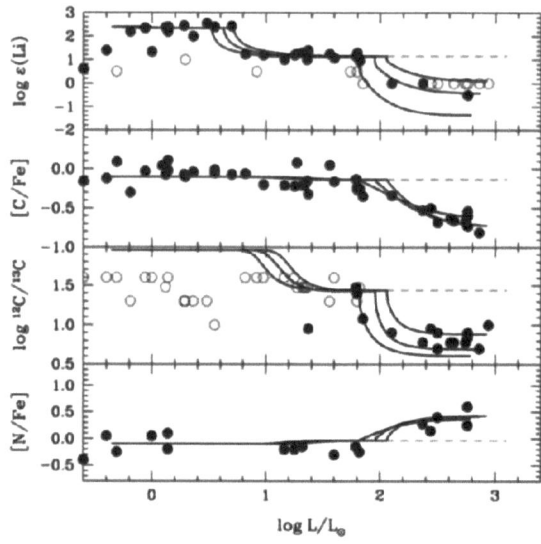

Fig. 2. Evolution of field star surface abundances by [17] and theoretical models imposing extra-mixing below the convective envelope by Denissenkov & VandenBerg [12]

Smith and Martell [23] recently have compiled similar data concerning the evolution of carbon along the RGBs of several globular clusters (M92, M13, M3, . . .), showing not only that the same process is operating in cluster as in field stars, but also that its effectiveness is very similar in both environments. Finally, since [15] the same evolution of $^{12}C/^{13}C$ has been known in open clusters, too, and most recently, it was discovered in the Small Magellanic Cloud [22], such that it is save to conclude that extra-mixing is taking place in most, if not all low-mass stars independent of composition and environment, and therefore truly can be termed 'canonical' extra-mixing.

Open clusters may in the future prove to be quite important in understanding the physics of the extra-mixing because they cover a wide range of ages and therefore stellar masses on the RGB.

3   The Primordial Complication

In addition to the CN-anomalies, similar (anti-)correlations for Na and O, and Mg and Al have been known for some time to occur in globular cluster giants, though not in field stars (see [21] for a compilation). Initially thought to be a purely evolutionary effect, too, we now have mounting evidence that it is a primordial feature, present already in unevolved stars (e.g. [16] for NGC 5762, [4] for 47 Tuc). We therefore will not discuss this anticorrelation further in this context, although J. Cohen (this meeting) reports about a possible evolutionary contribution in the O-Na-anomaly in M13, the cluster with the most expressed element variations.

A complication, however, arises from the fact that these anomalies are also due to proton nucleosynthesis (though operating at higher temperatures) and thus that some primordial CNO-abundance anomalies should be expected, too. In fact, 47 Tuc has been for a long time the example for a CN/CH dichotomy all along the cluster sequence, and by now, CN-variations have been found – mainly by Cohen, Briley and coworkers – in unevolved stars of other clusters as well (47 Tuc: [3,2], M71: [1], M5: [9], . . .). This complicates the task to identify the purely evolutionary effect due to extra-mixing, which is needed to develop and test physical models.

4   Modeling Extra-Mixing

From the observational fact that the extra-mixing sets in at the bump, and from relating the observed CN-anomalies with the structure in the hydrogen shell, as shown in Fig. 3 for a star evolved just past the bump, two quantities for modeling the deep mixing with a parametric ad hoc model can be deduced: starting epoch and mixing depth. The latter is found to be around $\delta m \approx 0.15$ (see Fig. 3). The third one, the mixing speed or, equivalently, a diffusion constant D_{mix} can be estimated from theories about rotation-induced mixing; this results in values of $\approx 10^8 \cdots 10^{10}$ cm^2/s.

There is an increasing level of self-consistency in the models. While the original ones (see [25,5,13]) where done by post-processing of canonical background models, using a diffusive scheme for the extra-mixing and an extended nuclear network, today all effects are calculated simultaneously within a single stellar evolution code (see, e.g., [12]). The old approach, useful for exploratory work, is no longer necessary since both the observations [23] and the parametric models demonstrated that the parameters are very well constrained to $D_{\mathrm{mix}} \approx 4 \cdot 10^8$, $\delta m_{\mathrm{mix}} \approx 0.15$. The success of the parametric approach has been demonstrated in many papers, as in [12].

Recent research now concentrates on the more physical models involving theories of rotation. The long-term aim of these attempts are to provide fully self-consistent models which include stellar evolution, rotation, transport of angular momentum and of chemical species. The key players in this field are P. Denissenkov [11,12] and C. Charbonnel and coworkers (for their approach, see the contributions by Charbonnel and Palacios in this volume). Both groups employ the theoretical description of rotation by Zahn and Maeder [27,19].

While this theory has been shown to be very successful in the case of massive stars (Maeder and Meynet, this volume) and Population 1 low-mass stars (Charbonnel, this volume), full and self-consistent application in the case of globular

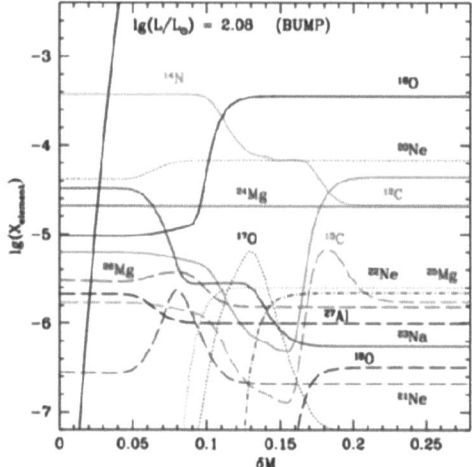

Fig. 3. Abundances within the hydrogen-burning shell as function of relative mass δm, defined as ranging from 0 to 1 between the bottom of the shell and the convective envelope. From observational evidence a penetration of extra-mixing down to $\delta m \approx 0.15$ has been inferred

cluster stars has not lead to the desired extra-mixing so far.[1] To obtain mixing, [12] for example, had to increase D_{mix} by a factor of seven. This problem might be linked to an insufficient treatment of angular momentum transport.

The inclusion of rotation and rotation-induced mixing into the stellar evolution codes is by far not trivial and the individual treatment might lead to different results. One example is the so-called Lithium-flash, found by [20], but not confirmed by [10]. Here, the back-reaction of enhanced nuclear energy production (by $^7\mathrm{Li}(p,\gamma)$) to the local angular momentum appears to be the point of disagreement. Both groups, however, agree that enhanced mixing due to enhanced rotation rates, is necessary to produce for a limited time $^7\mathrm{Li}$ overabundances in red giants as have been observed (see de la Reza, this volume). In [20] the Lithium-flash is providing this, while Denissenkov and Weiss [14] deposit angular momentum by the capture of a planet companion to achieve this. As in [12] this leads to an increase in D_{mix} up to values of $10^{11}\,\mathrm{cm}^2/\mathrm{s}$, which also is sufficiently high to mix shell layers hot enough for the NeNa-cycle and therefore to produce the ONa-anomaly. [12] argue that this "enhanced extra-mixing" could be one source for the observed primordial ONa-anomalies, if the mass lost by upper RGB stars of slightly higher mass than today's values (say, around $1.2\,M_\odot$) has been incorporated into the red giants (of $\approx 0.8\,M_\odot$) observed today.

---

[1] See also Chaname, this meeting, for a similar approach and result: the rotation speeds required for extra-mixing in GC stars result in too much mixing for open cluster objects

5   Conclusions

Extra-mixing has been shown to be a generic feature during the first red giant evolution of low-mass stars. So far, models based on rotation as the basic physical process have not succeeded in reproducing the mixing, but a lot of progress has been made in the last few years. Although red giant evolution seems to be unaffected by extra-mixing (but see Palacios' contribution concerning the effect on the nature and structure of the bump), it nevertheless can have far-reaching consequences, as is demonstrated by the case of ^{3}He. According to canonical evolution low-mass stars are strong ^{3}He-producers, but extra-mixing counterbalances this [6,26], leading to a better agreement with observations of the galactic ^{3}He evolution (see also D. Galli, this volume). An urgent question remains that with respect to the primordial abundance anomalies and whether they could be linked to (enhanced) extra-mixing in a previous generation of stars.

References

1. M.M. Briley, J.G. Cohen: AJ **122**, 242 (2001)
2. M.M. Briley, D. Harbeck, G.H. Smith, E.K. Grebel: AJ **127**, 1588 (2004)
3. R.D. Cannon, B.F.W. Croke, R.A. Bell, J.E. Hesser, R.A. Stathakis: MNRAS **298**, 601 (1998)
4. E. Carretta, R.G. Gratton, A. Bragaglia, P. Bonifacio, L. Pasquini: A&A **416**, 925 (2004)
5. R.M. Cavallo, A.V. Sweigart, R.A. Bell: ApJ **464**, L79 (1996)
6. C. Charbonnel: ApJL **453**, L41 (1995)
7. C. Charbonnel, J.A. Brown, G. Wallerstein: A&A **332**, 204 (1998)
8. D.D. Clayton: *Principles of stellar evolution and nucleosynthesis*, 2nd edn. (Univ. of Chicago Press, Chicago 1983)
9. J.G. Cohen, M.M. Briley, P.B. Stetson: AJ **123**, 2525 (2002)
10. P.A. Denissenkov, F. Herwig: ApJ **612**, 1081 (2004)
11. P.A. Denissenkov, C.A. Tout: MNRAS **316**, 395 (2000)
12. P.A. Denissenkov, D.A. VandenBerg: ApJ **593**, 509 (2003)
13. P.A. Denissenkov, A. Weiss: A&A **308**, 773 (1996)
14. P.A. Denissenkov, A. Weiss: A&A Letters **348**, L49 (2000)
15. K.K. Gilroy: ApJ **347**, 835 (1989)
16. R.G. Gratton, P. Bonifacio, A. Bragaglia, et al.: A&A **369**, 87 (2001)
17. R.G. Gratton, C. Sneden, E. Carretta, A. Bragaglia: A&A, **354**, 169 (2000)
18. I. Iben, Jr.: ApJ **142**, 1447 (1965)
19. A. Maeder, J.-P. Zahn: A&A, **334**, 1000 (1998)
20. A. Palacios, C. Charbonnel, M. Forestini: A&A **375**, L9 (2001)
21. S.V. Ramírez, J.G. Cohen: AJ **123**, 3277 (2002)
22. V.V. Smith, K.H. Hinkle, K. Cunha, et al.: AJ **124**, 3241 (2002)
23. G.H. Smith, S.L. Martell: PASP **115**, 1211 (2003)
24. A.V. Sweigart, K.G. Mengel: ApJ **229**, 624 (1979)
25. G.J. Wasserburg, A.I. Boothroyd, I.J. Sackmann: ApJL **447**, 37 (1995)
26. A. Weiss, J. Wagenhuber, P.A. Denissenkov: A&A **313**, 581 (1996)
27. J.-P. Zahn: A&A **265**, 115 (1992)
28. M. Zoccali, S. Cassisi, G. Piotto, M. Salaris: ApJL, **518**, L49 (1999)

Rotation-Induced Mixing in Red Giant Stars

A. Palacios[1], C. Charbonnel[2,3], S. Talon[4], and L. Siess[1]

[1] IAA-Université Libre de Bruxelles, CP-226 Bd du Triomphe, B-1050 Brussels, Belgium
[2] Geneva Observatory, 51 chemin des Maillettes, CH-1290 Sauverny, Switzerland
[3] LA2T-OMP, UMR 5572, 14 av. E. Belin, F-31400 Toulouse, France
[4] Département de Physique, Université de Montréal, Montréal PQ H3C 3J7, Canada

1 Introduction

Red giant stars, both in the field and in globular clusters, present abundance anomalies that can not be explained by standard stellar evolution models. Some of these peculiarities, such as the decline of $^{12}C/^{13}C$, and that of Li and ^{12}C surface abundances for stars more luminous than the *bump*, clearly point towards the existence of extra-mixing processes at play inside the stars, the nature of which remains unclear. Rotation has often been invoked as a possible source for mixing inside Red Giant Branch (RGB) stars ([8],[1],[2]). In this framework, we present the first fully consistent computations of rotating low mass and low metallicity stars from the Zero Age Main Sequence (ZAMS) to the upper RGB.

2 Physics of the Models

We present three models of a 0.85 $M_\odot$, $Z = 10^{-5}$ star. Model **A** is a standard non-rotating model. Model **B** is a slowly rotating model with $v_{ZAMS} = 5$ km.s^{-1}, undergoing no magnetic braking, and for which we assumed a solid body rotating convective envelope (CE) ($\Omega_{CE} = $ cst) during the whole evolution. Model **C** is an initially rapidly rotating model with $v_{ZAMS} = 110$ km.s^{-1}, which undergoes magnetic braking ($v_{TO} = 3$ km.s^{-1}), and for which we assumed uniform specific angular momentum in the CE ($\Omega_{CE} \propto r^{-2}$). In our rotating models, we compute the transport of both angular momentum and chemicals by meridional circulation and shear-induced turbulence from the ZAMS on, according to [5]. We use the new prescription for the horizontal turbulent viscosity given by [6].

3 Main Results

Rotational mixing does not affect significantly the stellar structure (L_{bump} is the same in models **A**, **B** and **C**), but leads to larger abundance variations on the lower RGB associated with a deeper first dredge-up (Fig. 1b).

When solid body rotation is assumed in the CE (model **B**), the degree of differential rotation at its base is too low to trigger efficient shear-induced turbulence between the outer part of the hydrogen burning shell (HBS) and the CE (solid lines in Fig. 1a). On the contrary, in our model **C** the differential rotation

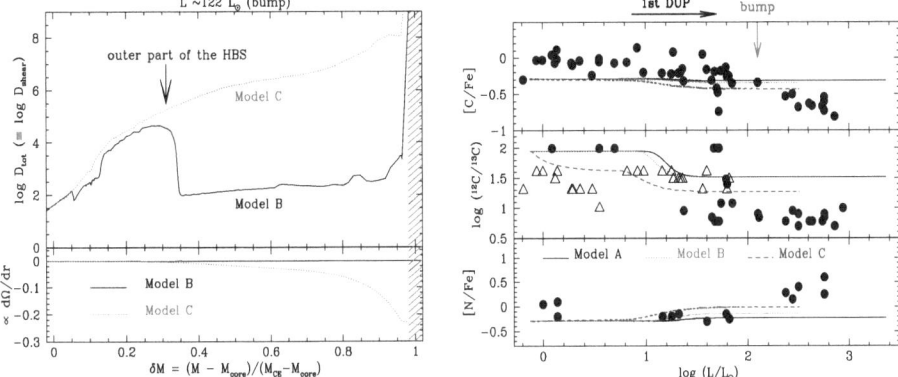

Fig. 1. (a) (*left*) Profiles at the bump, of the total diffusion coefficient (top) and of the degree of differential rotation (bottom) for model **B** (*solid lines*) and model **C** (*dotted lines*). Hatched regions correspond to the CE. **(b)** (*right*) Comparison of our models with observations ([4]). Triangles are lower limits. Dots are actual values.

of the CE ensures the onset of turbulence in the contiguous radiative region, and the CE is connected to the outer HBS through shear mixing (dotted lines in Fig. 1a). This confirms the conjecture by [8], [7] and [2] that the shear-mixing efficiency is enhanced in models with differentially rotating CE.

In our most favorable case (model **C**), the maximum value of the diffusion coefficient in the outer part of the HBS (where abundances of Li, C and N present large variations) is 10^5 cm^2.s^{-1} (Fig. 1a), far from the 4 10^8 cm^2.s^{-1} value that seems to be necessary to reproduce the observations according to [3]. As a result none of our rotating models can reproduce the observed patterns emphasized by [4] (Fig. 1b).

Rotation remains the best candidate for extra-mixing in RGB stars. The present modelling of the rotational mixing is however still incomplete, and agreement between self-consistent models and observations might be achieved by improving the description of the hydrodynamics related to rotation.

References

1. C. Charbonnel: ApJL **453**, L41 (1995)
2. P.A. Denissenkov, C.A. Tout: MNRAS **399**, 603 (2000)
3. P.A. Denissenkov, D.A. VandenBerg: ApJ **593**, 509 (2003)
4. R. G. Gratton , C. Sneden , E. Carretta , A. Bragaglia: A&A **354**, 169 (2000)
5. A. Maeder, J.-P. Zahn: A&A **334**, 1000 (1998)
6. S. Mathis, A. Palacios, J.-P. Zahn: A&A **425**, 243 (2004)
7. A. Sills, M. Pinsonneault: ApJ **540**, 489 (2000)
8. A.V. Sweigart, J.G. Mengel: ApJ **229**, 624 (1979)

3D Hydrodynamical Simulations of Convection in Red-Giants Stellar Atmospheres

R. Collet[1], M. Asplund[2], and R. Trampedach[2]

[1] Department of Astronomy & Space Physics, Uppsala University, BOX 515, SE-751 20 Uppsala, Sweden
[2] Research School of Astronomy & Astrophysics, Australian National University Mount Stromlo Observatory, Cotter Road, Weston ACT 2611, Australia

Abstract. We present preliminary results of 3D hydrodynamical simulations of surface convection in red giants stars. We investigate the main differences between static 1D and 3D time-dependent model stellar atmospheres of red giants for a range of metallicities between solar and [Fe/H] = −3 focusing in particular on the impact of 3D spectral line formation on the derivation of stellar abundances.

1 Introduction

Classical models of stellar atmospheres generally rely on a hydrostatic 1D geometry stratification and a simplified treatment of convective energy transport. In the case of late-type stars though the convective zone reaches and sensibly influences the regions ultimately responsible for the emitted stellar flux and therefore the use of 1D model atmospheres for stellar abundance analysis might lead to systematic errors. Recent 3D time-dependent hydrodynamical simulations of stellar convection ([3], [1]) have in fact indicated that the deviations of the 3D structure from the 1D stratification can produce significant differences in the strength of the synthetic spectral lines and eventually affect the determination of stellar abundances.

In this contribution we present preliminary estimates of the effect of convection in 3D model stellar atmospheres of red giants on the formation of spectral lines and on the derivation of chemical abundances in stars.

2 Calculations

We use a 3D, time-dependent, compressible, explicit, radiative-hydrodynamics code by [5] to simulate convection in the atmospheres of red giants stars with effective temperature $T_{\rm eff} \approx 4800$ K, surface gravity $\log g = 2.2$ and metallicity ranging from solar down to [Fe/H] = −3. The equations for conservation of mass, momentum and energy and for radiative transfer are discretized and solved on a Eulerian mesh with gridsize $100 \times 100 \times 125$. The physical domain of the simulations is set to be large enough to cover about ten granules in the horizontal plane and ten pressure scales in the vertical direction. The adopted boundary conditions and input physics are the same as in [3].

We then use a Feautrier scheme [4] to perform spectral line formation calculations in local thermodynamic equilibrium approximation (LTE) for the species indicated in table 1. At this stage we consider only rays in the vertical direction and a single snapshot per 3D simulation. Abundance corrections are computed differentially by comparing the predictions from 3D models with the ones from 1D MARCS model stellar atmospheres ([2]) generated for the same stellar parameters (a microturbulence $\xi = 2.0$ km s^{-1} is applied to calculations with 1D models).

3 Results

The results of the differential study are presented in table 1. Similarly as found by [3] for dwarf and subgiant stars, 3D model atmospheres of metal-poor red giants appear to be significantly cooler at the surface than their 1D counterparts. Consequently, at a given abundance, the populations of neutral atoms and molecules (e.g. Fe I, Ca I, OH) in these upper layers tend to be enhanced in 3D models comparing to 1D, leading to negative abundance corrections.

Table 1. $[X/H]_{3D}$-$[X/H]_{1D}$ abundance corrections for various chemical species.

Metallicity	Fe I	Fe II	Ca I	Mg I	[O I]	O I	OH	CH	NH
+0	~ 0.0	+0.05	~ 0.0	~ 0.0	~ 0.0	+0.05	+0.01	+0.02	+0.02
−2	−0.2	−0.04	−0.1	−0.05	−0.2	−0.04	−0.25	−0.25	−0.2
−3	−0.2	+0.01	−0.1	−0.05	−0.2	+0.01	−0.5	−0.5	−0.6

4 Conclusions

Preliminary simulations of convection in red giants' stellar atmospheres indicate that cooler surface layers are expected in 3D models than in 1D. Therefore, in the LTE case, for lines forming in those layers corrections to abundances derived with 1D analysis have to be applied in the same direction as for dwarfs and subgiants. The magnitude of the corrections though appears to be lower for red giants.

References

1. M. Asplund & A. E. García Pérez: A&A **372**, 601 (2001)
2. M. Asplund, B. Gustafsson, D. Kiselman, & K. Eriksson: A&A **318**, 521 (1997)
3. M. Asplund, Å. Nordlund, R. Trampedach, & R. F. Stein: A&A **346**, L17 (1999)
4. P. Feautrier: C. R. Acad. Sci. Paris **258**, 3189 (1964)
5. R. F. Stein & Å. Nordlund: ApJ, **499**, 914 (1998)

Rotational Mixing in Massive Stars and Its Many Consequences

A. Maeder, G. Meynet, R. Hirschi, and S. Ekström

Geneva Observatory, CH–1290 Sauverny, Switzerland

Abstract. Rotation plays a major role in massive star evolution. Rotation produces a significant mixing and enhances the mass loss. All model outputs are influenced. We show how the chemical yields are modified by rotation. Below 30 $M_\odot$, mixing increases the yields of the α–elements and above 30 $M_\odot$ rotational mass loss dominates and enhances the yield in helium. Primary ^{14}N is produced at very low metallicities.

1   Introduction and Basic Physics

Massive stars play a key role in the spectral evolution of galaxies, they are also the progenitors of Wolf–Rayet (WR) stars, supernovae and γ-ray bursts. They are the main agents of nucleosynthesis driving the chemical evolution of galaxies. The relative numbers of the various kinds of massive stars (blue, red supergiants, WR stars), their properties and nucleosynthesis very much depend on mass loss and rotation, as well as on the interaction of these two effects.

Great care has to be given to the physics of rotation and to the treatment of its interaction with mass loss. For differentially rotating stars, the structure equations need to be written differently [9] than for solid body rotation. For the transport of the chemical elements and angular momentum, we consider the effects of shear mixing, meridional circulation, horizontal turbulence and in the advanced stages the dynamical shear is also included. Caution has to be given that advection and diffusion are not the same physical effect.

Mass loss in rotating star is asymmetric. Very hot star have a dominant polar wind. Stars with T_{eff} below about 24 000 K, due to their larger opacities, may have an equatorial ejection forming a disc. Polar ejection removes little angular momentum, while equatorial ejection removes a lot. It is thus also important to consider the wind asymmetries in massive rotating stars. Also, rotation produces a general enhancement of the mass loss rates [7].

2   General Results

Grids of models have been made [11] at $Z = 0.020$, 0.004 and 10^{-5}. Fig. 1 illustrates the HR diagram for non–rotating and rotating stars at solar metallicity. The rotating models have an initial velocity v_{ini} of 300 km/s, which gives an average velocity of 220 km/s during the MS phase, corresponding to observations. We notice the following points:

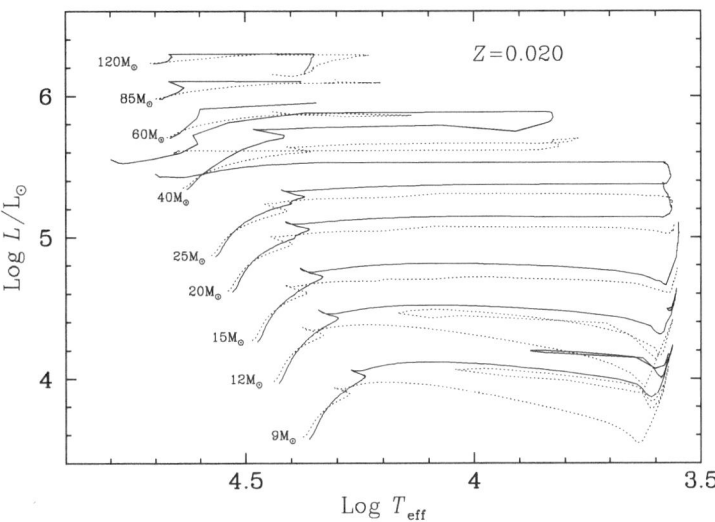

Fig. 1. HR diagram of massive stars with Z=0.02, for rotating stars with $v_{ini} = 300$ km/s (continuous lines) and for non–rotating stars (dotted lines). The initial rotation $v_{ini} = 300$ km/s corresponds to the average observed velocity of about 220 km/s for OB stars.

- Rotation increases the MS lifetime with respect to non–rotating models (up to about 40 %).
- The values assigned from isochrones with an average rotation velocity typically lead to ages 25% larger than without rotation.
- Rotation strongly affects the lifetimes as blue and red supergiants (RSG). In particular in the SMC, the high observed number of RSG can only be accounted for by rotating models [8].
- Steeper gradients of internal rotation Ω are built at lower Z, and this favours mixing. There are 2 reasons for the steeper Ω–gradients. One is the higher compactness of the star at lower Z. The second one is more subtle. At lower Z, the density of the outer layers is higher, thus the Gratton–Öpik term is less important. This produces less outward transport of angular momentum and contributes to steepen the Ω–gradient.
- At lower Z, rotating stars more easily reach break–up velocities and may stay at break-up for a substantial fraction of the MS phase.
- The various filiation sequences for massive stars can be described.

3   Evolution of the Surface Abundances

As a result of the rotational mixing, products of the CNO processing are reaching the stellar surface during MS evolution and produce N/C enhancements, as

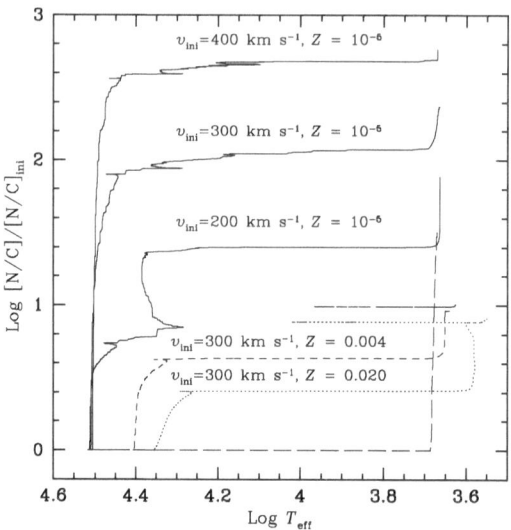

Fig. 2. Evolution of the N/C ratio in number for 9 (dashed and dotted lines) and 40 $M_\odot$ (continuous line) of different Z and v_{ini}. Zero rotation for 9 and 40 $M_\odot$ gives no enrichment until the red giant stage.

observed [2,3,6] since long in OB–stars. A N–enrichment progressively occurs during the MS phase, then it keeps about constant during the crossing of the HR and rises up again due to the convective dredge–up in red supergiants.

Fig. 2 shows the evolution of the N/C ratios in models of rotating stars with 9 $M_\odot$ for initial metallicities $Z = 0.02$ and 0.004. At zero rotation, there is no enrichment during the MS phase (except at $Z=0.02$ for M ≥ 60 $M_\odot$ due to very high mass loss). At 9 $M_\odot$ for solar Z and an initial rotation of 300 km s^{-1}, the N/C ratio increases by about 0.4 dex during the MS phase. For masses below about 30 $M_\odot$, the *relative* values of the N/C ratios increase with decreasing Z, in particular N/C would increase by two orders of a magnitude for $Z = 10^{-5}$. This results from the steeper Ω–gradients and greater compactness at lower Z. In general, large N/C enhancements are accompanied by small enrichments in helium, typically of a few hundredths. The larger N/C enrichments at lower Z have been nicely confirmed by abundances determinations in the SMC [14]. They found relative N/H enrichments for A–type supergiants in the SMC reaching an order of magnitude or more, while in the Milky Way the enrichments are only by a factor 2 to 3.

Fig. 2 also compares the relative N/C enrichments for an initial 40 $M_\odot$ star. For high mass stars (≥ 30 $M_\odot$), higher Z models have the larger enrichments, i.e. a situation opposite to that described above. The reason is that mass loss effects dominate over mixing. Higher Z stars have a higher mass loss, which

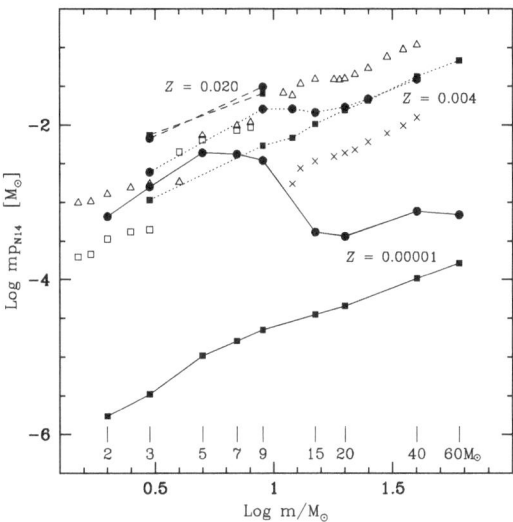

Fig. 3. Variation as a function of the initial mass of the stellar yields in ^{14}N for different Z and rotation velocities. At $Z = 10^{-5}$, the lowest continuous line is for $v_{\mathrm{ini}} = 0$, the upper line for 300 km/s. At higher Z, rotation makes little differences, see also [10].

peels off the stars and makes the products of CNO process visible at the stellar surface.

For Wolf–Rayet stars of type WN, rotation makes smoother changes of abundances, due to internal mixing. For WN stars, the transition phase, with still H present, becomes longer due to rotation and this increases the late WN phase (WNL) where H is usually present. The CNO abundances at the end of the WN phase are the same for rotating and non–rotating models, because they are model independent and determined just by CNO nuclear equilibrium. Indeed, CNO abundances in WN stars provide a unique test of the physics of the CNO cycle.

For the WC phase, the milder composition gradients, when revealed at the surface, make smoother transitions. This produces on the average lower C/He ratios, in very good agreement with observations [1]. The abundances in WC stars are not equilibrium values, but are products of the partial He–burning, thus they are model dependent and offer a most interesting test on the quality of stellar models.

The rare WO stars, characterized by a high O/C ratio represent a more advanced stage of nuclear processing. Curiously enough, such stars which may be the progenitors of supernovae SNIc are found only at lower Z. The physical reasons of that have been explained [13]: lower Z stars become WC stars (if they do it) only very late in evolution, i.e. with a high O/C ratio. At the opposite, at higher Z WC stars occur at an early stage of He–processing, i.e. with a low

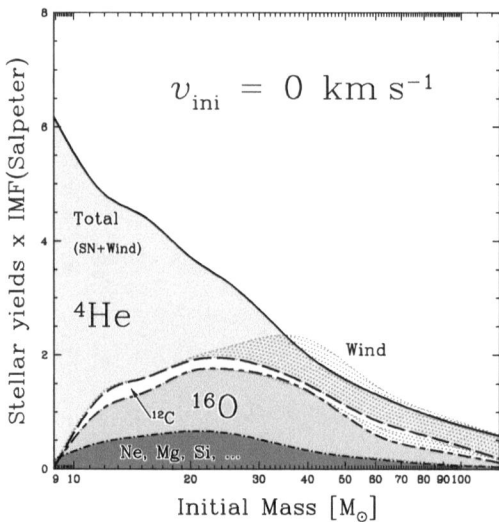

Fig. 4. The yields for models without rotation x IMF (initial mass function), from [5].

O/C ratio. The observed and predicted variations with Z of the number ratio of SNIb/c and SNII shows an interesting agreement [11]. This is noticeable because of the possible connection WO stars - SNIb/c - GRBs (γ Ray Bursts). In this context, we consider WO stars as good candidate for GRB progenitors.

4   Chemical Yields from Massive Stars

Let us first make some comments on the yield in nitrogen. For metallicities lower than that of the SMC, the internal mixing in intermediate and massive stars is sufficient to bring new C from the He–burning core to the H–shell where it is then turned to N. This nitrogen is called "primary" as it does not result from CNO elements initially present. Fig. 3 shows the yields in N as a function of the initial mass for $Z = 0.02$, 0.004 and 10^{-5}. At $Z = 0.02$ and $Z = 0.004$ the level of N production is just that resulting from the initial CNO elements, even if at $Z = 0.004$ the N–enrichment is somehow larger than at $Z = 0.02$. However, at $Z = 10^{-5}$ there is an overproduction of N by 1 or 2 orders of magnitude. This primary N is mainly produced in intermediate mass stars, but the contribution of massive stars is also significant.

The model evolution of rotating stars has been pursued up to the presupernova stage [5], since we know that nucleosynthesis is also influenced by rotation [4]. Figs. 4 and 5 show the chemical yields from models without and with rotation. These figures shows these yields multiplied by the initial mass function (IMF). The main conclusion is that below an initial mass of 30 $M_\odot$, the cores are larger and thus the production of α–elements is enhanced. Above 30 $M_\odot$,

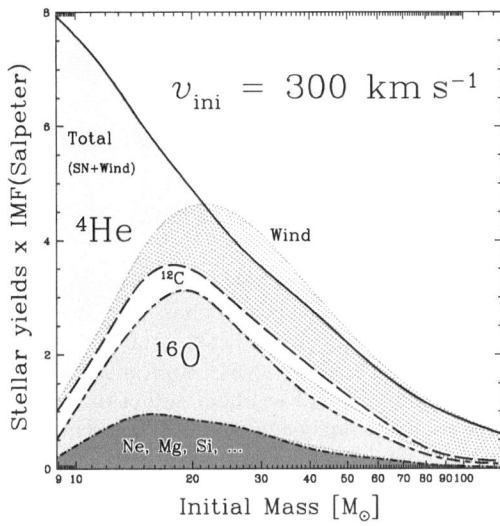

Fig. 5. The yields x IMF for models with $v_{ini} = 300$ km/s, from [5].

mass loss is the dominant effect and more He is ejected before being further processed, while the size of the core is only slightly reduced. When we account for the weighting by the IMF, the production of oxygen and of α–elements is globally enhanced, while the effect on the He–production in massive stars remains limited. It will be interesting to explore the consequences of these new yields for the chemical evolution of the Galaxy.

References

1. P.A. Crowther, L.J. Smith, A.J. Willis: A&A **304**, 269 (1995)
2. D.R. Gies, D.L. Lambert: ApJ **387**, 673 (1992)
3. A. Herrero, R.P. Kudritzki, J.M. Vilchez et al: A&A **261**, 209 (1992)
4. A. Heger, N. Langer, S.E. Woosley: ApJ **528**, 368 (2000)
5. R. Hirschi, G. Meynet, A. Maeder: A&A in press, (2004)
6. L.S. Lyubimkov: Ap&SS **243**, 329 (1996)
7. A. Maeder, G. Meynet: A&A **361**, 159 (2000)
8. A. Maeder, G. Meynet: A&A **373**, 555 (2001)
9. G. Meynet, A. Maeder: A&A **321**, 465 (1997)
10. G. Meynet, A. Maeder: A&A **390**, 561 (2002)
11. G. Meynet, A. Maeder: A&A in press, (2004)
12. N. Prantzos, S. Boissier: A&A **406**, 259
13. L.F. Smith, A. Maeder: A&A **241**, 77 (1991)
14. K. Venn, N. Przybilla: in **CNO in the Universe**, Eds. C. Charbonnel et al., ASP Conf. Ser. 304, p. 20 (2003)

Massive Rotating Stars at Very Low Metallicity

G. Meynet, S. Ekström, and A. Maeder

Geneva Observatory
CH–1290 Sauverny
Switzerland

Abstract. Massive rotating stars reach the break-up limit during the MS phase. Then they spend a great part of their core He-burning phase in the red part of the HR diagram and have their surface strongly enriched in heavy elements. All these effects enhance the mass loss by stellar winds. Typically 60 $M_\odot$ stars with $v_{\mathrm{ini}}/v_{\mathrm{crit}}$ equal to 0.55 and 0.7 at $Z = 10^{-8}$ and 10^{-5} respectively lose about half of their initial mass during their lifetimes. The wind material is strongly enriched in CNO elements. Composition of the wind/supernova ejecta presents striking similarities with the abundances observed at the surface of C-rich extremely metal poor halo stars.

Effects of Rotation at Very Low Z

Non–rotating models at very low metallicity are believed to present the following differences with respect to normal metal rich stars:

- The process of star formation in metal free gas might lead to the formation of much more massive stars than at solar metallicity (see e.g. the review [8]).
- The absence of metals implies that all stars will begin to burn their hydrogen in their core through the pp chains.
- When metallicity decreases, the opacity decreases and the stars are more compact. For instance the radius of a 20 $M_\odot$ star on the ZAMS is decreased by about a factor 4 when its metallicity passes from 0.020 in mass fraction to 0. Due to the decrease of the opacities, the radiative driven stellar winds are also weaker [5].

Now what changes at very low Z for rotating models ?

- At low Z, little angular momentum is removed by the stellar winds, since little mass is lost through stellar winds. On the other hand, during the Main Sequence phase (whatever the metallicity), internal transport mechanisms bring angular momentum from the inner regions to the surface, accelerating the outer layers. As a consequence of the above two effects, at low Z, the surface velocity reaches more easily the critical velocity. Near the break-up limit, the star loses mass at a high rate.
- The velocity of the meridional currents decreases when the metallicity decreases (see the contribution by Maeder et al. in this volume). As a consequence the inner regions of the star keeps a higher angular momentum. A high content of angular momentum in the core at the pre–supernova stage might lead to non spherical explosions or to the production of γ–ray burst events.

- The gradients of the angular velocity inside the stars are steeper at low metallicity. This favours the mixing of the chemical elements which is mainly due to shear turbulence. Thus for a given initial mass and velocity the composition of the surface of the star changes more rapidly when the metallicity decreases. This is true for initial masses which suffer little mass loss by stellar winds during the MS phase ($M < 25 M_\odot$). Above this limit, mass loss at high Z becomes an effective way to modify the surface abundances.
- The yields in CNO elements are increased by rotation [3]. At low metallicity, rotating stars become important sources of primary nitrogen [7]. As a numerical example, the quantity of primary nitrogen in pop III 60 $M_\odot$ models with v_{ini}/v_{crit} equal to 0.5 is 5.1 $10^{-2} M_\odot$ (end of the core He–burning phase), while in the non–rotating model this quantity amounts to 2.4 $10^{-7} M_\odot$. The observations of high N/O ratios at the surface of halo stars (see [1] and [4]) requires amounts of primary nitrogen production by massive stars of the order of magnitude obtained in the present rotating models (see the paper by Chiappini in the present volume). The quantities of ^{13}C are also strongly affected by rotation. In 60 $M_\odot$ models, at the the end of the core He-burning phase, the quantities of this element are 1.8 $10^{-8} M_\odot$ and 1.4 $10^{-2} M_\odot$ in the non–rotating and rotating model respectively.
- As explained in [6], rotation favours the evolution to the red part of the HR diagram after the Main Sequence phase. During the Red Supergiant Stage a deep outer convective envelope appears which dredges up CNO processed material and He–burning products. Thus the surface metallicity increases. For instance our rotating 60 $M_\odot$ with $Z = 10^{-5}$ and $v_{ini}/v_{crit} = 0.7$ presents at the end of its evolution a surface metallicity which is about 1400 times the initial metallicity. Likely this will increase the opacity and thus the mass loss rates.

Allowing for all the effects discussed above, we obtain that rotating 60 $M_\odot$ stellar models with v_{ini}/v_{crit} equal to 0.55 and 0.7 at $Z = 10^{-8}$ and 10^{-5} respectively lose about half of their initial mass during their lifetime (mass loss rates are supposed to increase with the square root of the surface metallicity). Very interestingly, these stars eject through their winds material which is strongly enriched in ^{12}C, ^{13}C, ^{14}N, ^{15}N, ^{16}O, ^{17}O, and to a lesser extent in ^{22}Ne, ^{23}Na, ^{26}Mg and ^{27}Al. Striking similarities are found with the abundance pattern observed at the surface of C-rich extremely metal poor halo stars as e.g. the Christlieb star [2].

References

1. R. Cayrel et al.: A&A **416**, 1117 (2004)
2. N. Christlieb et al.: Nature **419**, 904 (2002)
3. R. Hirschi, G. Meynet, A. Maeder: A&A **425**, 649 (2004)
4. G. Israelian et al.: A&A **421**, 649 (2004)
5. R. Kudritzki: ApJ **577**, 389 (2002)
6. A. Maeder, G. Meynet: A&A **373**, 555 (2001)
7. G. Meynet, A. Maeder: A&A **390**, 561 (2002)
8. F. Palla: MmSAI Supplement **3**, p.52 (2003)

The r-Process Yields in Massive Stars

S. Wanajo[1], Y. Ishimaru[2], K. Nomoto[3], and T.C. Beers[4]

[1] Department of Physics, Sophia University, 7-1 Kioi-cho, Chiyoda-ku, Tokyo, 102-8554, Japan
[2] Department of Physics and Graduate School of Humanities and Sciences, Ochanomizu University, 2-1-1 Otsuka, Bunkyo-ku, Tokyo 112-8610, Japan
[3] Department of Astronomy, School of Science, University of Tokyo, Bunkyo-ku, Tokyo, 113-0033, Japan
[4] Department of Physics/Astronomy, Michigan State University, E. Lansing, MI 48824, USA

1 Introduction

The astrophysical origin of the rapid neutron-capture (r-process) species has been a long-standing mystery. Even the most promising, "neutrino wind" scenario, encounters some difficulties [5]. Recent chemical evolution studies imply the dominant source of r-process elements to be the low-mass end of the supernova mass range, such as stars of $8 - 10 M_{\odot}$ [1,2]. The purpose of this study is to investigate conditions necessary for the production of r-process nuclei obtained in purely hydrodynamical models of prompt explosions of collapsing O-Ne-Mg cores, and to explore some of the consequences if those conditions are met (see [7,8] for more detail).

2 Prompt Explosion

A pre-supernova model of a $9 M_{\odot}$ star is taken from Nomoto [3], which forms a 1.38 $M_{\odot}$ O-Ne-Mg core. We link this core to a one-dimensional implicit Lagrangian hydrodynamic code with Newtonian gravity. The equation of state of nuclear matter (EOS) is taken from Shen et al. [4]. We find that a very weak explosion results, where no r-processing is expected. In order to examine the possible operation of the r-process in the explosion of this model, we artificially obtain an explosion with a typical energy of $\sim 10^{51}$ ergs by application of a multiplicative factor ($= 1.6$) to the shock-heating term in the energy equation.

3 The r-Process

The yields of r-process nucleosynthesis species are obtained by application of an extensive nuclear reaction network code that consists of ~ 4000 species. The mass-integrated abundances from the surface (zone 1) to the zones 83 (a), 92 (b), 95 (c), 98 (d), 105 (e), and 132 (f) are compared with the solar r-process abundances in Fig. 1. A solar r-process pattern for $A \gtrsim 130$ is naturally reproduced in cases c-f, while cases a-b fail to reproduce the third abundance

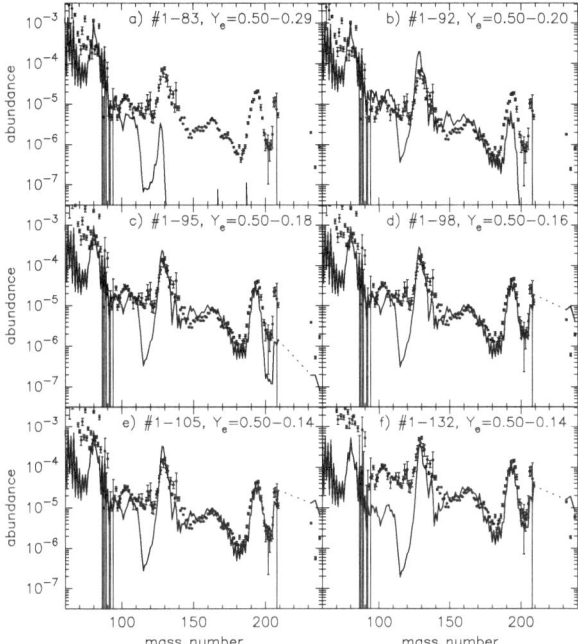

Fig. 1. Mass-averaged *r*-process abundances (line) as a function of mass number, which are compared with the solar *r*-process abundances (points).

peak. This implies that the region with $Y_e < 0.20$ must be ejected to account for production of the third *r*-process peak. Furthermore, to account for the solar level of thorium ($A = 232$) and uranium ($A = 235, 238$) production, the region with rather low Y_e (< 0.18) must be ejected.

Fig. 1 implies that the production of thorium and uranium differs from model to model, even though the abundance pattern seems to be *universal* between the second and third *r*-process peaks. Thus, the use of Th/Eu as a cosmochronometer should be regarded with caution, at least until the possible variations can be better quantified; U/Th might be a far more reliable chronometer [6,7].

References

1. Ishimaru, Y. & Wanajo, S. 1999, ApJ, 511, L33
2. Ishimaru, Y., Wanajo, S., Aoki, W., & Ryan, S. G. 2004, ApJ, 600, L47
3. Nomoto, K. 1984, ApJ, 277, 791
4. Shen, H., Toki, H., Oyamatsu, K., & Sumiyoshi, K. 1998, Nucl. Phys. A, 637, 435
5. Wanajo, S., Kajino, T., Mathews, G. J., & Otsuki, K. 2001, ApJ, 554, 578
6. Wanajo, S., Itoh, N., Ishimaru, Y., Nozawa, S., & Beers, T. C. 2002, ApJ, 577, 853
7. Wanajo, S., Tamamura, M., Itoh, N., Nomoto, K., Ishimaru, Y., Beers, T. C., & Nozawa, S. 2003, ApJ, 593, 968
8. Wanajo, S., Goriely, S., Samyn, M., & Itoh, N. 2004, ApJ, 606, 1057

Inhomogeneous Enrichment of r-Process Elements in Our Galaxy

Y. Ishimaru[1], S. Wanajo[2], W. Aoki[3], S.G. Ryan[4], and N. Prantzos[5]

[1] Department of Physics, Ochanomizu University, Bunkyo, Tokyo 102-8554, Japan
[2] Department of Physics, Sophia University, Chiyoda, Tokyo 102-8554, Japan
[3] National Astronomical Observatory of Japan, Mitaka, Tokyo 181-8588, Japan
[4] Department of Physics and Astronomy, The Open University, Milton Keynes, UK
[5] Institut d'Astrophysique de Paris, 98 bis, Boulevard Arago, 75014, Paris, France

Abstract. Observed large scatters in abundances of neutron-capture elements in metal-poor stars suggest that they are enriched a single or a few supernovae. Comparing predictions by an inhomogeneous chemical evolution model and new observational results with *Subaru* HDS, we attempt to constrain the origins of r-process elements.

1 Introduction

Observations of metal-poor stars reveal large dispersions in r-process elements [1]. This can be interpreted as a result of incomplete mixing of the interstellar medium (ISM) at the beginning of the Galaxy. If metal-poor stars contain products of a single or a few supernovae (SNe), distributions of stellar chemical compositions must reflect variation in SN yields. Thus, r-process yields may be highly dependent on the masses of SN progenitors. However, no consensus about the origins of r-process elements has been achieved, although a few scenarios show some promise[2,3]. In this study, we discuss the enrichment of Sr, Pd, Eu, and Ba, and attempt to constrain the origin of r-process.

2 Enrichment of Europium

We investigate the enrichment of Eu, a representative element of r-process, using a chemical evolution model, in which inhomogeneous gas mixing is taken into account[4]. When the r-process site is assumed as SNe from $8 - 10 M_\odot$ stars, significant numbers of stars having [Eu/Fe] < 0 are predicted at [Fe/H] < -3, owing to the delayed production of Eu (Fig. 1). In contrast, if r-process occurs in massive stars such as $\geq 30 M_\odot$, most of stars will have [Eu/Fe]> 0. Therefore, the lower limit of [Eu/Fe] at [Fe/H]< -3 is required to distinguish these predictions.

We selected three stars with [Fe/H] < -3, which were known to have [Ba/Fe] ~ -1, typical for their metallicities, and estimate Eu abundance using *Subaru* HDS. As shown in Fig. 1, our data add the lowest detections of Eu, at [Fe/H] < -3. The three stars and most others are located between the 50% confidence lines for this case. However, if Eu comes from more massive stars, these stars are located outside the 90% confidence region. We suggest, therefore, the r-process site is most likely to be SNe from low-mass progenitors such as $8 - 10 M_\odot$ stars.

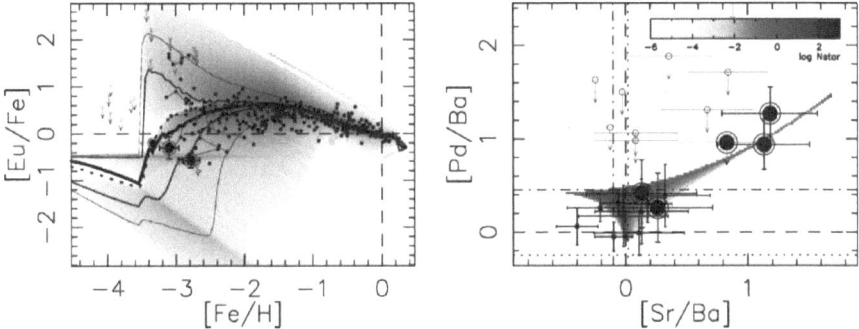

Fig. 1. *(left panel)* [Eu/Fe] as a function of [Fe/H]. Gray-scale indicates predicted distribution of stellar fraction. The r-process site is assumed to be SNe of $8 - 10 M_\odot$. The average stellar distributions are indicated by thick-solid lines with the 50% (solid lines) and 90% confidence intervals (thin-solid lines). The current observational data are given by large circles, with other previous data (small circles).

Fig. 2. *(right panel)* Same with Fig. 1 but for [Pd/Ba] vs. [Sr/Ba]. Weak r-process fraction for Sr, Pd, and Ba are respectively assumed as 60%, 10%, and 1%.

3   Origin of Palladium and "Weak" r-Process

Although Ba and heavier elements seem to fit the solar r-process pattern, lighter elements show wide varieties[5]. In particular, a large dispersion has been found in [Sr/Ba] at low metallicity[1], suggesting that lighter elements such as Sr does not come from a universal process, which produces Ba and Eu, but from "weak" r-process. An inhomogeneous chemical evolution model suggests that the dispersions in [Sr/Ba] are well-explained, when weak r-process produces $\sim 60\%$ of Sr but only $\sim 1\%$ of Ba in metal-poor stars. Furthermore, intermediate mass elements such as Pd must provide clues to understand the weak r-process yield.

Fig. 2 shows [Pd/Ba] as a function of [Sr/Ba] with our new data by *Subaru* HDS. By definition, [Sr/Ba] should increase with the fractional contribution of weak r-process to the stellar abundances. If Pd originates from weak r like Sr, [Pd/Ba] must show a correlation with the slope of unity to [Sr/Ba]. If Pd comes from main r like Ba, [Pd/Ba] must be constant. New data show a mild correlation with the slope less than unity, suggesting that the weak r-process fraction for Pd takes intermediate value between those of Sr and Ba; $\sim 10\%$. Therefore, this implies that the weak r-process yield decreases gradually from Sr to Ba.

References

1. e.g., Honda, S., Aoki, W., Kajino, T., et al., 2004, ApJ, 607, 474
2. Woosley, S. E., Wilson, J. R., Mathews, G. J., et al. 1994, ApJ, 433, 229
3. Wanajo, S., Tamamura, M., Itoh, N., et al. 2003, ApJ, 593, 968
4. Ishimaru, Y., Wanajo, S., Aoki, W., & Ryan S. G., 2004, ApJ, 600, L47
5. e.g., Sneden, C., Cowan, J. J., Lawler, J. E., et al. 2003, ApJ, 591, 936

The Weak sr(p)−Process in Massive Stars

M. Pignatari[1], R. Gallino[1], A. Heger[2], S.E. Woosley[3], and F. Käppeler[4]

[1] Dipartimento di Fisica Generale, Universitá di Torino, Via P. Giuria 1,
   10125 Torino, Italy
[2] Los Alamos National Laboratories, Los Alamos, NM 87545, USA
[3] Department of Astronomy and Astrophysics, University of California,
   Santa Cruz, CA 95064, USA
[4] Forschungszentrum Karlsruhe, Institut für Kernphysik, Postfach 3640,
   D-76021 Karlsruhe, Germany

Neutron captures in massive stars are mostly driven by the ^{22}Ne$(\alpha,n)^{25}$Mg reaction. A large abundance of ^{22}Ne results from the previous conversion of original CNO nuclei into ^{14}N during H burning followed by ^{14}N$(\alpha,\gamma)^{18}$F$(\beta^+)^{18}$O$(\alpha,\gamma)^{22}$Ne during the early phases of He burning. The β^+-decay by ^{18}F makes the neutron excess that allows the neutron source for the s process. A number of works in the past followed the s process during core He burning, where the average neutron density barely achieves 10^6 n/cm^3 ([2] and references therein). The classical analysis of the weak s-component cannot help much to constrain the expected abundance distribution below A $\sim$ 90. It is not even clear whether a single neutron exposure or an exponential distribution of exposures is best suited to reproduce the σN_s distribution. Note that the r-process theory fails to provide consistent contributions in the same atomic mass region. On the other hand, only preliminary studies on neutron captures during the subsequent convective shell C burning have been pursued so far. This process extends over most of the previous core He burning region. The release of α particles and the much higher temperature makes the residual ^{22}Ne to be severely consumed and to release neutrons in a much shorter time. Here the neutron density reaches quite high values, of up to a few 10^{12} n/cm^3, no more reconcilable with the classical concept of the s process. A large number of branchings at unstable nuclides along the neutron capture path are now open. Full stellar evolutionary calculations have been obtained by ([3], [5]) with a complete network extending from Fe to Bi for models of initial mass 15 and 25 $M_\odot$ and solar composition. This allows us to examine the nucleosynthesis of the heavy isotopes both in pre-explosive and explosive conditions. The final yields of all isotopes from ^{62}Ni to ^{87}Rb are mostly produced in hydrostatic conditions, by convective shell carbon burning and by previous core He burning. Exception is ^{64}Zn, which is produced in the neutrino winds [5]. In the innermost regions, photodisintegrations destroy the heavy isotopes beyond the iron peak. The production of p-only nuclei results from the previous neutron-rich seeds in a narrow region where explosive nucleosynthesis marginally affects oxygen burning. Interesting explosive nucleosynthesis effects result at the base of the He-burning shell, in the outer part of the oxygen-rich region. The case of Cu isotopes is reported in Fig. 1 for a 25 $M_\odot$ of solar composition. The pure s-process composition at core He exhaustion is visible at M_r ∼ 6 $M_\odot$. The yields of Cu isotopes in massive stars were studied at different metallicities by [4], using a more limited network, and showed a secondary-like trend for galactic disk stars. According to [5] (their Fig. 27) the production factors of almost all isotopes from ^{62}Ni up to ^{87}Rb are in the average a factor two higher than the one of ^{16}O. This includes pairs of s-only and r-only isobars, like ^{76}Se and ^{76}Ge. All the p-only nuclides, the r-only and the s-only nuclides as well as those of mixed rs origin in this mass region are of secondary origin, ultimately depending on the ^{22}Ne abundance. The production efficiency in this atomic mass region is sensitive to the choice of the ^{22}Ne$(\alpha,n)^{25}$Mg reaction rate, which in [5] was chosen from the recent result by [1]. Note that the higher yields for the heavy isotopes, by a factor of two with respect to ^{16}O, can be reconciled with their secondary-like nature. We conclude that neutron captures in massive stars are mostly attributable to a metallicity-dependent neutron capture process that is fairly different in nature from both classical predictions of the s-process and the r-process, which we may call the 'weak sr(p)-process'.

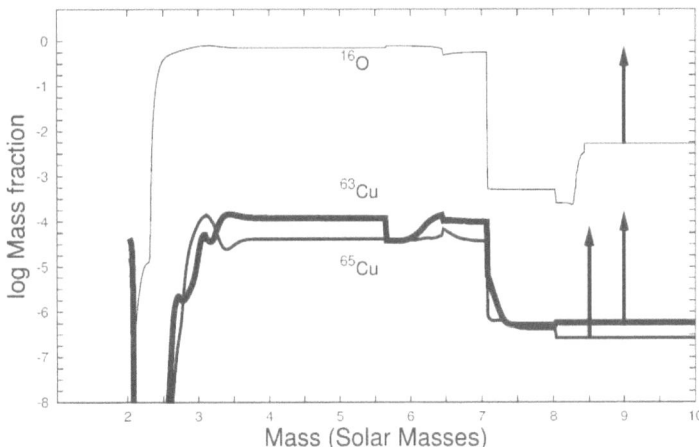

Fig. 1. Post−explosive yields versus mass of ^{63}Cu and ^{65}Cu for a 25 $M_\odot$ model of solar composition. The arrows indicate a production factor of 200 with respect to the initial composition. For ^{16}O the production factor is nearly 100.

References

1. M. Jaeger, R. Kunz, A. Mayer, J.W. Hammer, G. Staudt, K.-L. Kratz, B. Pfeiffer: Phys. Rev. Lett. **87**, 200501 (2001)
2. C.M. Raiteri, R. Gallino, M. Busso, D. Neuberger, F. Käppeler: ApJ **419**, 207 (1993)
3. T. Rauscher, A. Heger, R.D. Hoffman, S.E. Woosley: ApJ **576**, 323 (2002)
4. S.E. Woosley, T.A. Weaver: ApJ **101**, 181 (1995)
5. S.E. Woosley, A. Heger, T.A. Weaver: Rev. Mod. Phys. **74**, 1015 (2002)

Nucleosynthesis in Low Mass Very Metal Poor AGB Stars

A. Serenelli

Institute for Advanced Study, Einstein Drive, Princeton NJ 08540, USA

Calculations and Results

The evolution of a 1.5 $M_\odot$, Z= 10^{-5} stellar model has been followed starting at the ZAMS up to the thermally pulsating asymptotic giant branch (TP-AGB) phase. Calculations were done using the LPCODE [1], to which some changes were done. The most important and relevant to this work is the incorporation of a full nuclear network from H to Po, comprising about 525 isotopes and 910 nuclear reactions, appropriate for the computation of the s-process occurring in AGB stars. Convection is treated according to the mixing length theory (λ_{MLT} = 1.7) and convective mixing as a diffusive process. Diffusive overshooting is also included according to [2] and the free parameter f adopted is 0.015. Mass loss is given by the Reimers formula, with the parameter $\eta = 1$.

Starting from the ZAMS up to the beginning of the TP-AGB phase, the evolution of the 1.5 $M_\odot$, Z= 10^{-5} stellar model shows no qualitatively differences with that of more metal-abundant counterparts. However, in the first (full) thermal pulse, the entropy barrier between the He-convective shell (HECS) and the surrounding H-rich layers is overcome by the energy release in the helium convective shell [3–5]. In this way, the flash-driven convective shell (FDCS) penetrates into the H-rich layers and hydrogen is transported by convection into the He-burning shell (He-shell flash mixing [HESFM]), producing a H-flash. We focus our discussion on the HESFM and its consequences for the s-process that takes place in this kind of stars.

In the present calculations, the HESFM event starts when the He-convective shell extends from 0.5025 $M_\odot$ to 0.5425 $M_\odot$ and the hydrogen burning luminosity is $L_\mathrm{H}/L_\odot$=0.85. In about 0.35 yr, the hydrogen flash develops (mainly due to proton captures by ^{12}C) and L_H reaches its maximum value $L_\mathrm{H}/L_\odot$=10.5. At that point, the convective shell splits at 0.516 $M_\odot$ and an inner He-burning shell and an outer H-burning convective shell (HCS) develops, separated by a thin radiative region (its width is $\approx 5 \times 10^{-4}$ $M_\odot$). By the time the convective shell splits, the barion fraction of ^{13}C is $X(^{13}\mathrm{C})$= 0.0013. Four regions of interest can then be identified: the overshooting region below the HECS (OVHECS), the HECS, the radiative region between the HECS and the HCS (ICS), and the HCS. The temperature in the HECS is high enough so the reaction $^{13}\mathrm{C}(\alpha, n)^{16}\mathrm{O}$ is fully operative and s-processing occurs in the HECS. The neutron flux is very high (the neutron barion fraction reaches values of up to $X_n = 10^{-15}$) but it is diluted in the convective shell. As a result, s-process elements up to the second peak are produced here. The temperature in the OVHECS and the ICS is, at this point, too low for ^{13}C to be burnt by α captures. No s-processing ever occurs in the HCS.

After the flash, the HCS eventually merges with the convective envelope and the surface composition is enhanced in CNO elements (the star being now a carbon star, as $X(C)/X(O) = 4.8$). When the model evolves to the next helium flash, the temperature at the former OVHECS (between 0.5058 M$_\odot$ and 0.5076) increases and at some point activates the ^{13}C$(\alpha, n)^{16}$O reaction under radiative conditions and so the neutron flux is not diluted. Second and third peak elements are almost equally overproduced. Finally, the ^{13}C$(\alpha, n)^{16}$O reaction also activates in the former ICS. Here $X(^{13}$C$) \approx 0.003$ and the extremely high neutron flux leads to an overproduction of third peak elements more two orders of magnitude larger than those of the first and second peak. Dredge-up events occur in a few following thermal pulses (and stop as the stellar mass reduces) and in this way s-processed material reaches the stellar surface. The ^{13}C-pocket formed during these events is small and the s-processing that occurs there is of little effect on the abundance pattern formed as a result of the HEFM.

Finally we present a comparison of the final surface composition of our model (after ten thermal pulses) and the abundances of LP 625-44 [6] and LP 625-44 [7]. Model abundances are normalized to Ba.

Acknowledgements. AMS was partially supported by the W. M. Keck foundation through a grant to the IAS and the NSF (grant PHY-0070928). Calculations were done with the William Scheide computer cluster at the IAS.

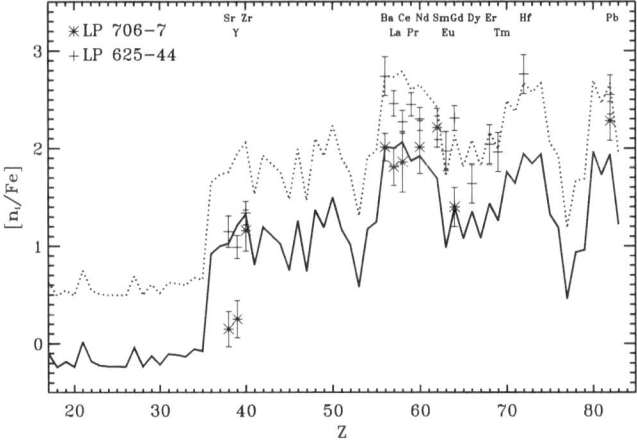

References

1. L. G. Althaus, et al.: A&A **404**, 594 (2003)
2. F. Herwig, et al.: A&A, **324**, L81 (1997)
3. S. Cassisi, V. Castellani, A. Tornambé: ApJ, **459**, 298 (1996)
4. M. Y. Fujimoto, Y. Ikeda, I. Iben: ApJ, **529** L25 (2000)
5. N. Iwamoto, K. Toshitaka, G. J. Mathews, W. Aoki: ApJ, **602**, 377 (2004)
6. W. Aoki, et al.: ApJ, **536**, L97 (2000)
7. W. Aoki, et al.: ApJ, **461**, 346 (2001)

A New Class of Type Ia Supernovae Inferred from Abundance Patterns of Halo Stars and High-z Galaxies

T. Tsujimoto

National Astronomical Observatory, Mitaka-shi, Tokyo 181-8588, Japan

Abstract. The recent discovery of a new population of stars exhibiting unusual elemental abundance patterns characterized by enhanced Ti to Ga elements and low α-elements suggests the contribution of a new class of supernovae, probably a kind of Type Ia supernovae associated with close binary evolution. The role of these supernovae in chemical evolution is negligible in normal galaxies that undergo moderate star formation such as our own. Thus, while the frequency of occurrence would be too low to detect in low-redshift galaxies, it may represent a prominent population in high-redshift objects such as early epoch massive elliptical galaxies and QSOs. The chemical contributor of this proposed type of supernovae in combination with recognized supernovae is compatible with the recent observational features in the distant universe.

1 Fossil Imprints in Metal-Poor Stars in the Solar Neighborhood

Recently, [1] found two stars belonging to a new class of low-α stars that exhibit enhanced abundances of elements from Ti to Ga, features never seen in other low-α stars. The deficiency of α-elements is significant in these stars, with characteristics such as [Mg/Fe] $= -0.64$ and [Si/Fe] $= -0.97$, whereas other low-α stars are only mildly deficient with [α/Fe]~ 0 at most. Such elemental abundance features are reminiscent of the nucleosynthesis of Type Ia supernovae (SNe Ia). Recent studies on the chemical compositions of metal-poor stars have revealed that these stars might have inherited the abundance pattern of the ejecta of the preceding single SN ([2]). If this is the case, these two stars may have been born from the ejecta of some kind of SN Ia, which would have had an explosion timescale comparable to those of massive stars ending in Type II SNe (SNe II). Otherwise, as in the general cases of normal SNe Ia, the fossil imprint of nucleosynthesis in SNe Ia will not be retained in stellar abundances due to mixing with the products from SNe II within the ejecta.

The merging of double white dwarfs as the end result of a close binary consisting initially of two massive stars of $\sim 8M_\odot$ could produce SNe Ia with an explosion timescale of the order of 10^6–10^7 yrs ([3]). Such SNe Ia are, if indeed they exist, expected to be locally rare events. However, the situation could be significantly different in other galaxies because of the potential for rates of close binary system formation in high-density environments, as suggested by the recent globular cluster (GC) study ([4]), which revealed with high confidence that the number of close binaries in GCs increases with the stellar encounter rate of the cluster.

2   Chemical Evolution of a Galaxy at $z = 2.626$

One signature of these potential SNe Ia can be seen in the high-z absorption system of QSOs. Recently, [5] observed elemental abundances in a damped Lyman alpha (DLA) system at $z = 2.626$. They concluded that this DLA galaxy is the progenitor of a massive elliptical galaxy based on the [α/Zn] ratios, which are suggestive of an SN II origin and imply that the potential age of this galaxy is of the order of several 10^8 yrs. However, this conclusion presents the puzzling problem of how the Ge abundance could become so enhanced in such a young galaxy. Incidentally, among the enhanced elements for the two solar neighborhood stars in question, a remarkable enhancement is seen in the Ga abundance. Since Ga and Ge may be produced in a similar fashion, the Ge abundance in this galaxy could be derived from the proposed new class of SNe Ia. This consideration is appropriate from the viewpoint that in the formation of massive elliptical galaxies, the gas density is expected to be high, leading to a large fraction of close binary systems.

3   Chemical Evolution of QSOs at $z \sim 6$

The high metallicities and high infrared luminosities of high-z QSOs are expected to be associated with a massive starburst. Therefore, an extremely high frequency of these potential SNe Ia occurrence would be expected in these QSOs because of the high gas density that triggered the starburst. The essential signature of enrichment by the proposed SNe Ia is a high Fe abundance with a very short timescale. In fact, accumulated observational results of the distant universe have revealed very early chemical enrichment in high-redshift QSOs. [6] have shown that QSOs at $z \sim 6.3$ exhibit FeII/MgII ratios about twice as high as those of low-z QSOs. At such an early stage, it appears to be impossible for normal SNe Ia to enrich the interstellar matter at a high Fe abundance. The results obtained by chemical evolution models incorporating the contribution from the new class of SNe Ia have been presented in [7].

References

1. I.I. Ivans et al.: ApJ **592**, 906 (2003)
2. T. Shigeyama, T. Tsujimoto: ApJ **507**, L135 (1998)
3. I.Jr. Iben, A.V. Tutukov: ApJS **54**, 335 (1984)
4. D. Pooley et al.: ApJ **591**, L131 (2003)
5. J.X. Prochaska, J.C. Howk, A.M. Wolfe: Nature **423**, 57 (2003)
6. R. Maiolino, Y. Juarez, R. Mujica, N.M. Nagar, E. Oliva: ApJ **596**, L155 (2003)
7. T. Tsujimoto: ApJ **611**, L17 (2004)

The Influence of the Treatment of Convection on the AGB Stars Yields

P. Ventura and F. D'Antona

Osservatorio Astronomico di Roma
Via Frascati 33, 00040 MontePorzio Catone (RM), Italy

1 Introduction

The interest of the astrophysical community on the evolution of the intermediate mass stars (IMS) raised in the last decades, as they have been suggested as possible responsible of the chemical anomalies which are observed in Giant and TO stars within Globular Clusters (see e.g. Gratton et al. 2004).

The idea behind this hypothesis is that an early generation of AGBs might have evolved and polluted the interstellar medium with nuclearly processed material, so that the new generations of stars were born from an already contaminated gas (D'Antona et al. 1983; Cottrell & Da Costa 1981).

We focus our attention on the role which the treatment of convection may play on the physical and chemical evolution of massive AGBs, and how their chemical yields depend on the efficiency of the convective model.

2 Results and Discussion

We compare the evolutions of 3 stellar models with initial mass $5M_\odot$ for a metallicity typical of those GCs like M3, M13, whose stars exhibit the quoted anomalies, i.e. $Z = 0.001$.

The three models were calculated with the same chemical and physical inputs with the only exception of convection, for which we adopted the Full Spectrum of Turbulence convective model (FST, Canuto & Mazzitelli 1991), and the MLT model (Vitense 1953) with two values of the free parameter connected to the mixing length: $\alpha = 1.7$ (the standard value, used to reproduce the evolution of the Sun) and $\alpha = 2.1$.

The left panel of fig. 1 shows the evolution with time of the luminosity of the three models. We see in the FST model a rapid increase of the luminosity starting from the first TPs, which is associated with the larger efficiency (hence, lower overadiabaticity) of the convective model, which makes the inner border of the convective envelope to be closer to the CNO burning shell peak. In this situation we have a larger extra-luminosity due to the burning of H-rich material in a high temperature region, with a consequent increase of the stellar luminosity.

On the chemical point of view, the larger luminosity of the FST model leads to a larger mass loss, which, in turn, reduces the whole AGB life-time and the number of 3rd dredge-up episodes. The consequence is that while in the MLT

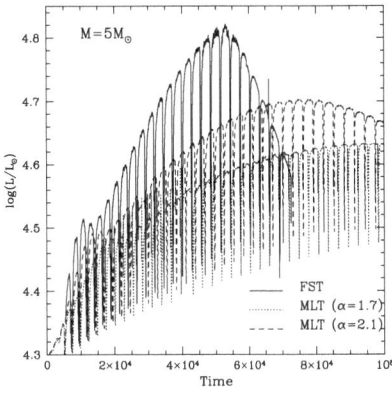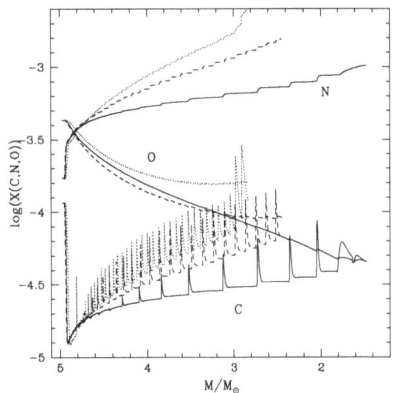

Fig. 1. Left: The variation with time of the luminosity of three models of initial mass $5M_\odot$ calculated with different prescriptions for the treatment of convection; times are counted from the beginning of the AGB phase. **Right**: The variation with the total mass of the surface abundances of the CNO elements for the same models shown in the left panel.

models (particularly in the MLT17 case) we expect that the expelled material on shows a great increase of the C+N+O abundance, in the FST model this latter quantity is approximately constant (see the right panel of fig. 1).

The models differ particularly in what concerns the Oxygen abundance of the ejecta: in the FST case Oxygen is destroyed due to full CNO burning at the base of the envelope, therefore an ^{16}O poor yield is expected. On the other hand, in the MLT models oxygen is not heavily destroyed, and it is also carried to the surface of the star in the latest AGB phases due to a deep 3rd dredge-up; therefore, in this case we expect an oxygen content of the ejecta which is close to the initial value.

The sodium yields are also different: in the FST model sodium is first created and the destroyed, when the Ne-Na cycle becomes fully operative. In the MLT models the 3rd dredge-up becomes so efficient that Sodium is eventually produced due to ^{22}Ne burning: in the MLT framework the AGB ejecta are expected to have great mass fractions of sodium.

References

1. V.M. Canuto, I. Mazzitelli 1991, ApJ, 370, 295
2. P.L. Cottrell, G.S. Da Costa 1981, ApJ, 245, 79
3. F. D'Antona, R. Gratton, A. Chieffi 1983, Mem.S.A.It., 54, 173
4. R. Gratton, C. Sneden, E. Carretta 2004, ARA&A, 42, 385
5. E. Vitense 1953, Zs.Ap., 32, 135

Part VI

Implications for BBN and Galaxy Formation and Evolution

BBN and the Primordial Abundances

G. Steigman

The Ohio State University, Columbus OH 43210, USA

Abstract. The relic abundances of the light elements synthesized during the first few minutes of the evolution of the Universe provide unique probes of cosmology and the building blocks for stellar and galactic chemical evolution, while also enabling constraints on the baryon (nucleon) density and on models of particle physics beyond the standard model. Recent WMAP analyses of the CBR temperature fluctuation spectrum, combined with other, relevant, observational data, has yielded very tight constraints on the baryon density, permitting a detailed, quantitative confrontation of the predictions of Big Bang Nucleosynthesis with the post-BBN abundances inferred from observational data. The current status of this comparison is presented, with an emphasis on the challenges to astronomy, astrophysics, particle physics, and cosmology it identifies.

1 Introduction and Overview

Our Universe is observed to be expanding and filled with radiation (the Cosmic Background Radiation: CBR), along with "ordinary" matter (baryons $\equiv$ nucleons). It is well known that if this evolution is followed backwards in time, then there was an epoch during its early evolution when the Universe was a "Primordial Nuclear Reactor", synthesizing in astrophysically interesting abundances the light nuclides D, ^{3}He, ^{4}He, and ^{7}Li. Discussion of BBN can start when the Universe was some tens of milliseconds old and the temperature (thermal energies) was of order a few MeV. At that time there were no complex nuclei, only neutrons and protons. Since there are nearly ten orders of magnitude more CBR photons in the Universe than nucleons, photodissociation ensures that at such high temperatures the abundances of complex nuclei are negligibly small. However, as the Universe expands (and the weak interactions transmute neutrons and protons), collisions among nucleons begin to build the light nuclides when the temperature drops below ~ 100 MeV, and the Universe is a couple of minutes old. Very quickly, almost all neutrons available are incorporated in the most tightly bound of the light nuclides, ^{4}He. As a result, the ^{4}He primordial abundance (mass fraction: Y_P) is a sensitive probe of the competition between the weak interaction rates and the universal expansion rate (the Hubble parameter: H); ^{4}He is a cosmological chronometer. The reactions building ^{4}He are not rate (nuclear reaction rate) limited and, therefore, Y_P is only weakly (logarithmically) sensitive to the baryon density. In contrast, the BBN abundances of the other light nuclides (D, ^{3}He, ^{7}Li) are rate limited and these do depend sensitively (to lesser or greater degrees) on the baryon density; D, ^{3}He, and ^{7}Li are all potential baryometers.

The relic abundances of the light nuclides predicted by BBN in the "standard" model of cosmology (SBBN) depend only on one free parameter: the nucleon density. There are, therefore, two complementary approaches to testing SBBN. On the one hand, the primordial abundances inferred from observational data should be consistent with the SBBN predictions **for a unique value (or range) of the nucleon density**. On the other hand, if there is a non-BBN constraint on the range allowed for the baryon density, this should lead to SBBN-predicted abundances in agreement with the observational data. Is this the case? If not, why not? That is, if there are conflicts between predictions and observations, is the "blame" to be laid at the feet of the observers (inaccurate data and/or unidentified systematic errors?), or the astrophysicists (incorrect models for analysing the data and/or extrapolating from abundances determined at present ("here and now") to their primordial ("there and then") values), or are the standard models of particle physics and/or cosmology in need of revision?

1.1   The Status Quo: Observations Confront SBBN

Space limitations prevent my presenting a full-fledged review of the history of the observational programs along with the evolution of the comparisons between theory and data. For some recent reviews of mine, see [1] and further references therein. Instead, an overview is presented which highlights the challenges to SBBN. The remainder of this article is devoted to some of the key issues associated with each of the light nuclide relic abundances.

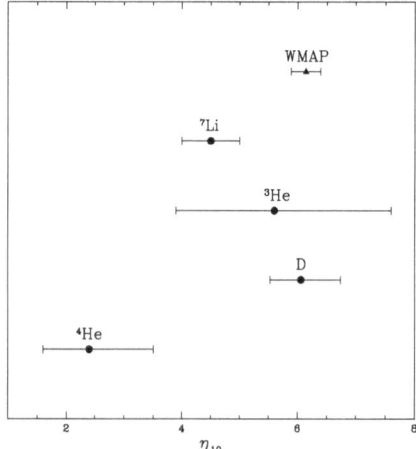

Fig. 1. The baryon density parameter, η_{10}, inferred from SBBN and the relic abundances of D, ^{3}He, ^{4}He, and ^{7}Li (filled circles), along with the non-BBN determination from WMAP (filled triangle). See the text for details.

If the standard model (SBBN) is the correct choice and if the primordial abundances inferred from the data were free of systematic errors, then the baryon densities determined from D, ^{3}He, ^{4}He, and ^{7}Li should agree among themselves and with that inferred from the CBR (WMAP) and other, non-BBN, cosmological data. In Figure 1 are plotted the various values of the baryon density parameter[1] determined by SBBN and the adopted primordial abundances and, also, from the WMAP-team analysis [2]. As may be seen from Fig. 1, the SBBN D abundance is in excellent agreement with the WMAP-inferred baryon density. However, neither ^{7}Li nor ^{4}He agree with them or, even with each other. While ^{3}He is consistent with SBBN deuterium and with WMAP (and is not in disagreement with ^{7}Li), the very large uncertainty in its primordial abundance, combined with its relative insensitivity to the baryon density, render it – at present – an even less sensitive baryometer than is ^{4}He.

In the next section each light nuclide is considered in turn, its post-BBN evolution briefly reviewed along with identification of a few of the potential challenges to accurately inferring the primordial abundances from the observational data. Then, having established that the current data – taken at face value – are not entirely consistent with SBBN, I investigate whether changes in the early universe expansion rate can reconcile them.

2   Primordial Abundances – Evolution, Uncertainties, Systematics

While Figure 1 suggests some problems with SBBN, the optimist may prefer to conclude that observations have provided impressive confirmation of the standard cosmological model. After all, if the standard model – or something very much like it – were not correct, there'd be no good reason why the abundances of four light nuclides, which range over some nine orders of magnitude, should be just such that the baryon densities inferred from each of them lie within a factor of three of each other, in nearly perfect agreement with that derived independently from non-BBN data. Only recently, when cosmology has entered an era of great precision, has it become important to distinguish accuracy from precision and to revisit the path from precise astronomical observations to accurate abundances. For each of the light nuclides of interest here, an all too abbreviated overview of the current uncertainties is presented, with the intentional goal of creating controversy in order to stimulate future observations and theoretical analyses.

2.1   Deuterium

Deuterium is the baryometer of choice. During its post-BBN evolution, as gas is cycled through stars, D is only destroyed (setting aside rare astronomical events

---

[1] After $e^\pm$ annihilation during the early evolution of the Universe, the ratio of baryons to photons is, to a very good approximation, preserved down to the present. The baryon density parameter is defined to be this ratio (at present): $\eta \equiv n_N/n_\gamma$; $\eta_{10} \equiv 10^{10}\eta$.

where, far from equilibrium, tiny amounts of D may be synthesized). Therefore, observations of D anywhere, at any time (*e.g.*, the solar system or the local ISM), provide a *lower* bound to its primordial abundance. As a result, it is expected that observations of D in regions which have experienced minimal stellar evolution (*e.g.*, the high redshift and/or low metallicity QSO Absorption Line Systems: QSOALS) should provide a good estimate of the relic abundance of deuterium. Kirkman *et al.* [3] have gathered the extant data; see [3] for details and related references. In Figure 2 are shown the QSOALS deuterium abundances as a function of metallicity; for reference, solar system and ISM D abundances are also shown.

While the observers are to be commended for their heroic work in identifying and analysing the tiny fraction of QSOALS which can be used to infer a low-metallicity, high redshift, D abundance, there are several unsettling aspects of the data displayed in Figure 1. Perhaps most noticeable is the paucity of data points. Without at all minimizing the difficulty of finding and analysing such systems, it is very nearly a sin to claim that the primordial abundance of a cosmologically key light nuclide is determined by five data points. If, however, the data points were in agreement within the estimated statistical errors, this might be less disturbing. It is clear from Fig. 1 that this is not the case. For these five data points the χ^2 about the mean is $\gtrsim 16$, suggesting either that the uncertainties have been underestimated, or that some of these data may be contaminated by unidentified systematic errors. While the dispersion may simply be due to the small number of data points, it might be significant that

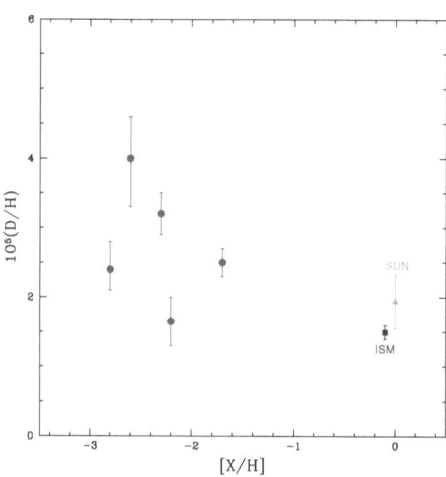

Fig. 2. The deuterium abundance (by number relative to hydrogen), $y_D \equiv 10^5(D/H)$, derived from high redshift, low metallicity QSOALS [3] (filled circles). The metallicity is on a log scale relative to solar; depending on the line-of-sight, X may be oxygen or silicon. Also shown is the solar system abundance (filled triangle) and that from observations of the local ISM (filled square).

the three QSOALS with the lower D/H ratios are LLS, while the two highest D/H determinations are from DLAs.

In the absence of evidence for changing or eliminating any of the current D abundance determinations, it is not unreasonable to follow the advice of [3] and adopt the weighted mean as an estimate of the primordial D abundance. Here, too, there is a (minor) problem. Kirkman *et al.* [3] advocate finding the mean of $\log(D/H)$ to determine $y_D \equiv 10^5 (D/H)$; $y_D \equiv 10^{(5 + <\log(D/H)>)} = 2.78$. In contrast, if the weighted mean of the five D/H determinations is used, $y_D = 2.60$. While this difference is well within the dispersion, it reflects the fact that in determining the mean of $\log(D/H)$, PKS1937 with $y_D = 3.24$ dominates, while for the mean of D/H, HS0105 with $y_D = 2.54$ dominates. In what follows I adopt $y_D = 2.6 \pm 0.4$, where the error, following [3], is derived from the dispersion in D/H determinations. The corresponding BBN (SBBN) prediction for the baryon abundance, $\eta_{10}(D) \approx 6.1^{+0.7}_{-0.5}$, is shown in Figure 1.

2.2   Helium-3

In my talk at this meeting I actually avoided discussion of ^{3}He, until it was forced upon me during the question session. In part, this was because this subject was ably covered in Tom Bania's talk, in Dana Balser's poster, and in Bob Rood's conference summary. In part, however, it was because, in my opinion, for both observational and theoretical reasons ^{3}He has more to teach us about stellar and Galactic evolution than about BBN. ^{3}He is a less sensitive baryometer than is D since $(D/H)_{BBN} \propto \eta^{-1.6}$, while $(^3He/H)_{BBN} \propto \eta^{-0.6}$. Even more important

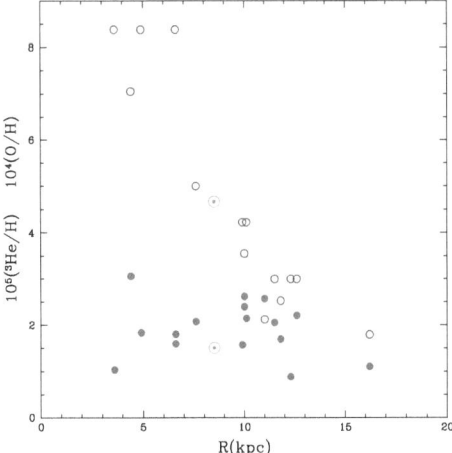

Fig. 3. The ^{3}He abundances (by number relative to hydrogen), $y_3 \equiv 10^5 (^3He/H)$, derived from Galactic H II regions [4], as a function of galactocentric distance (filled circles). Also shown for comparison is the solar system abundance (solar symbol). The open circles are the oxygen abundances for the same H II regions (and for the Sun).

is that, in contrast to the expected, monotonic, post-BBN evolution of D, the post-BBN evolution of ^{3}He is quite complicated, with competition between stellar production, destruction, and survival. For years, indeed decades, it had been anticipated that net stellar production would "win" and the abundance of ^{3}He would increase with time (and with metallicity); see, *e.g.*, [5]. Unfortunately, observations of ^{3}He are restricted to the solar system and the Galaxy. Nonetheless, since there is a clear galactic gradient of metallicity (see Fig. 3), a gradient in ^{3}He abundance would also be expected. If net production "wins", then ^{3}He should be highest in the inner galaxy; the opposite if net destruction dominates. The absence of any statistically significant gradient in the Bania, Rood, Balser (BRB) data [4] (see Fig. 3), points to an extremely delicate balance (cancellation) between production and destruction. This suggests that the mean ^{3}He abundance in the Galaxy ($y_3 \approx 1.9$) might provide a reasonable estimate of the primordial abundance. However, BRB recommend that the ^{3}He abundance determined in the most distant (from the Galactic center), most metal poor Galactic H II region yields an upper limit, $y_3 \lesssim 1.1 \pm 0.2$, to the primordial abundance. The estimate of $\eta_{10}(^3\mathrm{He}) \approx 5.6^{+2.0}_{-1.7}$ shown in Figure 1 is based on this choice. Had I adopted the mean value of $y_3 = 1.9$, I would have inferred $\eta_{10}(^3\mathrm{He}) \approx 2.3$. While the former choice is in excellent agreement with deuterium (and with ^{7}Li and with the WMAP result [2]), the very large uncertainty renders ^{3}He an insensitive baryometer; the latter option would be consistent with ^{4}He, but not with D (or with ^{7}Li or WMAP).

2.3   Helium-4

Helium-4 is the textbook example of a relic nuclide whose abundance is known precisely but, likely, inaccurately. To be of value in testing SBBN as well as in constraining non-standard models, Y_P should be determined to 0.001 or better. The largest uncertainty in the SBBN prediction is from the very small error in the neutron lifetime ($\tau_n = 885.7 \pm 0.7$ s). For the WMAP estimate of the baryon density, including its uncertainty, the SBBN-predicted primordial abundance is $Y_P = 0.2482 \pm 0.0007$, as shown in Figure 4. Also shown in Figure 4 is a record of Y_P determinations, from observations of extragalactic, low metallicity, H II regions, covering the period from the late 1970s to the present (2004). During this time it has been well known, but often ignored, that there are a variety of systematic uncertainties which might dominate the Y_P determinations. In the hope of accounting for these systematic errors (rather than constraining them by observations), the error estimates for Y_P have often been inflated beyond the purely statistical uncertainties. Thus, until the late 1990s, the typical error estimate for Y_P was 0.005. For example, summarizing the status as of 2000, Olive, Steigman, and Walker [7] suggested that the data available at that time were consistent with $Y_P = 0.238 \pm 0.005$; see Fig. 4. However, if ^{4}He were used as an SBBN baryometer (not recommended!), the error in the baryon density would have been $\sim 50\%$. Of course, as the number of H II regions observed increased, largely due to the work of Izotov & Thuan [6], the statistical errors decreased. For example, from observations of 82 extragalactic H II regions, in their 2004 paper Izotov & Thuan quote [6] $Y_P = 0.2429 \pm 0.0009$; this data point is shown in Fig. 4. In contrast, a very recent, detailed study of the effects of *some* of the identified systematic uncertainties by Olive & Skillman [8] suggests the true errors are likely larger than this, by at least an order of magnitude.

As may be seen from Fig. 4, none of the Y_P estimates agree with the SBBN prediction, all being low by roughly 2-σ. Indeed, from their sample of 82 data points Izotov & Thuan [6] derive such a small uncertainty that their central value is low by nearly 6-σ. In their analysis, Izotov & Thuan commit the cardinal sin of examining their data and then, *a posteriori*, choosing a subsample of 7 H II regions to derive their favored estimate of $Y_P = 0.2421 \pm 0.0021$. One wonders what they may have found from a random series of choosing 7 of 82 data points. In any case, this estimate also falls short of the SBBN prediction by nearly 3-σ. Using this suspect subsample, Olive & Skillman [8] do find consistency with SBBN once they have corrected for the systematic errors they've chosen to include. However, one they have ignored, the ionization correction factor [9], almost certainly would have the effect of reducing their central value and increasing their error estimate. For the entire Izotov & Thuan sample, Olive & Skillman find a very large range for Y_P, from 0.232 to 0.258 (or, $Y_P \approx 0.245 \pm 0.013$, entirely consistent with SBBN)). If the average correction for ionization suggested by Gruenwald, Steigman, and Viegas [9] for the somewhat smaller 1998 Izotov & Thuan data set is applied to their 2004 compilation, it would suggest the Olive & Skillman central value be reduced and their error increased: $Y_P \approx 0.239 \pm 0.015$.

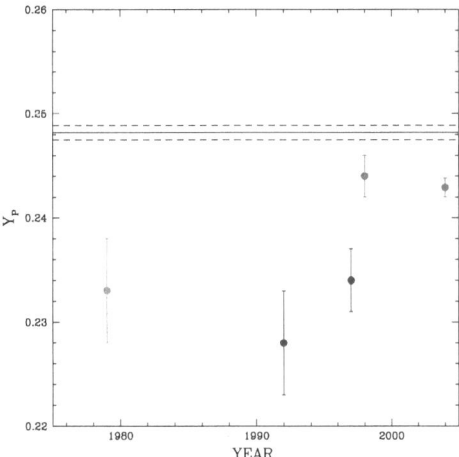

Fig. 4. A summary of the time evolution of primordial ^{4}He abundance determinations (mass fraction Y_P) from observations of metal-poor, extragalactic H II regions (see the text for references). The solid horizontal line is the SBBN-predicted ^{4}He abundance expected for the WMAP (and/or D) inferred baryon density. The two dashed lines show the 1σ uncertainty in this prediction.

If, indeed, the true uncertainty in Y_P is really this large, it opens the possibility that there might be alternate approaches to Y_P which are competitive with the extragalactic, low-metallicity H II regions and, even more important, complementary in that they are subject to completely independent sets of systematic errors. One such example, with a venerable history of its own, is to use effect of the initial helium abundance on the evolution of low-mass Pop II stars, employing the R-parameter [10]. Recently, Cassisi, Salaris, and Irwin [11] have attempted this using observations of a large sample of Galactic Globular Clusters (GGC) and new stellar models. While they claim an extraordinarily accurate determination of Y_P (0.243 ± 0.006), this does not seem to be supported by the data they present since, for the lowest metallicity GGCs, Y ranges from $\lesssim 0.19$ to $\gtrsim 0.27$. Nonetheless, if there is the possibility that the R-parameter method might achieve theoretical and observational uncertainties $\lesssim 0.01$, it is certainly an approach worth pursuing.

An alternate approach, subject to large theoretical uncertainties, would be to attempt to use chemical evolution models, tied to the solar helium and heavy element abundances, to extrapolate back to the relic abundance of ^{4}He. While at present this approach appears to be limited by the theoretical uncertainties (*e.g.*, metallicity dependent stellar winds and stellar yields), the following example may serve as a stimulus (or challenge) to those who might believe they can do better. In a recent paper employing new yields, Chiappini, Matteucci, and Meynet [12] find $\Delta Y \equiv Y_\odot - Y_P \approx 0.018 \pm 0.006$. Using the recent Bahcall & Pinsonneault [13] estimate of $Y_\odot \approx 0.270 \pm 0.004$, this leads to a primordial estimate of $Y_P \approx 0.252 \pm 0.007$. Although this result is consistent with the SBBN prediction, it would be entirely premature to declare victory on the basis of such a crude estimate. The possible lesson illustrated by this example is that the **error** associated with such an approach might not be uncompetitive with those from the standard H II region analyses.

2.4   Lithium-7

As with ^{4}He, the recent history of the comparison between the SBBN predictions and the observational data leading to the relic abundance of ^{7}Li is one of conflict, with the spectre of systematic errors looming large. ^{7}Li, along with ^{6}Li, ^{9}Be, ^{10}B, and ^{11}B, is produced in the Galaxy by cosmic ray spallation/fusion reactions. Furthermore, observations of super-lithium rich red giants provide evidence that (some) stars are net producers of lithium. Therefore, to probe the BBN yield of ^{7}Li, it is necessary to restrict attention to the most metal-poor halo stars (the "Spite plateau"). Using a specially selected data set of the lowest metallicity halo stars, Ryan *et al.* [14] claim evidence for a 0.3 dex increase in the lithium abundance ($[\text{Li}] \equiv 12 + \log(\text{Li/H})$) for $-3.5 \leq [\text{Fe/H}] \leq -1$, and they derive a primordial abundance of $[\text{Li}]_P \approx 2.0 - 2.1$. This value is low compared to the estimate of Thorburn (1994) [15], who found $[\text{Li}]_P \approx 2.25 \pm 0.10$. In the steps from the observed equivalent widths to the derived abundances, the stellar temperature plays a key role. When using the infrared flux method effective temperatures, studies of halo and Galactic Globular Cluster stars [16] suggest a higher abundance: $[\mathrm{Li}]_\mathrm{P} = 2.24 \pm 0.01$. Very recently, Melendez & Ramirez [17] have reanalyzed 62 halo dwarfs using an improved infrared flux method effective temperature scale. They fail to find the [Li] vs. [Fe/H] gradient claimed by Ryan *et al.* [14] and they confirm the higher lithium abundance, finding $[\mathrm{Li}]_\mathrm{P} = 2.37 \pm 0.05$. As shown in Figure 1, if this were the true primordial abundance of $^7\mathrm{Li}$, then $\eta_{10}(\mathrm{Li}) = 4.5 \pm 0.4$. **All** of these observational determinations of primordial lithium are significantly lower than the SBBN expectation of $[\mathrm{Li}]_\mathrm{P} = 2.65^{+0.05}_{-0.06}$ for the WMAP baryon density.

As with $^4\mathrm{He}$, the culprit may be the astrophysics rather than the cosmology. Since the low metallicity, dwarf, halo stars used to constrain primordial lithium are the oldest in the Galaxy, they have had the most time to modify (by dilution and/or destruction) their surface abundances. While mixing of the stellar surface material with the interior would destroy or dilute the prestellar lithium abundance, the very small dispersion in [Li] among the low metallicity halo stars suggests this effect may not be large enough to bridge the ≈ 0.3 dex gap between the observed and WMAP/SBBN-predicted abundances ($[\mathrm{Li}]_\mathrm{P}^{obs} \approx 2.37$ versus $[\mathrm{Li}]_\mathrm{P}^{pred} \approx 2.65$); see, *e.g.*, [18] and further references therein.

3   Non-Standard BBN

The path from acquiring observational data to deriving primordial abundances is long and complex and littered with pitfalls. While the predicted and observed relic abundances are in rough qualitative agreement, at present there exist some

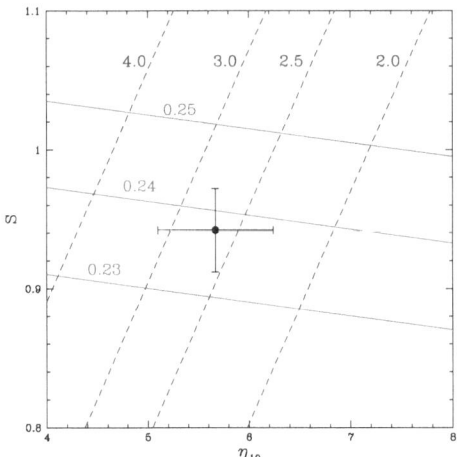

Fig. 5. Isoabundance curves for $^4\mathrm{He}$ (solid) and D (dashed) in the baryon abundance (η_{10}) – expansion rate factor (S) plane. The labels on the $^4\mathrm{He}$ curves are for Y_P, while those on the D curves are for $y_\mathrm{D} \equiv 10^5(\mathrm{D/H})$. The filled circle with error bars corresponds to the adopted values of the D and $^4\mathrm{He}$ primordial abundances (see the text).

potential discrepancies. These may well be due to the data, the data analysis, or the extrapolations from here and now to there and then. But, there is also the possibility that these challenges to SBBN are providing hints of new physics beyond the standard models of particle physics and/or cosmology. Once this option is entertained, the possibilities are limited only by the imagination and creativity of physicists and cosmologists. Many such models have already been proposed and studied. One option is to discard most or all of standard physics and start afresh. A more conservative (not in the pejorative sense!) approach is to recognize that the standard model (SBBN) does quite well and to look for small variations on the same theme. Here, I'll adopt the latter strategy and explore one such option which can resolve some, but not all, of the conflicts: a non-standard, early universe expansion rate.

There are many different extensions of the standard model of particle physics which result in modifications of the early universe expansion rate (the time – temperature relation). For example, additional particles will increase the energy density (at fixed temperature), resulting in a faster expansion. In such situations it is convenient to relate the extra energy density to that which would have been contributed by an additional neutrino with the ordinary weak interactions [19]. Just prior to $e^{\pm}$ annihilation, this may be written as

$$\frac{\rho'}{\rho} \equiv 1 + \frac{7\Delta N_{\nu}}{43}. \tag{1}$$

Since the expansion rate (the Hubble parameter) depends on the square root of the combination of G (Newton's constant) and the density, the expansion rate factor (S) is related to ΔN_{ν} by,

$$S \equiv \frac{H'}{H} = (1 + \frac{7\Delta N_{\nu}}{43})^{1/2} = (\frac{G'}{G})^{1/2}. \tag{2}$$

Thus, while adding new particles (increasing the energy density) results in a speed-up in the expansion rate ($\Delta N_{\nu} > 0$; $S > 1$), changing the effective Newton's constant may either increase or decrease the expansion rate. Another example of new physics which can alter the expansion rate is the late decay of a very massive particle which reheats the universe, but to a "low" reheat temperature [20]. If the reheat temperature is too low ($\lesssim 7$ MeV) the neutrinos will fail to be fully populated, resulting in $\Delta N_{\nu} < 0$ and $S < 1$. Finally, in some higher dimensional extensions of the standard model of particle physics additional terms appear in the 3+1 dimensional Friedman equation whose behavior mimics that of "radiation", resulting in an effective ΔN_{ν} which could be either positive or negative [21].

Thus, a nonstandard expansion rate ($S \neq 1$) is a well-motivated, one parameter modification of SBBN which has the potential to resolve *some* of its challenges. A slower expansion would leave more time for neutrons to become protons and a lower neutron abundance at BBN would result in a smaller Y_P (good!). Since ^{4}He is the most sensitive chronometer, the effect on its abundance is most significant. However, a modified expansion rate would also affect the predicted abundances of the other light nuclides as well. A slowdown will result in more destruction of D and ^{3}He and, for $\eta_{10} \gtrsim 4$, in production of more ^{7}Be (which becomes ^{7}Li via electron capture). In Figure 5 are shown the quite accurate approximations to the isoabundance curves for D and Y_P in the $S - \eta_{10}$ plane, from the recent work of Kneller and myself [22]. Also shown in Fig. 5 is the location in this plane corresponding to the adopted D and ^{4}He abundances (including their uncertainties). Not surprisingly, it is possible to adjust these two parameters (S and η) to fit the relic abundances of these two nuclides. Note, however, that the best fit corresponds to a *slower* than standard expansion rate ($S \approx 0.94 \pm 0.03$; $\Delta N_\nu \approx -0.70 \pm 0.35$). While this combination of parameters is consistent with WMAP (see, *e.g.*, Barger et al. [23] and further references therein), it does not resolve the conflict with ^{7}Li. Although a slowdown in the expansion rate does have the effect of increasing ^{7}Li, this is compensated by the somewhat lower baryon density, which has the opposite effect. The result is that for these choices of S and η, which do resolve the conflicts between D and ^{4}He (and WMAP and ^{4}He), the predicted primordial abundance of ^{7}Li is still $[\text{Li}]_P \approx 2.62 \pm 0.10$.

4   Summary and Conclusions

Four light nuclides (D, ^{3}He, ^{4}He, ^{7}Li) are predicted to emerge from the early evolution of the universe in abundances large enough to be observed at present. In SBBN there is only one parameter, the baryon density parameter η, which determines the relic abundances of these nuclides. The current observational data identifies a range in this parameter of about a factor of 2-3 for which the SBBN predictions are in agreement with the primordial abundances inferred from current data. This range of η is also consistent with independent, non-BBN estimates [2]. Tests of the standard model of cosmology at 20 minutes (BBN) and nearly 400 thousand years later (WMAP) agree. While this is a great triumph for the standard models of cosmology and of particle physics, the agreement is not perfect and, if the uncertainty estimates are taken seriously, there are some challenges to this standard model. In this talk, to an audience of astronomers, I have emphasized the observational uncertainties in the hope of helping to stimulate further observational (and theoretical) work. Will more and better data resolve these apparent conflicts? Or, will we be pointed to new physics beyond the current standard models? Only time will tell.

Acknowledgments

It is with great pleasure that I express my thanks to the organizers of this very interactive and stimulating meeting, especially Luca Pasquini and Sofia Randich for their tireless efforts to facilitate my participation and to make it so enjoyable. The research described here has been supported at The Ohio State University by the US Department of Energy (DE-FG02-91ER40690).

References

1. G. Steigman: 'Primordial Alchemy: From The Big Bang To The Present Universe'. In: *Cosmochemistry: The Melting Pot of the Elements, XIII Canary Islands Winter School of Astrophysics, Tenerife, Canary Islands, Spain, November 19 – 30, 2001*, ed. by C. Esteban, R.J. García López, A. Herrero, & F. Sánchez (Cambridge University Press, Cambridge 2004) p. 1; G. Steigman: 'Big Bang Nucleosynthesis: Probing The First 20 Minutes'. In: *Measuring and Modeling the Universe", Carnegie Observatories Astrophysics Series, Vol. 2*, ed. by W.L. Freedman (Cambridge: Cambridge University Press 2004) p. 169
2. D.N. Spergel, et al.: ApJS, **148**, 175 (2003)
3. D. Kirkman, D. Tytler, N. Suzuki, J. O'Meara, D. Lubin: ApJS, **149**, 1 (2003)
4. T.M. Bania, R.T. Rood, D.S. Balser: Nature, **415**, 54 (2002)
5. R.T. Rood, G. Steigman, B.M. Tinsley, ApJ, **207**, L57 (1976)
6. J. Lequeux, M. Peimbert, J.F. Rayo, A. Serrano, S. Torres-Peimbert: A&A, **80**, 155 (1979); B.E.J. Pagel, E.A. Simonson, R.J. Terlevich, M.G. Edmunds: MNRAS **255**, 325 (1992); K.A. Olive, E.D. Skillman, G. Steigman: ApJ, **489**, 1006 (1997); Y.I. Izotov, T.X. Thuan: ApJ, **500**, 188 (1998); *ibid*, ApJ, **602**, 200 (2004)
7. K.A. Olive, G. Steigman, T.P. Walker: Phys. Rep. **333**, 389 (2000)
8. K.A. Olive, E.D. Skillman: astro-ph/0405588 (2004)
9. S.M. Viegas, R. Gruenwald, G. Steigman: ApJ, **531**, 813 (2000); R. Gruenwald, G. Steigman, S.M. Viegas: ApJ, **567**, 931 (2002); D. Sauer, K. Jedamzik: A&A, **381**, 361 (2002)
10. I. Iben: Nature **220**, 143 (1968)
11. S. Cassisi, M. Salaris, A.W. Irwin: ApJ, **588**, 862 (2003)
12. C. Chiappini, F. Matteucci, G. Meynet: A&A, **410**, 257 (2003)
13. J.N. Bahcall, M.H. Pinsonneault: Phys. Rev. Lett. **92**, 121301 (2004)
14. S.G. Ryan, J.E. Norris, T.C. Beers: ApJ, **523**, 654 (1999); S.G. Ryan, T.C. Beers, K.A. Olive, B.D. Fields, J.E. Norris: ApJ, **530**, L57 (2000)
15. J.A. Thorburn: ApJ, **421**, 318 (1994)
16. P. Bonifacio, P. Molaro: MNRAS, **285**, 847 (1997); P. Bonifacio, P. Molaro, L. Pasquini: MNRAS, **292**, L1 (1997)
17. J. Melendez, I. Ramirez: ApJL, **615**, L33 (2004)
18. M.H. Pinsonneault, G. Steigman, T.P. Walker, V.K. Narayanan: ApJ, **574**, 398 (2002)
19. G. Steigman, D.N. Schramm, J.E. Gunn: Phys. Lett. B **66**, 202 (1977)
20. M. Kawasaki, K. Kohri, N. Sugiyama: Phys. Rev. Lett. **82**, 4168 (1999); S. Hannestad: Phys. Rev. D **70**, 043506 (2004)
21. L. Randall, R. Sundrum: Phys. Rev. Lett. **83**, 3370 (1998); *ibid*, Phys. Rev. Lett. **83**, 4690 (1998)
22. J.P. Kneller, G. Steigman: New J. Phys. **6**, 117 (2004)
23. V. Barger, J.P. Kneller, H.-S. Lee, D. Marfatia, G. Steigman: Phys. Lett. B **566**, 8 (2003)

The Cosmic Saga of ^{3}He

D. Galli

INAF – Osservatorio Astrofisico di Arcetri, Largo E. Fermi 5, 50125 Firenze, Italy

Abstract. We recall the emergence of the "^{3}He problem", its currently accepted solution, and we summarize the presently available constraints on models of stellar nucleosynthesis and studies of Galactic chemical evolution from observations of the He isotopic ratio in the Galaxy.

1 In the Beginning Was Tralphium

The isotope ^{3}He probably first entered the astrophysical arena in 1949 with the (unpublished) calculations of Fermi & Turkevich concerning the chemical evolution of the first half-hour of the Universe. The names "tralphium" and "tralpha particles" invented by George Gamow for this isotope and its nuclei, have survived only in his humorous description of nucleogenesis: *And God said: "Let there be mass three." And there was mass three. And God saw tritium and tralphium, and they were good".* And so on to transuranium elements, with Fred Hoyle's help to bridge the gap at mass five (Kragh 1996). The rough estimate of by Fermi & Turkevich (^{3}He $\simeq 10^{-2}$ by mass) was later refined by more detailed calculations, like e.g. those of Wagoner, Fowler, & Hoyle (1967) who showed that ^{3}He could be produced at levels comparable to its terrestrial abundance ($\sim 5 \times 10^{-5}$ by mass) during the evolution of a "universal fireball or a supermassive object", or, as we say today, in the big bang . Thus, at least in principle, the abundance of ^{3}He could be used (together with D, ^{4}He and Li) to test theoretical predictions, and, in particular, to constrain the baryon density of the Universe. Having gained the special status of "cosmic baryometer" and caught the attention of cosmologists, the interest in ^{3}He spread rapidly in the astronomical community.

2 Trouble Ahead

At around the same time, Iben (1967) and Truran & Cameron (1971) showed that ordinary stars produce ^{3}He in the ashes of hydrogen burning by p–p cycle on the main sequence. They found that the stellar production of ^{3}He roughly scales as M^{-2}, where M is the mass of the star, indicating that low-mass stars ($M \simeq 1$–$3\,M_\odot$) are the dominant site of ^{3}He production in the Galaxy. Problems followed soon, when Rood, Steigman, & Tinsley (1976) incorporated the stellar production of ^{3}He in simple models of Galactic chemical evolution, and found the predicted present-day abundances to be larger by orders of magnitude than the value measured in samples of gas-rich meteorites, representative of interstellar medium abundances at the time of formation of the Sun. The paper by Rood, Steigman, & Tinsley (1976) marked the first appearance of the "^{3}He problem". However, additional observations of ^{3}He in the Galaxy were needed to confirm the extent of the discrepancy.

3   Observing ^{3}He

Radioastronomers first learned of ^{3}He in 1955 at the fourth I.A.U. Symposium in Jodrell Bank, when the frequency of the hyperfine ^{3}He$^+$ line at 8.666 GHz (3.46 cm) was included by Charles Townes in a list of "radio-frequency lines of interest to astronomy" (Townes 1957). The line was (probably) detected for the first time only twenty years later, by Rood, Wilson & Steigman (1979) in W51, opening the way to the determination of the ^{3}He abundance in the interstellar gas of our Galaxy via direct (although technically challenging) radioastronomical observations. In the last two decades, a considerable collection of ^{3}He$^+$ abundance determinations has been assembled in H\textsc{ii} regions and planetary nebulae. The relevance of these results will be discussed in Sect. 4 and 5 respectively.

For many years, meteorites have provided the only means to determine the abundance of ^{3}He in protosolar material. The values obtained by mass spectroscopy techniques in the so-called "planetary" component of gas-rich meteorites have been critically examined by Geiss (1993) and Galli et al. (1995). The latter recommend the value ^{3}He/^{4}He= $(1.5 \pm 0.1) \times 10^{-4}$. The meteoritic value has been confirmed by *in situ* measurement of the He isotopic ratio in the atmosphere of Jupiter by the Galileo Probe Mass Spectrometer. The isotopic ratio obtained in this way, ^{3}He/^{4}He= $(1.66 \pm 0.04) \times 10^{-4}$ (Mahaffy et al. 1998), is slightly larger than, but consistent with, the ratio measured in meteorites, reflecting possible fractionation in the protosolar gas in favor of the the heavier isotope, or differential depletion in Jupiter's atmosphere.

The He isotopic ratio in the present day local ISM (inside and beyond the heliosphere at 3–5 AU from the Sun) has been determined by two recent space experiments, and the two results agree within the uncertainties. Helium atoms entering the solar system from the surrounding interstellar cloud and ionized deep inside the heliosphere (the so-called "pick-up" ions), analyzed by the Solar Wind Ion Composition Spectrometer on the Ulysses spacecraft, show an isotopic ratio ^{3}He/^{4}He= $(2.5^{+0.7}_{-0.6}) \times 10^{-4}$, with the uncertainty resulting almost entirely from statistical error (Gloecker & Geiss 1998). In the Collisa experiment on the Russian space station Mir, on the other hand, samples of the local *neutral* ISM were collected on thin metal foils exposed to the flux of interstellar particles, and later analyzed in terrestrial laboratories. The He isotopic ratio measured in this way is ^{3}He/^{4}He = $(1.7 \pm 0.8) \times 10^{-4}$ (Salerno et al. 2003).

4   The Age of Reason

The old problems have now largely been overcome, and new ones have appeared. As for the cosmological implications, thanks to the continuing effort of Rood and collaborators over more than two decades to determine ^{3}He abundances in HII regions (see contribution by Bania et al. in these proceedings), the usefulness of this isotope as a cosmic baryometer has now been fully established. The trend (or better, the absence of a trend) of ^{3}He vs. metallicity in a sample of about 40 HII regions reveals the existence of a "^{3}He plateau" at ^{3}He/H= $(1.1\pm0.2)\times10^{-5}$, similar in many ways to the celebrated "Li plateau". The resulting baryon-to-photon ratio $\eta_{10} = 5.4^{+2.2}_{-1.2}$ (Bania, Rood & Balser 2002) is in agreement with other independent determinations of this fundamental cosmological parameter. After fifty years, the program started by Fermi & Turkevich's theoretical prediction of "tralphium" production in the early universe seems to have reached its fulfillment.

As for the discrepancy between observed abundances of ^{3}He and the predictions of models of Galactic chemical evolution, the natural explanation of the problem was found by Charbonnel (1995) and Hogan (1995) in the existence of a non-standard mixing mechanism acting in low-mass stars during the red-giant branch evolution or later, leading to a substantial (or complete) destruction of all their freshly produced ^{3}He. In this way, the "^{3}He problem" was reduced to "just another" isotopic anomaly, similar to those commonly observed in the atmospheres of giant stars for elements like carbon and oxygen, as originally suggested by Rood, Bania & Wilson (1984) almost ten years earlier. The characteristics of this mixing mechanism, and the attempts to identify a physical mechanism responsible for its occurrence (rotation?) have been nicely reviewed by Charbonnel (1998), and will not be repeated here. For an impressive demonstration of the effects of extra-mixing on the carbon isotopic ratio in globular cluster stars, the reader should look at Fig. 2 of Shetrone (2003).

Fig. 1 (adapted from Fig. 4 of Romano et al. 2003) shows the evolution of ^{3}He/H in the solar neighborhood, computed with the model of Tosi (2000) assuming the standard (without extra-mixing) stellar yields of Dearborn, Steigman, & Tosi (1996) and the extra-mixing yields of Boothroyd & Sackmann (1999) for 90% ans 10% of stars with $M < 2.5$ $M_\odot$ (see Galli et al. 1997 and Romano et al. 2003). Symbols and errorbars show the ^{3}He/H value measured in: meteorites (Galli et al. 1995); Jupiter's atmosphere (Mahaffy et al. 1998); the local ionized ISM (Gloecker & Geiss 1998); the local neutral ISM (Salerno et al. 2003); the sample of "simple" HII regions (Balser et al. 2002). The primordial abundance of ^{3}He corresponding to the baryon-to-photon ratio determined by WMAP (Spergel et al. 2003) is indicated by an arrow at $t = 0$. Taken together, the observational data support the hypothesis that negligible changes of the abundance of ^{3}He have occurred in the Galaxy during the past 4.55 Gyr. The failure of the standard ^{3}He yields to account for the measured abundances is not a peculiarity of the particular Galactic model shown in Fig. 1, as the interested reader may see in Fig. 6 of Tosi (1998). It should be noted however that the discrepancy with the observational data is rather model dependent.

It is evident from Fig. 1 that consistency with the observed abundance of ^{3}He in the Galaxy is achieved only if the fraction of low-mass stars ($M < 2.5$ $M_\odot$) undergoing extra-mixing is larger than $\sim$ 90%, assuming the ^{3}He yields of

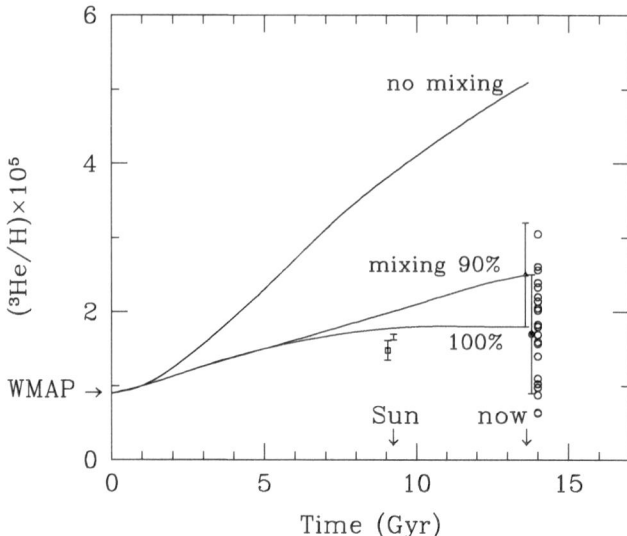

Fig. 1. Evolution of ^{3}He/H in the solar neighborhood, computed without extra-mixing (*upper curve*) and with extra-mixing in 90% or 100% of stars $M < 2.5\ M_\odot$ (*lower curves*). The two arrows indicate the present epoch (assuming a Galactic age of 13.7 Gyr) and the time of formation of the solar system 4.55 Gyr ago. Symbols and errorbars show the ^{3}He/H value measured in: meteorites (*empty squares*); Jupiter's atmosphere (*errorbar*); the local ionized ISM (*filled triangle*); the local neutral ISM (*filled circle*); the sample of "simple" HII regions (*empty circles*). Data points have been slightly displaced for clarity. The He isotopic ratios has been converted into abundances relative to hydrogen assuming a universal ratio He/H= 0.1. See text for references.

Boothroyd & Sackmann (1999). Thus, to solve the ^{3}He problem in terms of extra-mixing in low-mass stars, the vast majority of them (90%–100%) must be affected by this phenomenon (Galli et al. 1997). The same conclusion has been reached independently by Charbonnel & do Nascimento (1998) on the basis of the statistics of carbon isotopic ratios in a sample of red-giant stars with accurate Hipparcos parallaxes.

5   A Final Touch: Planetary Nebulae

The most direct, model independent, way to test the validity of the mixing solution is to measure the ^{3}He abundance in the ejecta of low-mass stars, i.e. in planetary nebulae (PNe). The search for ^{3}He in the ejecta of PNe via the 8.667 GHz spin-flip transition of ^{3}He$^+$, painstakingly carried out by Rood and coworkers at the Green Bank radiotelescope since 1992 (see summary of results in Balser et al. 1997), has produced so far *one solid detection* (NGC 3242, see Rood, Bania, & Wilson 1992; confirmed with the Effelsberg radiotelescope by

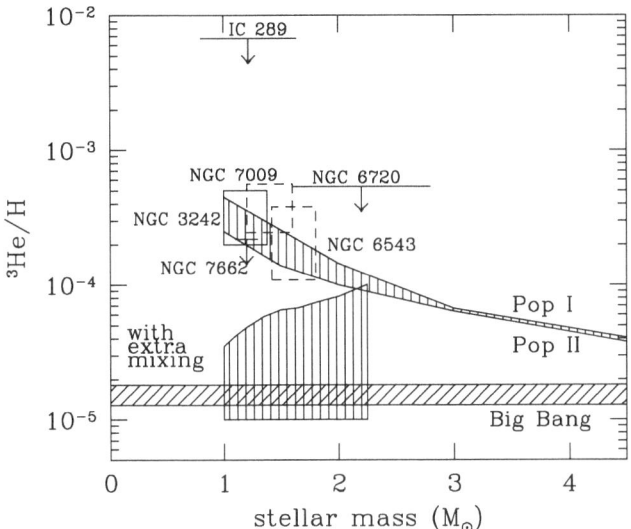

Fig. 2. Abundance of ^{3}He vs. main-sequence masses (determined by Galli et al. 1997) for the six PN of the sample of Balser et al. (1997) and Balser, Rood, & Bania (1999). The curves labeled "Pop I" and "Pop II" show the "standard" abundance of ^{3}He computed by Weiss, Wagenhuber, & Denissenkov (1996) for two metallicities. The curves labeled "with extra-mixing" show the results of stellar nucleosynthesis calculations with deep mixing by Boothroyd & Sackmann (1999) (upper curve) and the equilibrium value ^{3}He/H $= 10^{-5}$ for $M < 2.5$ $M_\odot$ (lower curve).

Balser, Rood, & Bania 1999), *two tentative detections* (IC 289, NGC 6720) and *three upper limits* (NGC 7662, NGC 6543, NGC 7009). One more detection has been recently obtained with the NRAO VLA in the PN J320 (Balser et al., these proceedings). All these objects can be placed in the progenitor mass–^{3}He diagram (see details of the procedure in Galli et al. 1997), and compared with the predictions of stellar models with and without extra-mixing (Fig. 2). Ironically enough, the position of the six PNe definitely supports the standard ^{3}He yields, in particular the (only) solid detection of the sample, NGC 3242. Although the statistical significance of the sample is questionable, and selection biases are certainly present, the only way to reconcile Fig. 1 with Fig. 2 is to conclude that most, if not all, the PNe shown in Fig. 2 belong to the 10% (or less) of low-mass stars which did *not* experience extra-mixing.

6   Conclusions

We have learned many things about Gamow's tralphium since 1949. A personal selection includes: (1) the abundance of ^{3}He has not changed significantly over ~ 14 Gyr of Galactic evolution, which is remarkable; (2) it has not changed not because nothing happened, but because two independent processes of opposite sign and equal magnitude were at work, which is *truly* remarkable; (3) one object does not make a statistically significant sample; (4) many objects do not make it either, if they are selected carefully enough.

Acknowledgements. This work is supported by a grant COFIN 2002027319 003. It is a pleasure to thank the members of the international "^{3}He community" for sharing their results and enthusiasm on our beloved isotope. Special thanks to Monica Tosi for providing an enlightening guidance to the intricacies of ^{3}He, and an exquisite company at many conferences on the light elements.

References

1. D. Balser, T.M. Bania, R.T. Rood, T.L. Wilson: Astrophys. J. **483**, 320 (1997)
2. D. Balser, R.T. Rood, T.M. Bania: Astrophys. J. **522**, 73 (1999)
3. T.M. Bania, R.T. Rood, D.S. Balser: Nature **415**, 54 (2002)
4. A.I. Boothroyd, I.-J. Sackmann: Astrophys. J. **510**, 232 (1999)
5. C. Charbonnel: Astrophys. J. **543**, L41 (1995)
6. C. Charbonnel: Space Sci. Rev. **84**, 199 (1998)
7. C. Charbonnel, J.D. Do Nascimento: Astron. Astrophys. **336**, 915 (1998)
8. D.S.P. Dearborn, G. Steigman, M. Tosi: Astrophys. J. **473**, 560 (1996)
9. D. Galli, F. Palla, Ferrini, F., Penco, U.: Astrophys. J. **443**, 536 (1995)
10. D. Galli, L. Stanghellini, M. Tosi, F. Palla: Astrophys. J. **477**, 218 (1997)
11. J. Geiss: In *Origin and evolution of the elements;* eds. N. Prantzos, E. Vangioni-Flam, M. Cassé (Cambridge University Press, 1993), p. 89
12. G. Gloecker, J. Geiss: Space Sci. Rev. **84**, 275 (1998)
13. C. Hogan: Astrophys. J. **441**, L17 (1995)
14. I. Iben: Astrophys. J. **147**, 650 (1967)
15. H. Kragh: *Cosmology and controversy* (Princeton University Press, 1996)
16. P.R. Mahaffy, T.M. Donahue, S.K. Atreya et al.: Space Sci. Rev. **84**, 251 (1998)
17. D. Romano, M. Tosi, F. Matteucci, C. Chiappini: M. N. R. A. S. **346**, 295 (2003)
18. R.T. Rood, G. Steigman, B.M. Tinsley: Astrophys. J. **207**, L57 (1976)
19. R.T. Rood, T.L. Wilson, G. Steigman: Astrophys. J. **227**, L97 (1979)
20. R.T. Rood, T.M. Bania, T.L. Wilson: Astrophys. J. **280**, 629 (1984)
21. R.T. Rood, T.M. Bania, T.L. Wilson: Nature **355**, 618 (1992)
22. E. Salerno, F. Bühler, P. Bochsler, et al.: Astrophys. J. **585**, 840 (2003)
23. M. D. Shetrone: Astrophys. J. **585**, L45 (2003)
24. D.N. Spergel, L. Verde, H.V. Peiris, et al.: Astrophys. J. Suppl. **148**, 175 (2003)
25. M. Tosi: Space Sci. Rev. **84**, 207 (1998)
26. M. Tosi: In *I.A.U. Symposium n. 198*, eds. L. da Silva, R. de Medeiros, M. Spite, p. 525 (Kluwer, 2000)
27. C. H. Townes: In *I.A.U. Symposium n. 4* ed. by H. C. Van de Hulst, p. 92 (Cambridge University Press, 1957)
28. J. W. Truran, A.G.W. Cameron: Astrophys. J. Suppl. **14**, 179 (1971)
29. R.V. Wagoner, W.A. Fowler, F. Hoyle: Astrophys. J. **148**, 3 (1967)
30. A. Weiss, J. Wagenhuber, P.A. Denissenkov: Astron. Astrophys. **313**, 581 (1996)

First Epoch Observations of ^{3}He
with the Green Bank Telescope

T.M. Bania[1], R.T. Rood[2], D.S. Balser[3], and C.Q. Campos[4]

[1] Institute for Astrophysical Research, Boston University, USA
[2] Astronomy Department, University of Virginia, USA
[3] National Radio Astronomy Observatory, USA
[4] Instituto de Astronomia, Geofisica e Ciencias Atmosfericas,
    Universidade de Sao Paulo, Brazil

Abstract. We have made the first observations of ^{3}He$^+$ using the new 100 m Green Bank Telescope (GBT) toward a sample of Galactic H II regions and planetary nebulae (PNe). The study of the origin and evolution of the elements is one of the cornerstones of modern astrophysics. The ^{3}He abundance is derived from measurements of the spin-flip line of ^{3}He$^+$ with a rest frequency of 8.665 GHz (3.45 cm). Potential ^{3}He sources include PNe and H II regions located throughout the Milky Way. ^{3}He abundances have a broad, multidisciplinary impact; they can serve as a critical probe of both stellar/galactic chemical evolution [1] and cosmology [2]. Our ^{3}He results to date raise specific questions which can only be addressed by telescopes like the GBT. Indeed, the GBT's design seems to be optimized for the extremely low line-to-continuum ratio spectroscopy the ^{3}He experiment requires.

We have been measuring the ^{3}He abundance in Galactic H II regions [3] and PNe [4] for many years. The H II region abundances reveal a "^{3}He Plateau"

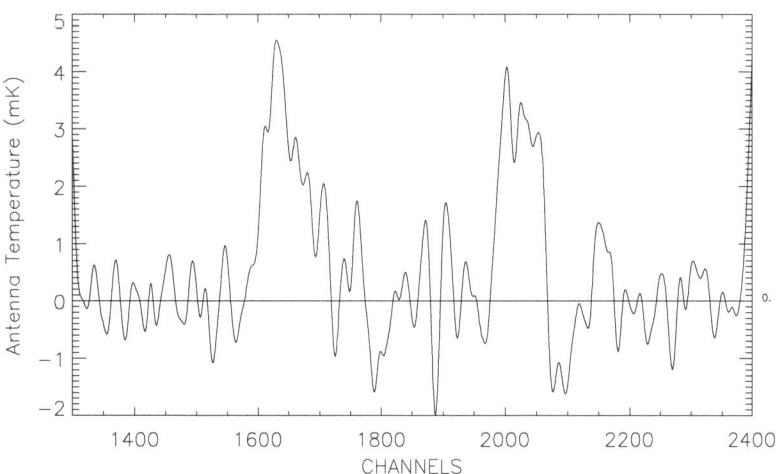

Fig. 1. GBT Spectrum for the S 209 H II Region. This 14.5 hr integration has been smoothed to a 5 km/sec resolution. The H171 η (*left*) and ^{3}He$^+$ (*right*) transitions are clearly detected

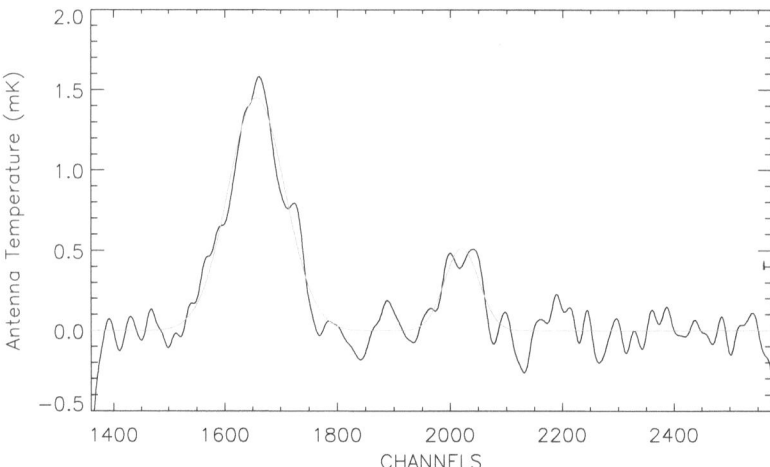

Fig. 2. GBT Composite Spectrum of Planetary Nebulae. This 125.7 receiver hour integration is the spectral average of 4 PNe: NGC 3242 + NGC 6543 + NGC 6826 + NGC 7009. Shown are the Gaussian fits to the H171 η (*left*) and ^{3}He$^+$ (*right*) transitions

which means that very little ^{3}He has been produced by chemical evolution, yet some PNe, NGC 3242 in particular [5], show abundances in accord with the enrichment predicted by standard stellar evolution. This PNe sample is very small, however, and is biased toward sources expected to be very ^{3}He rich. A larger PNe sample is needed. After much sturm und drang the GBT is now fully operational for the ^{3}He experiment. Here we show GBT ^{3}He$^+$ spectra for an H II region (Fig. 1) and a PNe composite source (Fig. 2). We do not yet have sufficient signal-to-noise on any single PN, but the composite is fully consistent with our MPIfR 100 m results: the Fig. 2 composite PNe has a ^{3}He abundance consistent with that predicted by standard stellar evolution. In sum, the current status of the ^{3}He experiment is:

- The GBT is now fully operational for ^{3}He$^+$ observations. The first GBT observing epoch was completed in June 2004.
- The GBT spectral baselines are of unprecedented quality.
- The composite PNe GBT spectrum is consistent with MPIfR survey results.
- We report a VLA ^{3}He$^+$ 4-σ detection for PN J320 [6].
- We will soon make ^{3}He$^+$ observations with the Arecibo Observatory 305 m.

References

1. D. Galli: 'The Cosmic Saga of ^{3}He'. In: this volume
2. G. Steigman: 'BBN in the post-WMAP era'. In: this volume
3. D. Balser, T. Bania, R. Rood, T. Wilson: Ap. J. **510**, 73 (1999)
4. D. Balser, T. Bania, R. Rood, T. Wilson: Ap. J. **483**, 320 (1997)
5. D. Balser, R. Rood, T. Bania: Ap. J. **522**, L73 (1999)
6. D. Balser, et al.: 'Observations of ^{3}He in Planetary Nebulae'. In: this volume

Li, Be, B and Cosmic Rays in the Galaxy

N. Prantzos

Institut d'Astrophysique de Paris, 98bis Bd Arago, Paris, France

Abstract. A short overview is presented of current issues concerning the production and evolution of Li, Be and B in the Milky Way. It is argued that the currently popular idea that Galactic Cosmic rays are accelerated inside metal-rich superbubbles (which leads "naturally" to the production of primary Be and B, as observed) encounters the same problems as the previously popular idea of supernovae accelerating their own ejecta. A major challenge to theories of light element production is presented by the recent (and still preliminary) data suggesting a surprisingly high and ~constant abundance of ^{6}Li in halo stars; attempts to explain such a "plateau" are critically examined.

1 The 1970s: Problems with Late ^{7}Li, ^{11}B

The idea that the light and fragile elements Li, Be and B are produced by the interaction of the energetic nuclei of galactic cosmic rays (CGR) with the nuclei of the interstellar medium (ISM) was introduced 35 years ago (Reeves et al. 1970, Meneguzzi et al. 1971). In those early works it was shown that, taking into account the relevant cross-sections and with plausible assumptions about the GCR properties (injected and propagated spectra, intensities etc.; see Fig. 1, *right column*) one may reproduce reasonably well the abundances of those light elements observed in meteorites *and* in GCR.

Two problems were identified with the GCR production, compared to meteoritic composition: the ^{7}Li/^{6}Li ratio (~2 in GCR but ~12 in meteorites) and the ^{11}B/^{10}B ratio (~2.5 in GCR but ~4 in meteorites). Modern solutions to those problems involve *stellar* production of ~70% of ^{7}Li (in the hot envelopes of AGB stars and/or novae) and of ~40% of ^{11}B (through ν-induced spallation of ^{12}C in SNII). In both cases, however, uncertainties in the yields are such that observations are used to constrain the yields of the candidate sources rather than to confirm the validity of the scenario.

2 The 1980s: the Li Plateau; Primordial, but Low or High?

One of the major cosmological developments of the 1980s was the discovery of the Li plateau in low metallicity halo stars (Spite and Spite 1982, see Fig. 1 *top left* panel). The unique behaviour of that element, i.e. the constancy of the Li/H ratio with metallicity, strongly suggests a primordial origin. The observed value has been extensively used (along with those of D and ^{4}He) to constrain the physics of primordial nucleosynthesis and, in particular, the baryonic density of the Universe (e;g. Steigman, this meeting). In particular, the difference between the observed plateau value (Ryan et al. 1999) and the Li abundance corresponding to the baryonic density derived from WMAP data (Fig. 1) is rather high ($\sim$0.5 dex) and points to a failure of our understanding, of either stellar atmospheres, primordial nucleosynthesis or Li depletion in stars (e.g. Lambert 2004 and references therein).

3   The 1990s: Problems with Early Be and B (Primaries!)

Observations of halo stars in the 90s revealed a linear relationship between Be/H (as well as B/H) and Fe/H. That was unexpected, since Be and B were thought to be produced as *secondaries*, by spallation of the increasingly abundant CNO nuclei. Only the Li isotopes, produced at low metallicities mostly by $\alpha + \alpha$ fusion reactions (Steigman and Walker 1992) are produced as primaries. The only way to produce primary Be and B is by assuming that GCR have always the same CNO content (Duncan et al. 1992). The most convincing argument in that respect is the "energetics argument" put forward by Ramaty et al. (1997): if SN are the main source of GCR energy, there is a limit to the amount of light elements produced per SN, which depends on GCR and ISM composition. If the metal content of both ISM and GCR becomes low, there is simply not enough energy in GCR to keep the Be and B yields constant (as required by observations). Since the ISM metallicity certainly increases with time, the only possibility to have $\sim$constant LiBeB yields is by assuming that GCR have a $\sim$constant metallicity.

A $\sim$constant abundance of C and O in GCR can naturally be understood if SN accelerate their own ejecta. However, the absence of unstable ^{59}Ni (decaying through e^--capture within 10^5 yr) from observed GCR suggests that acceleration occurs $>10^5$ yr after the explosion (Wiedenbeck et al. 1999) when SN ejecta are presumably diluted in the ISM. Higdon et al. (1998) suggested that GCR are accelerated in *superbubbles* (SB), enriched by the ejecta of many SN as to have a large and $\sim$constant metallicity. Since then, this became *by default*, the "standard" scenario for the production of primary Be and B by GCR, invoked in almost every work on that topic.

However, the SB scenario suffers from (at least) two problems. First, core collapse SN are observationally associated to HII regions (van Dyk et al. 1996) and it is well known that the metallicity of HII regions reflects the one of the *ambient ISM* (i.e. it can be very low, as in IZw18) rather than the one of SN. Moreover, Higdon et al. (1998) evaluated the time interval Δt between SN explosions in a SB to a comfortable $\Delta t \sim 3 \; 10^5$ yr, leaving enough time to ^{59}Ni to decay before the next SN explosion and subsequent acceleration. However, SB are constantly powered not only by SN but also by the strong *winds of massive stars* (with integrated energy and acceleration efficiency similar to the SN one, e.g. Parizot et al. 2004), which should continuously accelerate ^{59}Ni, as soon as

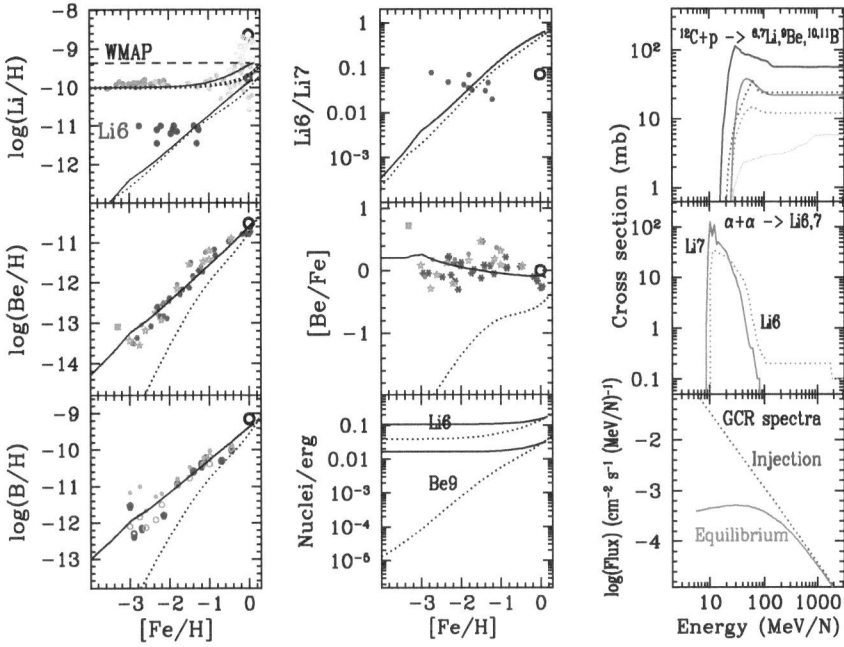

Fig. 1. *Left and middle columns*: Evolution of LiBeB with two different compositions assumed for GCR (1) $X_{GCR}(t) = f_{ENH} X_{ISM}(t)$ always (*dotted curves*) and (2) $X_{GCR} = f_{ENH} X_{\odot}$ always, (*solid curve*), i.e. metallicity dependent and independent compositions, respectively (X representing *mass fraction*); in both cases, the *enhancement factor* f_{ENH} is such as to match the present day observed GCR composition, which is enriched in C (by ~ 8), N (by ~ 2) and O (by ~ 5), i.e. the "metallicity" of the GCR fluid today is about 5 times solar in the solar vicinity. In both cases, the GCR injection spectrum is $Q(E,t) = S\ q(E)\ R_{SN}(t)$, where $R_{SN}(t)$ is the SN rate, $q(E)$ is displayed in *right bottom panel* and the normalisation factor S is such that the energy of all nuclei in GCR is $\sum_i X_{i,GCR} \int q(E)EdE = \epsilon_{CR} E_{KIN}$ (where the kinetic energy $E_{KIN} = 1.5\ 10^{51}$ ergs per SN and $\epsilon_{CR} \sim 0.1$-0.2 is a typical acceleration efficiency of SN). Model (2) produces naturally primary Be and B with reasonable (and metallicity $\sim$independent) values of ϵ_{CR}, while Model (1) requires metallicity dependent values which become unacceptably large at low metallicities (*middle bottom* panel). Despite its success with Be and B, Model (2) cannot produce the Li6 plateau. The evolution of C,N,O,Fe is followed with metallicity dependent yields, but no Li7 from Hot-bottom burning in AGB stars or B11 from ν-nucleosynthesis in massive stars is included. As a result, the abundances of Li7, B11 and Li6/Li7 at [Fe/H]=0 differ from their solar values (by factors 6, 2 and 6, respectively). Production cross-sections from ^{12}C+p (for all light isotopes) and from $\alpha + \alpha$ (for Li6, Li7), as well as GCR spectra at the source (injection) and after propagation (equilibrium) are displayed in the *right column*. The *equilibrium spectrum* should be folded with the corresponding cross-sections and abundances to calculate relevant production rates of LiBeB. Energies are in MeV/nucleon.

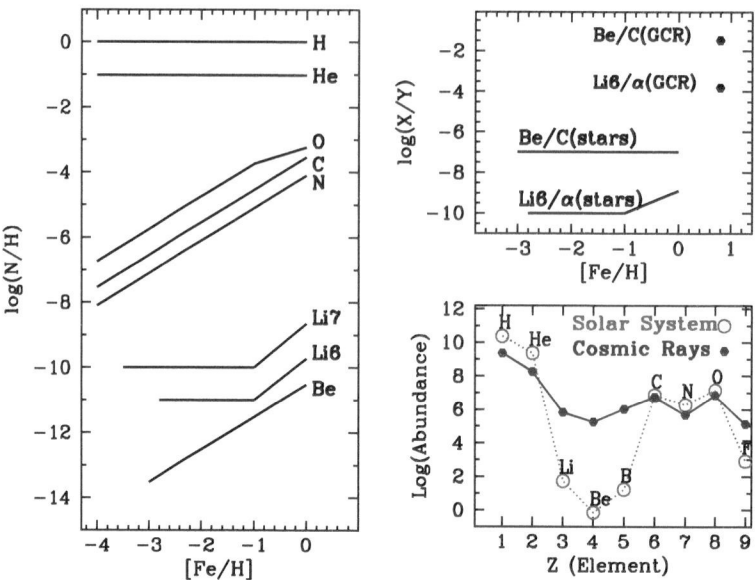

Fig. 2. *Left:* Schematic evolution of the light elements, from H to O, in the local Galaxy, according to observations (the evolution of B, similar to the one of Be, is not shown for clarity). *Right top:* Schematic evolution of abundance ratios Be/C and Li6/He (by number) in the local Galaxy. Those ratios are chosen because Be is produced mainly by C (and N,O, which display similar evolution) while Li6 is produced mainly by $\alpha + \alpha$ (especially in the early Galaxy, while production by CNO becomes important later). In both cases, the constancy of those ratios is reminiscent of an "equilibrium" process (with the production rate balanced by the destruction rate, as e.g. in the CNO cycle operating at equilibrium). In GCR, the light isotopes are certainly at equilibrium with their "father" nuclei, since the observed abundances (*Right bottom*) correspond indeed to such equilibrium values (ratios displayed at *Right top*); moreover, the ratio of $(X/Y)_{GCR}/(X/Y)_{ISM}$ is the same for Be/C and Li6/He (around 10^5-10^6). However, although it is easy to understand equilibrium abundances of light elements in GCR (where production and destruction by the abundant H and He of the ISM medium are rapid), it is difficult to conceive such an equilibrium situation for the nuclei residing in the ISM (where production and destruction by the rarefied H and He gas of the GCR are very slow).

it is ejected from SN explosions. Thus, SB suffer exactly from the same problem that plagued SN as accelerators of metal rich ejecta.

The problem of the acceleration site of GCR (so crucial for the observed linearity of Be and B vs Fe) has not found a satisfactory explanation yet.

Table 1. Problems with LiBeB and GCR

Problem	Suggested solution	Comments
Late Li: ^{7}**Li**/^{6}**Li** Solar value=12 but GCR=2	Late ^{7}Li (but not ^{6}Li) from AGB and novae	Plausible, but Li yields of AGB and rates/yields of novae VERY uncertain
Late B: ^{11}**B**/^{10}**B** Solar value=4 but GCR=2.5	40% of ^{11}B produced by ν-nucleosynthesis in SNII	Plausible, although ν-spectra are uncertain Produces *primary* ^{11}B
Early Be and B Observations: *primaries*, while "Standard" GCR produce *secondaries*	GCR metallicity always the same, originating in SN ejecta or in *Superbubbles* (SB)	Problem with absence of unstable ^{59}Ni in GCR; it becomes stable if directly accelerated in SN or continuously accelerated in SB by *stellar winds*
Early ^{6}**Li/H :** A too high "plateau"	1) *Primordial*, non-standard production in BBN 2) *Pregalactic* production during structure formation 3) *Equilibrium*: ^{6}Li/$\alpha \sim$const. Production=Destruction	Particles and cross-sections unknown CR energetics unknown; hard to explain "plateau" Requires too rapid reactions, incompatible with GCR densities

4   The 2000s: a ^{6}Li Plateau? Primordial or (Pre-)Galactic?

The recent report of a "plateau" for ^{6}Li/H in halo stars (Asplund et al. this meeting) gives a new twist to the LiBeB saga. The detected ^{6}Li/H (and corresponding ^{7}Li/^{6}Li) value at [Fe/H]=-2.8 is much larger than expected if GCR are the only source of the observed Be/H and ^{6}Li/H in that star (see Fig. 1). A few explanations have been proposed for such a high early amount of ^{6}Li:

1) Primordial, non-standard, production during Big Bang Nucleosynthesis (BBN): the decay/annihilation of some massive particle (e.g. neutralino) releases energetic nucleons/photons which produce ^{3}He or ^{3}H by spallation/photodisintegration of ^{4}He, while subsequent fusion reactions between ^{4}He and ^{3}He or ^{3}H cre-

ate ^{6}Li (e.g. Jedamzik 2004). Observations of ^{6}Li/H constrain then the masses/cross-sections/densities of the massive particle.

2) Pre-galactic, by fusion reactions of ^{4}He nuclei, accelerated by the energy released during structure formation (Suzuki and Inoue 2002); in that case, CR energetics are decoupled from SN energetics. In view of the many uncertainties related to the behaviour of the baryonic component during structure formation, the energetics of CR in that scenario are very poorly known/constrained at present[1]. Moreover, in order to explain the observed ^{6}Li "plateau" the effect must end (or drastically decrease) *before* the first stars form and explode releasing Fe, otherwise ^{6}Li/H would continue increasing at metallicities [Fe/H]>-3. But this runs against our current understanding of structure formation, which suggests that merging of sub-structures continues with mildly reduced intensity during a large fraction of the Galaxy's early life (e.g. Helmi et al. 2003).

3) A third possibility is suggested by the observed ∼constancy of ^{6}Li/He and Be/C with [Fe/H] (Fig. 3 *right top*), reminiscent of an *equilibrium* process. Indeed, the abundances of the LiBeB isotopes *in GCR* are much higher than the solar ones (*right bottom*) and are in equilibrium with those of the progenitor He,C,N,O nuclei, i.e. the ratio of daughter/progenitor abundances in GCR is roughly equal to the one of the corresponding production/destruction spallation cross-sections. The advantage of that idea is that it explains at one stroke both the primary Be and B (always at equilibrium with progenitor CNO nuclei) and the ^{6}Li plateau (with ^{6}Li at equilibrium with progenitor ^{4}He). However, equilibrium requires very fast production and destruction reactions (with timescales shorter than the evolutionary timescale of the system); this certainly happens inside GCR (with the fast GCR particles interacting with the numerous ISM nuclei) but the opposite does not hold[2].

References

1. Fields B, Prodanovic T. (2004), astro-ph/0407314
2. Helmi A., White S., Springel V. (2003) MNRAS 339, 834
3. Higdon J., Lingenfelter R., Ramaty R. (1998) ApJ 509, L33
4. Jedamzik K. (2004) PhysRevD 70, 0603524
5. Lambert D. (2004), astro-ph/0410418
6. Duncan D., Lambert D., Lemke M. (1992) ApJ 584, 595
7. Meneguzzi M., Audouze J., Reeves H. (1971) AA 15, 337
8. Parizot E., et al. (2004) AA 424, 747
9. Prantzos N., Cassé M. (1994) ApJS 92, 575
10. Prantzos N., Casse M., Vangioni-Flam E. (1993) ApJ 403, 630

---

[1] Fields and Prodanovic (2004) suggest that the associated production of γ-rays (from decaying pions, produced by energetic $p+p$ reactions) could contribute significantly to the extragalactic γ-ray background; see also Prantzos and Cassé (1994).

[2] The interaction timescale is: $\tau \sim (n~\sigma v)^{-1}$. For GCR ^{6}Li interacting with ISM protons of density $n_p \sim 1$ cm^{-3} one has $\tau \sim 10^7$ yr, but for ISM ^{6}Li interacting with GCR protons of $n_p \sim 10^{-9}$ cm^{-3} (corresponding to the observed energy density of 1 eV/cm^3) τ is a billion times larger ($\sigma \sim 400$ mb being the destruction cross-section).

11. Ramaty R., Kozlovsky B., Lingenfelter R., Reeves H. (1997) ApJ 488, 730
12. Reeves H., Fowler W., Hoyle F. (1970) Nature 226, 727
13. Ryan S., Norris J., Beers T. (1999) ApJ 523, 654
14. Spite F., Spite M. (1982) AA 115, 357
15. Steigman G., Walker T.(1992) ApJ 385, L13
16. Suzuki T., Inoue S. (2002) ApJ 573, 168
17. van Dyk S., Hamuy M., Filippenko A. (1996) AJ 111, 2017
18. Wiedenbeck M. et al. (1999) ApJ 523, L61

Abundance Ratios
and the Formation of the Milky Way

C. Chiappini

Osservatorio Astronomico di Trieste - OAT/INAF - Italy

1   The N/O Abundance Ratio Along the Galactic Disk

Although the data for the Milky Way are not yet conclusive about the existence of a N/O abundance gradient along the Galactic disk, such a gradient is clearly seen in other spiral galaxies [7]. In this work we computed the abundance gradient of N/O for the MW. In our formalist the MW formed out of two-infall episodes, one forming the halo/thick disk on a short timescale and a second one forming the thin disk on a much longer timescale. Moreover, in our framework the thin disk formed inside-out (see details in [3]). Here we show that a negative gradient is predicted for N/O once we adopt stellar yields where rotation is taken into account (see Fig.1 - thick line). We point out that although [5] did not include formally the hot bottom burning (HBB), models computed with their stellar yields are still compatible with the available observational data.

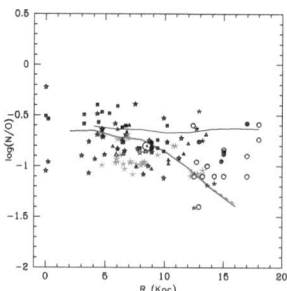

Fig. 1. Abundance gradient of N/O predicted by models adopting stellar yields where rotation is not taken into account (as model 7 of [3] - thin solid line) and the same models computed with MM02 yields ([2] - thick solid line). A model where we increased only the amount of primary N in massive stars for metallicities below $Z=10^{-5}$ overlaps with the thick solid line shown here [1]. This shows that the N/O gradient along the MW disk is affected mainly by the amount of nitrogen production in low and intermediate mass stars and not the primary N in massive stars. For the abundance data see [3] and references therein - asterisks are B stars (see Cunha, this conference).

The existence of abundance gradients of N/O in spiral galaxies imposes limits on the efficiency of HBB since for large efficiencies they would vanish [3,1].

2 Evolution of the Abundance Gradients

Our chemical evolution model for the MW predicts [4] that the abundance gradients for different elements were almost flat in the very early phases of the thin disk formation (Fig. 2). Moreover we predict little evolution of the abundance gradients in the last 5 to 8 Gyrs.

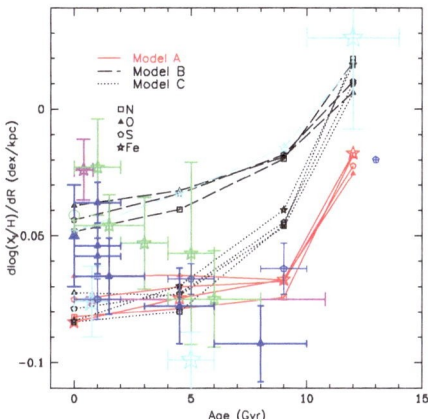

Fig. 2. dlog(X/H)/dR (dex/kpc) vs. Age (Gyrs). The models A, B and C are from [4] where a detailed description can be found. The several observational attempts to measure the evolution of abundance gradients in time are also shown. Different symbols indicate different elements as shown by the label in the figure. The abundances were obtained from (see Maciel, Friel and Cunha this conference): iron in open clusters (open small stars), iron from the Geneva-Copenhagen Survey for F & G stars [6] (large open stars), oxygen abundance gradient at present obtained both from HII regions and B stars (large triangle - the error bar shows the spread in the determinations from different authors), planetary nebula data for oxygen (triangle) and sulfur (pentagons).

Given the scatter in the data (Fig. 2) and the difficulties involved in each one of these determinations, it is clear that the observations available at the present moment are not inconsistent with the two main predictions of our chemical evolution models outlined above.

References

1. Chiappini, C., Matteucci, F. & Ballero, S. 2004, A&A (submitted)
2. Chiappini, C., Matteucci, F. & Meynet, G. 2003, A&A, 410, 257
3. Chiappini, C., Romano, D. & Matteucci, F. 2003, MNRAS, 339, 63
4. Chiappini, C., Matteucci, F. & Romano, D. 2001, ApJ, 554, 1044
5. Meynet, G. & Maeder, A. 2002, A&A, 390, 561
6. Nordstrom, B. et al. 2004, A&A, 418, 989 (Geneva-Copenhagen Survey)
7. Pilyugin, L. S., Vilchez, J. M. & Contini, T. 2004, A&A, 425, 849

Chemical and Photometric Evolution of NGC 6822 in a Cosmological Context

L. Carigi[1], P. Colín[2], and M. Peimbert[1]

[1] Instituto de Astronomía, UNAM, Apdo. Postal 70-264, México 04510 D.F., Mexico
[2] Centro de Radioastronomía y Astrofísica, UNAM, Apdo. Postal 72-3, 58089 Morelia, Michoacán, Mexico

NGC 6822 is an isolated irregular galaxy of the Local Group located at 495 kpc from our Galaxy and at 880 kpc from M31, there are no small galaxies in its neighbourhood, and it does not show tidal effects.

We present chemical evolution models for NGC 6822 computed with five fixed parameters, all constrained by observations, and only a free parameter, related with galactic winds. The fixed parameters are: i) the infall history that has produced NGC 6822 is derived from its rotation curve and a cosmological model; ii) the star formation history of the whole galaxy based on star formation histories for 8 zones inferred from H-R diagrams; iii) the IMF, the stellar yields, and the percentage of Type Ia SNe progenitors are the same than those that reproduce the chemical history of the Solar Vicinity and the Galactic disk.

We adopted as the present chemical composition the O/H, C/O, N/O, and Fe/H abundances from the H II region Hubble V (Peimbert et al. 2005) and from A-type supergiants (Venn et al. 2001). With these abundances and assuming the solar abundances by Asplund et al. (2005), we have determined its metallicity ($Z = 0.6 Z_\odot$). Since Venn et al. find no metallicity gradient, we have assumed that at present the ISM is well mixed. We have obtained the amount of gaseous mass, $M_{gas} = 2 \times 10^8 M_\odot$, based on: the H I measurement inside $r < 5$ kpc, the present helium abundance, and an estimation of $M(H_2)$.

From HST data Wyder (2001, 2003) obtained the SFH of 8 zones in NGC 6822. We have assumed that these zones are representative of the whole galaxy during its evolution. Using the evolutionary stellar population synthesis code by Bruzual & Charlot (2003, GALAXEV) with the metallicity evolution given by Wyder we have concluded that these eight HST fields produce 20% of the luminosity of the galaxy, therefore the SFH of the whole galaxy adopted is 5 times higher than the sum of the eight SFHs. Based on the total SFH from GALAXEV we find that the M_V, $U - B$, and $B - V$ values at present (-16.04, -0.22, and 0.34) are in good agreement with the observed ones (-16.1 ± 0.1, -0.15 ± 0.20, and 0.50 ± 0.06).

Based on the ΛCDM "concordance" cosmological model and adopting the maximum circular velocity as the rotational velocity at the last measured point, we have obtained the DM mass of the galaxy ($3.4 \times 10^{10} M_\odot$). We use the recipe for the halo mass assembly history described by van den Bosch (2002) and the effect that reionization has on the infalling gas (Kravtsov et al., 2004) to derive the rate at which gas is accreted by the galaxy.

The 'naive' model is based on the five fixed ingredients. This model predicts 3.7 times more gaseous mass than observed, therefore this model needs to reduce the accreted mass or lose gaseous mass by a chemically well mixed galactic wind. We computed a second model that considers an infall rate half as large than that suggested by the cosmological model adopted. This model reproduces the gaseous mass observed, but predicts an O/H value higher than observed by 0.18 dex. Therefore the second model requires the ejection to the IGM of 30 % of the SNe products. To be consistent with the cosmological model, another half of the baryonic mass should be either in the halo, or in in the outside of the disk, or has not been accreted yet by the galaxy.

In the figure we show the evolution of a third model, our best model, computed without changing any of our five fixed parameters and assuming outflows of well mixed material to the IGM only during the first 5 Gyr. This proposed outflow cannot be discarded from the SFH derived by Wyder (2001,2003).

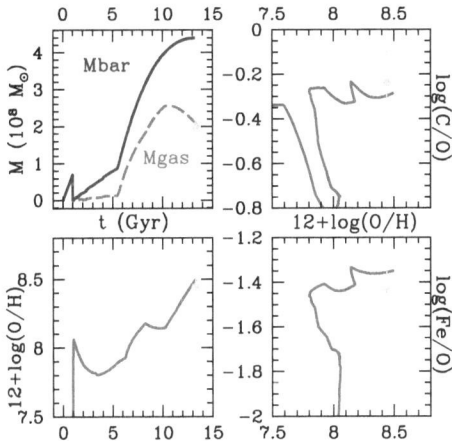

Fig. 1. Evolution of baryonic mass, gaseous mass, O/H, C/O and Fe/O predicted by a model that considers an early galactic wind. Observations from Peimbert et al. (2005) and Venn et al. (2001).

References

1. Asplund, M., Grevesse, N., & Sauval, A. J. 2005, astro-ph/0410214
2. Bruzual, G. & Charlot, S. 2003, MNRAS, 344, 1000
3. Kravtsov, A.V., Gnedin, O.Y., & Klypin, A.A. 2004, ApJ, 609, 482
4. Peimbert, A., Peimbert, M., & Ruiz, M. T. 2005 in preparation
5. van den Bosch, F. 2002, MNRAS, 331, 98
6. Venn, K. A. et al. 2001, ApJ, 547, 765
7. Wyder, T. K. 2001, AJ, 122, 2490
8. Wyder, T. K. 2003, AJ, 125, 3097

The Chemical Evolution of Dwarf Spheroidals

F. Matteucci[1] and G. Lanfranchi[2]

[1] Department of Astronomy, University of Trieste, Italy
[2] IAG-USP, University of Sao Paulo, Brazil

Abstract. We describe some models for the chemical evolution of dwarf spheroidals of the Local Group: Draco, Sagittarius, Sculptor, Sextan, Ursa Minor and Carina. These models assume one long or at maximum two episodes of star formation followed by a strong galactic wind. The efficiency of the star formation is lower than in the Milky Way and the star formation histories take into account the information derived from the observed color-magnitude diagrams of these galaxies. Under these hypotheses, we are able to well reproduce the observed [α/Fe] ratios in these galaxies and to predict their age-metallicity relations and stellar metallicity distributions.

1 The Chemical Model

We call Standard Chemical Model a general model devised for a typical dwarf spheroidal galaxy (for details see Lanfranchi & Matteucci, 2003). The main assumptions are:

- The basic equation of chemical evolution is:

$$\dot{G}_i = -\psi(t)X_i(t) + R_i(t) + (\dot{G}_i)_{inf} - (\dot{G}_i)_{out} \qquad (1)$$

- The SFR is assumed :

$$\psi(t) = \nu \cdot G(t) \qquad (2)$$

  where $\nu(Gyr^{-1})$ is the efficiency of star formation and $G(t) = \frac{M_{gas}}{M_{tot}}$.
- The star formation lasts for several Gyrs (7-8), then a powerful galactic wind is produced owing to the energy injected into the ISM by SN explosions. At this point the dwarf spheroidal (dSph) ejects most of its residual gas.
- The galactic wind rate is simply:

$$(\dot{G}_i)_{out} = w_i \cdot \psi(t) \qquad (3)$$

- The accretion rate is:

$$(\dot{G}_i)_{inf} = Ae^{-t/\tau} \qquad (4)$$

  where $\tau = 0.5$ Gyr for all galaxies.
- We have computed the chemical evolution of Sextan, Sculptor, Sagittarius, Draco, Ursa Minor & Carina just by changing ν and w_i and keeping $\tau = 0.5 Gyr$.
- We have followed the evolution of several chemical elements (H, D, He, C, N, O, Mg, Si, S, Ca, Fe).

- The energetics from type II and Ia SNe is taken into account for the development of a galactic wind (the star formation does not halt during the wind but lowers its intensity due to the gas loss).
- Heavy but diffuse halos of dark matter are considered, an initial mass of $\sim 10^8 M_\odot$ is assumed ($M_D/M_B = 10$).

1.1 Stellar Nucleosynthesis

We adopt the following nucleosynthesis prescriptions for stars of all masses and take into account the stellar lifetimes.

- Low and Intermediate mass stars ($0.8 \leq M/M_\odot \leq 8.0$): we have adopted the yields of van den Hoeck & Groenewegen (1997). They produce ^4He, C, N and s-process ($A > 90$) elements.
- Massive stars ($M \geq 10 M_\odot$): we have adopted Nomoto et al. (1997). They produce α-elements (O, Ne, Mg, Si, S, Ca), some Fe-peak elements, s-process elements ($A < 90$) and r-process elements.
- For the nucleosynthesis in Type Ia SNe, we have adopted Nomoto et al. (1997). These objects produce mainly Fe-peak elements. The Type Ia SN rate is computed according to the original prescriptions of Greggio & Renzini (1983).

2 Results

We have computed the chemical evolution of Sextan, Sculptor, Sagittarius, Draco Ursa Minor and Carina just by changing the efficiency of star formation ν and the wind efficiency w_i and by keeping $\tau = 0.5 Gyr$. The assumed star formation histories, based on the color-magnitude diagrams observed in these galaxies are shown in Table 1. In the first column is indicated the name of the galaxy, in column 2 the number of episodes of star formation, in column 3 the time at which each episode of star formation occurs. In column 4 the duration of each star formation episode, and finally in columns 5 and 6 the ranges of acceptable values for the star formation efficiency ν and the wind parameter w_i, respectively. In particular, ν and w_i have been chosen in order to best reproduce the observed [α/Fe] vs. [Fe/H] relations in each galaxy.

In figures 1 and 2 we show as examples the predicted and observed [α/Fe] ratios in Sextan and Carina. The data in the figures were collected from Smecker-Hane & McWilliam (1999), Bonifacio et al. (2000, 2004) and Shetrone et al. (2001,2003). Similar diagrams for the other dSphs can be found in Lanfranchi & Matteucci (2004).

3 Interpretation and Conclusions

The sharp decrease of the [α/Fe] ratios observed in dSphs is due to the combined effect of a slow star formation regime and a strong galactic wind. The

Table 1. Model parameters for several dSphs. Salpeter IMF. SF histories from Hernandez et al. (2000) and Dolphin (2002).

model	n	t(Gyr)	d(Gyr)	ν (Gyr^{-1})	w_i
Standard Model	1	1	8	0.01-1	5-15
Sextan	1	1	8	0.01-0.3	9-13
Sculptor	1	0	7	0.05-0.5	11-15
Sagittarius	1	0	13	1.0-5.0	9-13
Draco	1	6	4	0.005-0.1	6-10
Ursa Minor	1	0.0	3	0.05-0.5	8-12
Carina	2	6/10	3/3	0.02-0.4	7-11

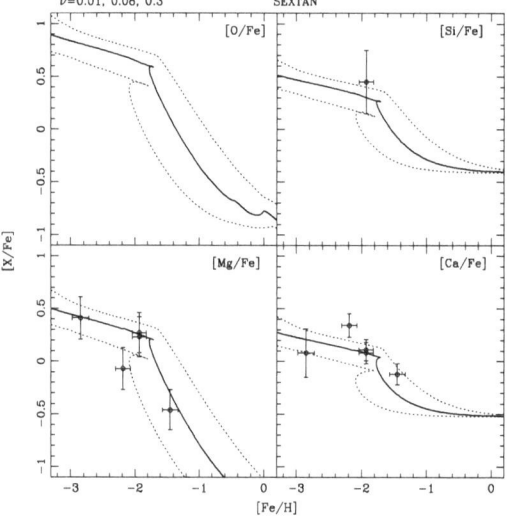

Fig. 1. Predicted and observed [α/Fe] ratios in Sextan. The curves refer to different star formation efficiencies (ν), indicated on the top left side of the Figure. The lowest curves refer to the lowest νs. The central line is the best fit to the data and the other two curves represent the maximum and minimum ν which can be acceptable. The wind parameter is assumed to be $w_i = 9$.

strong wind, in fact, lowers the star formation even more thus depressing the production of α-elements, whereas the Fe production is independent of the star formation rate and Fe continues to be produced and ejected even after star formation has stopped. We predict also the stellar metallicity distribution for dSphs and find that the peak in the distribution occurs always at a lower [Fe/H] than in the solar neighbourhood (roughly 1.5 dex below). The predicted metallicity distribution for Carina seems to be in reasonable agreement with recent data (Koch, this conference). The [α/Fe] ratios in the Milky Way show a different

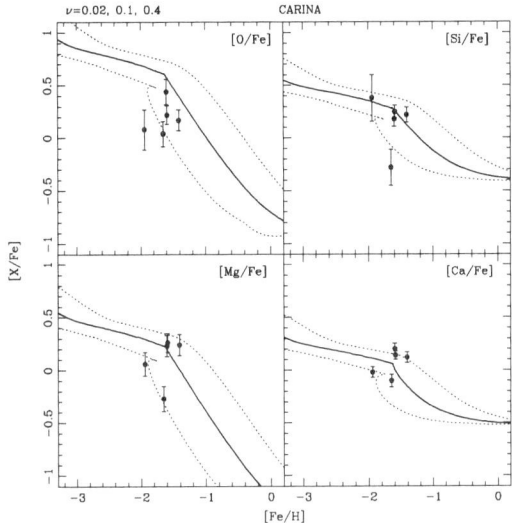

Fig. 2. The same as figure 1. Predicted and observed [α/Fe] ratios in Carina. The meaning of the curves are the same as in Figure 1 and the allowed star formation efficiencies are shown in the top left side of the Figure. The wind parameter is assumed to be $w_i = 7$.

behaviour than the ratios observed in dSphs (see Shetrone and Venn, this conference) : in particular, they are lower in dSphs at the same [Fe/H] than in the solar neighbourhood. This is suggesting that the dSphs and our Galaxy evolved as independent systems. As a consequence of that, we can conclude that it is unlikely that the dwarf spheroidals are the building blocks of the Milky Way and spiral galaxies in general.

References

1. Bonifacio P., Hill V., Molaro P., Pasquini L., Di Marcantonio P., Santin P., 2000, A&A, 359, 663
2. Bonifacio, P., Sbordone, L., Marconi, G., Pasquini, L., Hill, V. 2004, A&A, 414, 503
3. Dolphin A.E., MNRAS, 2002, 332, 91
4. Greggio, L., Renzini, A., 1983, A&A, 118, 217
5. Hernandez X., Gilmore G., Valls-Gabaud D., 2000, MNRAS, 317, 831
6. Lanfranchi, G., Matteucci, F., 2003, MNRAS, 345, 71
7. Lanfranchi, G., Matteucci, F., 2004, MNRAS, 351, 1338
8. Nomoto K., Hashimoto M., Tsujimoto T., Thielemann F.-K., Kishimoto N., Kubo Y., Nakasato N., 1997, Nucl. Phys. A, 616, 79
9. Shetrone M., Côté P., Sargent W.L.W., 2001, ApJ, 548, 592
10. Shetrone M., Venn K.A., Tolstoy E., Primas F., 2003, AJ, 125, 684

11. Smecker-Hane T., McWilliam A., 1999, in Hubeny I. et al., eds. Spectrophotometric Dating of Stars and Galaxies, ASP Conference Proceedings, Vol. 192, p.150
12. Van den Hoeck L.B., Groenwegen M.A.T., 1997, A&AS, 123 , 305

Chemical Evolution of Late-Type Dwarfs

D. Romano[1], M. Tosi[1], and F. Matteucci[2]

[1] INAF – Osservatorio Astronomico di Bologna,
   Via Ranzani 1, I-40127 Bologna, Italy
[2] Dipartimento di Astronomia, Università di Trieste,
   Via Tiepolo 11, I-34131 Trieste, Italy

Abstract. After reviewing some general properties of late-type dwarf galaxies, we present and discuss chemical evolution models for two windy starburst dwarfs: NGC 1569 and NGC 1705. Both have their recent star formation history and stellar initial mass function derived from color-magnitude diagram analysis. In order to reproduce the available observations, we must take into account the presence of dark matter halos and the development of late galactic winds. Two possible solutions are envisaged which might explain the origin of the different N/O ratios observed in NGC 1569 and NGC 1705: either *(i)* small variations in the stellar initial mass spectrum, leading to the production of different amounts of nitrogen and oxygen from intermediate- and high-mass stars; or *(ii)* the complex interplay among the mechanisms determining cooling and heating of the gas and its conversion into stars. Although the second is by far the most effective in determining the global properties of these galaxies, it still remains also the most elusive.

1 Introduction

Dwarf galaxies are common objects in the nearby universe. They present a well pronounced dichotomy between the two main classes of dwarf spheroidals (Matteucci, these proceedings) and dwarf irregulars (DIGs).

DIGs (and blue compact dwarfs, BCDs) have low mass, low metallicity, large gas content and mostly young stellar populations. Because of this, they were long considered the local analogues to primeval galaxies and to the building blocks out of which larger galaxies form in hierarchical formation models. Recent observational evidence points against their primordial nature. Indeed, old underlying stellar populations are routinely found in all studied DIGs and BCDs – even the most metal-poor ones [1] – which appear then to be at least as old as surveyed by the observations, thus suggesting that these galaxies are not truly young, but rather 'rejuvenated' objects. The scarce evolution suffered by DIGs and BCDs makes them interesting for a variety of cosmological and astrophysical problems: *(i)* by offering the possibility to approach the primordial helium abundance with the need for a minimum extrapolation to early conditions, they put stringent constraints on Big Bang nucleosynthesis theories; *(ii)* they offer information complementary to that available for the Milky Way on metal dependencies of stellar yields and stellar winds; *(iii)* they constitute ideal environments for studying the processes of star formation from metal-poor gas clouds and feedback from massive stars, etc.

Given their importance for our understanding of so many unsettled astronomical issues, DIGs and BCDs have been the subject of several extensive chemical evolution studies [2–8]. However, with a few exceptions [7,8], most of this work has dealt with these objects as *a class*, rather than *individually*. It has been demonstrated that differential galactic winds, i.e. winds releasing mostly metals in different proportions, can reproduce the observed spread in the log(N/O) vs. log(O/H)+12 diagram and the high $\Delta Y/\Delta O$ ratios observed in extragalactic H II regions [4,5]. Observational evidence for starburst-driven large-scale outflows has been found for a number of galaxies, *in primis* NGC 1569 [9] and NGC 1705 [10].

Nowadays, the star formation history (SFH), initial mass function (IMF) and detailed chemical properties have been determined for many dwarfs, both in the Local Group and outside it (e.g. Grebel, Shetrone, Tolstoy, these proceedings). This in principle allows us to base theories of late-type galaxy formation and evolution on firmer grounds, by reducing the free parameter space.

2 NGC 1569 and NGC 1705

Despite being differently classified, as a DIG NGC 1569 and as a BCD NGC 1705, these two galaxies share almost the same properties (see Table 1). They have fairly similar total mass, gaseous mass, metallicity and oxygen content (as measured from H II regions). However, they differ in their N/O abundance ratios, log(N/O) = -1.39 ± 0.05 for NGC 1569 and log(N/O) = -1.75 ± 0.06 for NGC 1705[1].

Table 1. Observational properties

Quantity	NGC 1569		NGC 1705	
	Observed	Ref.	Observed	Ref.
Distance (Mpc)	2.2 ± 0.6	[11]	5.1 ± 0.6	[16]
Gas mass ($M_\odot$)	$(1.5 \pm 0.3) \times 10^8$	[11]	1.7×10^8	[17][a]
Total mass ($M_\odot$)	3.3×10^8	[11]	3.4×10^8	[17][a]
Z	0.004	[12]	0.004	[18]
Log(O/H)+12	8.19–8.37	[13–15]	8.21 ± 0.05	[19]
Log(N/O)	-1.39 ± 0.05	[15]	-1.75 ± 0.06	[19]

[a] Gaseous and total masses were modified to reflect the distance used here.

Their relatively recent, detailed SFH has been derived, as well as the stellar IMF, from *HST* observations up to 1–2 Gyr ago for NGC 1569 [20,21] and up to

---

[1] A slightly higher value, -1.63 ± 0.07, is found if the anomalously low N abundance measured in region B4 is ignored [19].

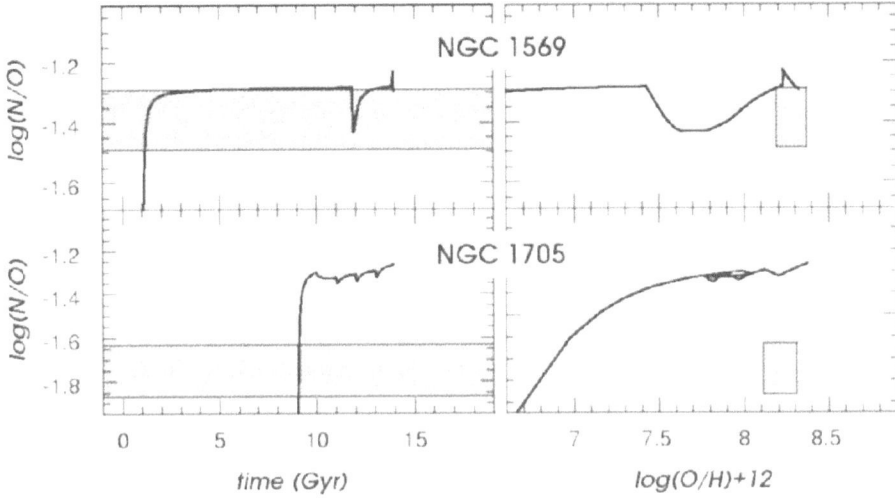

Fig. 1. Theoretical *(thick lines)* log(N/O) vs. time *(left panels)* and log(N/O) vs. log(O/H)+12 *(right panels)* compared to 2-σ observations *(dashed boxes)* for NGC 1569 *(top panels)* and NGC 1705 *(bottom panels)*

5 Gyr ago for NGC 1705 [22] by applying the synthetic color-magnitude diagram method [23]. It has been found that both galaxies have passed through a series of bursts of star formation at very high rates during the past few million years. The latest bursts likely triggered the galactic winds that are currently observed. The star formation preceding this recent, violent activity must have been less dramatic, with a low-level, almost continuous star formation being the preferred picture. Here we adopt the evolutionary scenario derived by [20,21] for NGC 1569 and by [22] for NGC 1705. The IMF too is that (Salpeter's) inferred from the observations, extended to the whole stellar mass range (0.1–100 $M_\odot$).

3   Model Results and Discussion

A full discussion of the model results obtained by exploring the whole parameter space will be presented elsewhere (Romano, Tosi & Matteucci, in preparation). Here we want to address a few 'hot' questions and focus on a couple of points.

In Fig. 1 we display the log(N/O) vs. time and log(N/O) vs. log(O/H)+12 behaviours predicted by two 'successful' models for NGC 1569 and NGC 1705. These models well reproduce several observational constraints – the present-day gaseous and total masses as well as the overall metallicity and oxygen content of the gas – by adopting the same prescriptions on the stellar nucleosynthesis, stellar IMF and galactic outflow onset conditions and efficiency. However, as can be seen from Fig. 1, the present-day N/O ratio is reproduced only for NGC 1569, whilst for NGC 1705 the theoretical N/O ratio during the last ~ 4 Gyr of galaxy's evolution is 0.3–0.4 dex higher than observed at the present time.

Changing the nucleosynthesis prescriptions does not help to solve the discrepancy. In fact, since there is no reason for the final products of stellar evolution to be different for two galaxies displaying the same metallicity, once the stellar nucleosynthesis is changed so as to reproduce the N/O ratio currently observed in NGC 1705, the agreement between model predictions and observations for NGC 1569 gets destroyed. Any IMF variation of high- and intermediate-mass stars is in turn constrained by *HST* photometry and does not prove useful in solving the discrepancy. Alternatively, one can imagine that the relative fractions of nitrogen and oxygen lost from the galaxies through the outflows must be different. In particular, by assuming a higher efficiency of nitrogen ejection with respect to oxygen for NGC 1705 than for NGC 1569, we are able to reproduce both the observed N/O ratios. However, the physical reasons for expecting such a behaviour haven't been assessed yet: is it due to differences in the type of supernovae triggering the wind, or to different mixing conditions of the ISM, or to other reasons? Detailed modelling of more galaxies with reasonably well-known SFH and stellar IMF, combined with energetic and hydrodynamical considerations, is needed in order to shed more light on this issue.

References

1. A. Aloisi, M. Tosi, L. Greggio: AJ **118**, 302 (1999)
2. F. Matteucci, C. Chiosi: A&A **123**, 121 (1983)
3. F. Matteucci, M. Tosi: MNRAS **217**, 391 (1985)
4. L.S. Pilyugin: A&A **277**, 42 (1993)
5. G. Marconi, F. Matteucci, M. Tosi: MNRAS **270**, 35 (1994)
6. F. Bradamante, F. Matteucci, A. D'Ercole: A&A **337**, 338 (1998)
7. L. Carigi, P. Colín, M. Peimbert: ApJ **514**, 787 (1999)
8. S. Recchi, F. Matteucci, A. D'Ercole: MNRAS **322**, 800 (2001)
9. C.L. Martin, H.A. Kobulnicky, T.M. Heckman: ApJ **574**, 663 (2001)
10. T.M. Heckman, K.R. Sembach et al.: ApJ **554**, 1021 (2002)
11. F.P. Israel: A&A **194**, 24 (1988)
12. R.M. González Delgado, C. Leitherer, T.M. Heckman, M. Cerviño: ApJ **483**, 705 (1997)
13. D. Calzetti, A.L. Kinney, T. Storchi-Bergmann: ApJ **429**, 582 (1994)
14. C.L. Martin: ApJ **491**, 561 (1997)
15. H.A. Kobulnicky, E.D. Skillman: ApJ **489**, 636 (1997)
16. M. Tosi, E. Sabbi et al.: AJ **122**, 1271 (2001)
17. G.R. Meurer, K.C. Freeman, M.A. Dopita, C. Cacciari: AJ **103**, 60 (1992)
18. T. Storchi-Bergmann, D. Calzetti, A.L. Kinney: ApJ **429**, 572 (1994)
19. H. Lee, E.D. Skillman: ApJ, in press (2004)
20. L. Greggio, M. Tosi et al.: ApJ **504**, 725 (1998)
21. L. Angeretti, M. Tosi et al.: AJ, submitted (2005)
22. F. Annibali, L. Greggio et al.: AJ **126**, 2752 (2003)
23. M. Tosi, L. Greggio, G. Marconi, P. Focardi: AJ **102**, 951 (1991)

N/O Abundance Ratios: Milky Way and DLAs

C. Chiappini[1], F. Matteucci[2], and G. Meynet[3]

[1] Osservatorio Astronomico di Trieste - OAT/INAF - Italy
[2] Dipartimento di Astronomia, Università di Trieste - Italy
[3] Geneva Observatory, Switzerland

Results and Discussion

It has been recently shown in Chiappini et al. [2] that models of chemical evolution computed with the Meynet & Maeder [4] yields for the whole range of masses, predict a slower increase of N with respect to what is obtained with other sets of stellar yields, thus leading to important implications for the interpretation of the DLAs abundance data. Thanks to the slower increase of N in time, the DLAs abundance patterns can be reproduced by "bursting models" and in this framework, the "low N/O" ($\simeq -2.2$ dex) and "high N/O" ($\simeq -1.6$ dex) groups of DLAs could be explained as systems which show differences in their star formation histories rather than an age difference (see [2] references therein). In fact, we were able to obtain models that show both a low log(N/O) and a [O/Fe]$\sim$0.2-0.3 dex during almost all their evolution, in agreement to what is observed in some DLAs. Alternative interpretations (see [2] for a discussion) of the "low N/O" DLAs suggested in the literature imply a high overabundance of α-elements in disagreement with observations. DLAs could also be identified with outer regions of spiral galaxies but in this case DLAs with low log(N/O) would be quite young systems (younger than $\sim$150 Myr) and no discontinuity in the log(N/O) vs. log(O/H) diagram would be expected [2].

However, as pointed out in [2], it remains to be seen to what extent the Meynet & Maeder [4] yields for N in the intermediate mass star range would increase once the hot bottom burning (HBB) is taken into account. Although Meynet & Maeder did not formally include the third dredge-up and HBB, they predict an important N production in low and intermediate mass stars, at low metallicities. In absence of a real quantitative assessment of the importance of the HBB it is interesting to study the importance of this new process, which produce "non-parametric" yields, independently of HBB.

In the massive range, the yields of Meynet & Maeder predict some primary N production. In [2] we showed that models for the MW computed with this new set of yields lead to a plateau in log(N/O), due to massive stars with initial rotational velocities of 300 kmsec^{-1}, at log(N/O) ~ -4. This value is below the value of -2.2 dex observed in some DLAs and hence we suggested that in these systems both, massive and intermediate mass stars, would be responsible for the N enrichment. This is at variance with recent claims that massive stars were the only ones to enrich systems which show a log(N/O)~ -2.2.

More stringent constraints on N nucleosynthesis come from the study of the N abundances in stars in the MW since they represent a true evolutionary

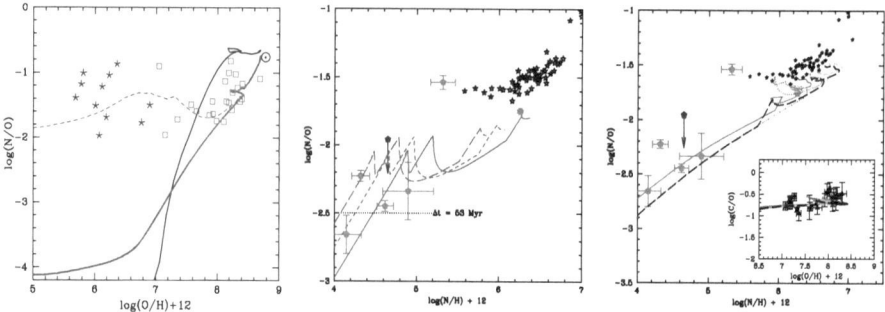

Fig. 1. Left panel: Log(N/O) vs. log(O/H)+12 diagram. The thick and thin solid lines show models of Chiappini et al. [2] for the Milky Way computed with stellar yields given in [4] and [5]+[6] respectively. The dashed-line shows the new model of Chiappini et al. [1] where we assumed "ad hoc" larger yields of N for massive stars of very low metallicity (the data plotted are from Spite - this conference - and Israelian - this conference); Middle panel: Log(N/O) vs. log(N/H)+12 diagram. From [2], "bursting models" of very short duration computed for DLAs (big symbols) with [4] stellar yields - these models are able to explain the "low N/O DLAs group" keeping α/iron ratios low; Right panel: Log(N/O) vs. log(N/H)+12 diagram. From [2], "bursting models" of longer duration computed with [4] stellar yields - these models are able to explain the locus of the blue compact galaxies (stars) and "high N/O DLAs". We also show the log(C/O) vs. log(O/H) plot predicted by the same models, compared with abundance data for blue compact galaxies (details on the abundance data can be find in [3]. In the latter 2 plots we show only DLAs with measured oxygen (see [2] for details).

sequence, where the stars with lower metallicity are the oldest ones. Moreover, the halo very metal poor stars play a fundamental role since at metallicities below [Fe/H] = −3, only Type II supernovae have had time to contribute to the interstellar medium enrichment from which these stars formed, thus offering a way to constrain the N production in massive stars at low metallicities. By the time we published [3] and [2] no conclusive data was available for nitrogen in metal poor halo stars. This situation has now dramatically changed (see also Spite and Israelian contributions to these proceedings and Fig. 1). The new data imply a large production of N in very metal poor massive stars as shown by Chiappini et al. [1] (this new model is plotted as a dashed curve in Fig. 1).

References

1. Chiappini, C., Matteucci, F. & Ballero, S. 2004, A&A (submitted)
2. Chiappini, C., Matteucci, F. & Meynet, G. 2003, A&A, 410, 257
3. Chiappini, C., Romano, D. & Matteucci, F. 2003, MNRAS, 339, 63
4. Meynet, G. & Maeder, A. 2002, A&A, 390, 561
5. van den Hoek, L. B. & Groenewegen, M. A. T. 1997, A&AS, 123, 305
6. Woosley, S. E. & Weaver, T. A. 1995, 101, 181

The Effects of Pop III Stars
on the Abundances of Spheroids

A. Pipino and F. Matteucci

Dipartimento di Astronomia, Universitá di Trieste, via G.B. Tiepolo 11,
34131 Trieste, Italy

Abstract. We have studied the effects of an hypothetical initial generation made only of very massive stars ($M > 100 M_\odot$, pair-creation SNe) on the chemical and photometric evolution of spheroidal systems. We found that the effects of Population III stars on the chemical enrichment is negligible if only one or two generations of such stars occurred, whereas they produce quite different results from the standard models if they continuously formed for a period not shorter than 0.1 Gyr. In this case, the results produced are at variance with the main observational constraints of ellipticals such as the average $[< \alpha/Fe >_*]$ ratio in stars and the color-magnitude diagram.

1 The Model

The photo-chemical code adopted here is described in full detail in Pipino & Matteucci (2004, PM04 hereafter). We limited our analysis to a $10^{11} M_\odot$ (baryonic mass) galaxy and considered the following cases. **a:** It is based on PM04 Model II (called here a1c). Simultaneous presence of both pair-creation SNe and stars of all masses until a threshold gas metallicity $Z_{thr} \sim 10^{-5} Z_\odot$ (Umeda & Nomoto, 2002, UN). **b:** As *a*, but without initial infall. **c:** Strongly bimodal star formation history: in the early stages, only very massive stars ($M > 100 M_\odot$). Different sets of pair creation SNe yields were taken into account. In particular we used Heger & Woosley (2002, HW02; models a2c, b2, c1, c2 and c3), UN (models a3c, b3, c5 and c6) and Ober et al. (1983, OFE83; models a4, b4, c7 and c8).

2 Results

In models a1c-a2c the predicted stellar distributions are almost indistinguishable (see Figure 1), except for the absence of stars with [Fe/H]< -3.0 in the case with Pop III stars. The reason for this resides in the fact that the Pop III phase is very short and at the same time the star formation rate is small at early stages. In the closed-box cases (models b1 and b2) the difference is more noticeable since at the beginning the star formation is quite high. In both models, in fact, no stars with metallicity lower than -3.0 and -2.0, respectively, are predicted. The distribution of stars with metallicity, in turn, influences the calculation of $[< Mg/H >_*]$ and $[< Fe/H >_*]$. For the a1c and a2c models there is very little difference in these average values, whereas for b1 and b2 model the $[< Mg/Fe >_*]$ varies from

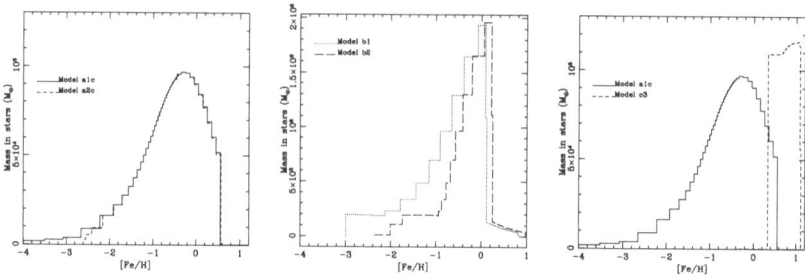

Fig. 1. Predicted distribution of stars as a function of metallicity for several models.

+0.507 to +0.32. The cases b2, b3 and b4, indicate that the yields of HW02 and UN are similar whereas the old yields by OFE83 produce quite different results, namely a predicted $[< Mg/Fe >_*] \sim 4.66$ as opposed to ~ 0.3 of the other two cases. Clearly this case should be discarded because it is at odd with observations ($[< Mg/Fe >_*] \sim$0.2-0.3, e.g. Thomas et al. 2002).

If, instead, the duration of the massive stars phase is as long as 0.1 Gyr (models c3, c4, c6 and c8) then we obtain a too high metallicity for the next stellar generations, with the consequence of obtaining too high metallicity indices and too red integrated colors. Therefore these models should be ruled out.

We refer the reader to Matteucci & Pipino (2005, in preparation) for a more detailed discussion on the topic.

3   Conclusions

Only one or two first stellar generations containing normal stars plus very massive stars included in the best model of PM04, produce negligible effects on the subsequent photo-chemical evolution, when either the yields of HW02 or those of UN are adopted. Therefore, these models are acceptable and we cannot assess Pop III existence nor disproved it.

Pop III stars in classic monolithic models for ellipticals (closed-box) produce larger, although still acceptable, $[< Mg/Fe >_*]$ ratios. However, the predicted integrated colors are too red. As a consequence these models should be rejected.

Therefore, from the chemical and photometric point of view there is no need to invoke the existence of Population III stars in spheroids.

References

1. Heger, A., Woosley, S.E.: 2002, ApJ **567**, 532 (HW02)
2. Ober, W.W., Fricke, K.J., El Eid, M.F.: 1983, A&A **119**, 61 (OFE83)
3. Pipino, A., Matteucci, F.: 2004, MNRAS **347**, 968 (PM04)
4. Schneider, R., Ferrara, A., Natarayan, P., Omukai, K.: 2002, ApJ **571**, 30
5. Thomas, D., Maraston, C., & Bender, R.: 2002, Ap&SS **281**, 371
6. Umeda, H., Nomoto, K.: 2002, ApJ **565**, 385 (UN)

Radio Recombination Line (RRL) Observations: Primordial Helium Determination and First Results at 408 MHz with the Croce del Nord Radiotelescope

A.P. Tsivilev[1], S. Cortiglioni[2], S. Poppi[3], and S. Montebugnoli[3]

[1] P.N. Lebedev Physical Institute of RAS, Moscow, Russia
[2] Istituto IASF, CNR, via Gobetti 101, 40129 Bologna, Italy
[3] Istituto IRA, CNR, via Gobetti 101, 40129 Bologna, Italy

Abstract. New results of the Primordial Helium abundance (Y_p) measurement by radio recombination line (RRL) observations from five galactic HII regions are presented. The RRL observations were carried out with two telescopes: RT32 (22.4 and 8.3 GHz, Medicina, Italy) and RT22 (36.5 and 22.4 GHz, Pushchino, Russia). The results of the first run of the low frequency RRL observations (408 MHz) with the Croce del Nord radiotelescope (Medicina Observatory, Italy) are also presented.

Primordial Helium Abundance

Table 1. Measured Y_p values and the correction parameters

Source	y^+	$\Delta y(IS)$	T_{eff}	y	metallicity	Y_p
	%	%	K	%	%	%
OrionA[a]			~ 37000	10.2(0.8)	1.12(.22)	26.1(2.0)
W3A[a]	9.9 (0.5)	-0.6	45000	9.3(0.5)	distant	26.1(1.6)
M17[a]	11.1(1.1)	-0.7	45000	10.4(1.1)	1.83(.18)	24.7(2.7)
NGC7538	7.7(1.0)[b]	+1.95	~ 37000[b]	9.65(1.8)	distant	26.8(4.0)[b]
W48	9.6(1.3)	0.0		9.6(1.3)	1.83(.19)[c]	23.3(3.3)

a) - data of the H,He RRL mapping was also used; b) - it will be further checked; distant - source at galactic distance outer than solar position; c) - Z (metallicity) value is from Galaxy Z model [5]; for other Z see [6]

Where $\Delta y(IS)$ is the correction for the ionization structure [6] by model calculations, depending on a star effective temperature (T_{eff}) and dust; $y^+=$ N(He$^+$)/N(H$^+$), y = N(He)/N(H). Correction for a stellar nucleosynthesis He production was either using $Y \sim Z$ linear dependence with the slope value of [3] or for distant source $\Delta y = -(0.5\pm0.5)\%$ being accepted as half of [2] calculation.

The obtained average value: $Yp = (25.74 \pm 1.0)\%$ *) (by mass) seems to be higher than the optical data, allowing the presence of 1-2 unknown light particles, and to be in agreement with the results from CMB anisotropy experiments [1].

RRL at 408 MHz

Table 2. Line parameters and electron temperature

Source	Int. time	Line	Amplitude	Line width	V_{lsr}	Te(LTE)
	h		$T_l/T_c * 10^{-3}$	km/s	km/s	K
W51	13.0	H252α	1.9($\pm$0.18)	36.0($\pm$4.2)	52.0($\pm$2.0)	8400($\pm$940)
		C252α	0.96($\pm$0.39)	7.9($\pm$3.8)	57.4($\pm$2.0)	
Rosette	17.1	H252α	2.4($\pm$0.86)	32.4($\pm$13.4)	18.9($\pm$5.5)	7800($\pm$3700)
Nebula		C252α	3.8($\pm$1.1)	19.4($\pm$6.6)	15.0($\pm$3.0)	
W3A	27.9	H252α	3.5($\pm$0.6)	11.6($\pm$2.3)	-39.2($\pm$1.0)	

The obtained T_e for W51 and Rosette Nebulae are similar to the T_e obtained at higher frequencies, probably evidencing the absence of strong T_e gradient inside the sources. The Hydrogen line width for W3A is narrower than usual, likely it's from partly ionized component, from which the narrow lines were observed at higher radio frequencies for this source.

Moreover, the extragalactic object 3c123 was observed to look for the H,C236α in absorption, and the upper level (3σ) for T_l/T_c absorption of 1.5% is achieved. So far, that is rougher than Dr. Matveenko has observed (private communication, 27.04.04, at conference in Pushchino, Russia) the H78a absorption line toward 3c345 quasar (RT100, Bonn) with, probably, amplitude of 0.5% (T_l/T_c).

Note: *) - Y_p may be further recalculated according to the new slope value [4]

References

1. A. Coc et al.: ApJ , **600**, 544 (2004).
2. F.R.S. Hoyle, R.J. Teyler: Nature, **203**, 1108 (1964).
3. Yu.I. Izotov and T.X. Thuan, ApJ, **500**, 188 (1998).
4. Yu.I. Izotov and T.X. Thuan, ApJ, **602** 200 (2004).
5. P.A. Shaver et al., MNRAS, **204**, 53 (1983).
6. A.P. Tsivilev, S. Cortiglioni, R.L. Sorochenko et al., Gravitation & Cosmology, Supplement, **8**, 122 (2002).

Multipopulation Models for Galactic Star Formation and Chemical Evolution. Effects of Stochastic Accretion and Collisional Stripping (Especially for Deuterium)

G. Valle[1], S.N. Shore[1], and D. Galli[2]

[1] Pisa University, Physics Department, Italy
[2] Arcetri Astronomical Observatory, Florence, Italy

This poster illustrates the results of ongoing simulations – within our previously developed modeling framework – allowing for stochastic accretion within closed and open systems. All models include three "zones" (halo, thick disk, and thin disk) and are initialized with only the halo, the abundances for which are assumed to be primordial. The time history of star formation is computed directly from the model equations rather than being assumed *a priori* without instantaneous recycling [1]. The yields and deterministic rates are the same as used in our previous work [2]. Several types of sample simulations have been performed (details in preparation): (a) closed system with bursty mass loss from the halo and the thick disk with independent Gaussian distributions and means equal to our previously adopted *standard* model; (b) assuming the standard deterministic *internal* rates we examine the effects of random environmental (extragalactic) accretion, e.g. by infall of high velocity clouds directly to the thin disk; (c) schematic collision event in which all *diffuse* gas is continuously removed from the system beginning at some time t_{start} and lasting for an interval Δt after which stripping is turned off and the system reconstructs through only internal processes [3]. For the open models, the mass of the galaxy has been explicitly

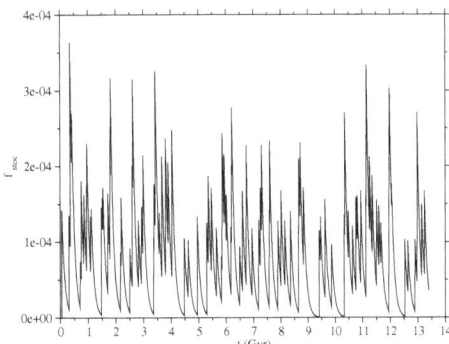

Fig. 1. Single realization illustrating the time history for the events for the model shown in Fig. 2. The added mass always goes directly to the *thin* disk.

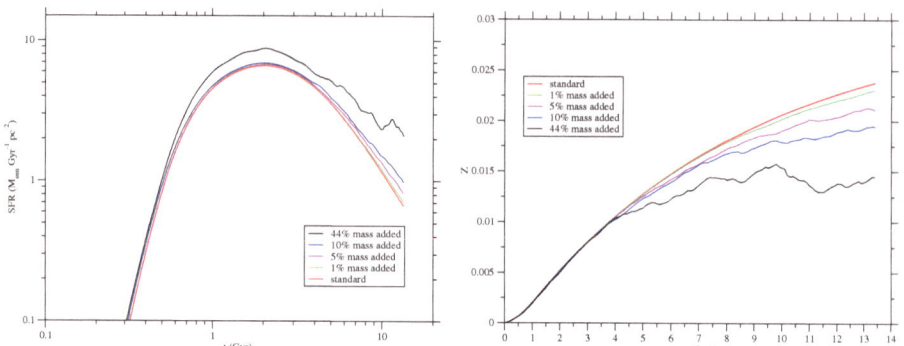

Fig. 2. Stochastic accretion models for an open system. The infalling gas is assumed to be extragalactic material with standard Big Bang nucleosynthetic abundances ($X_0 = 0.758, Y_0 = 0.242$, ^{2}D$=6.5\times10^{-5}$, SBBN) and zero metals. (a) Star formation rate vs. time for the thin disk. From the top to the bottom the curves refer to: 44%, 10%, 5%, 1% and no mass added. (b) Metallicity vs. time for the thin disk. From the top to the bottom the curves refer to: standard case (no mass added), 1%, 5%, 10%, 44% of mass added. The metallicity evolution curve illustrates the relatively weak dilution effects that are offset by continuing star formation. Details for the Deuterium abundances are shown in Fig. 3

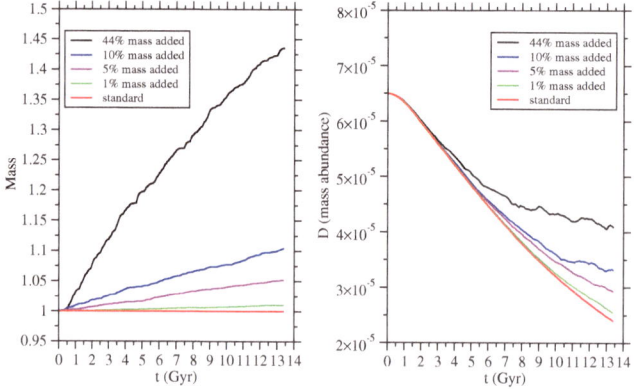

Fig. 3. ^{2}D evolution for the open system described in Fig. 2. The temporal development of total galactic mass (a) and ^{2}D thin disk abundance (by mass) (b) for a set of open stochastic accretion models with primordial gas accreted. From the top to the bottom the curves refer to: 44%, 10%, 5%, 1% and no mass added. As shown, the early time development of deuterium is dominated by astration, while the later history (post-star forming peak) is controlled by the infall rate and feedback to the maintenance of the star formation.

tracked with time (see Fig. 3). We present here the case (b) with stochastic accretion of primordial intergalactic gas directly to thin disk.

Collisions are, in effect, environmental stripping events seen in "fast forward". For example (Fig. 4), we have calculated simulated collisions with the

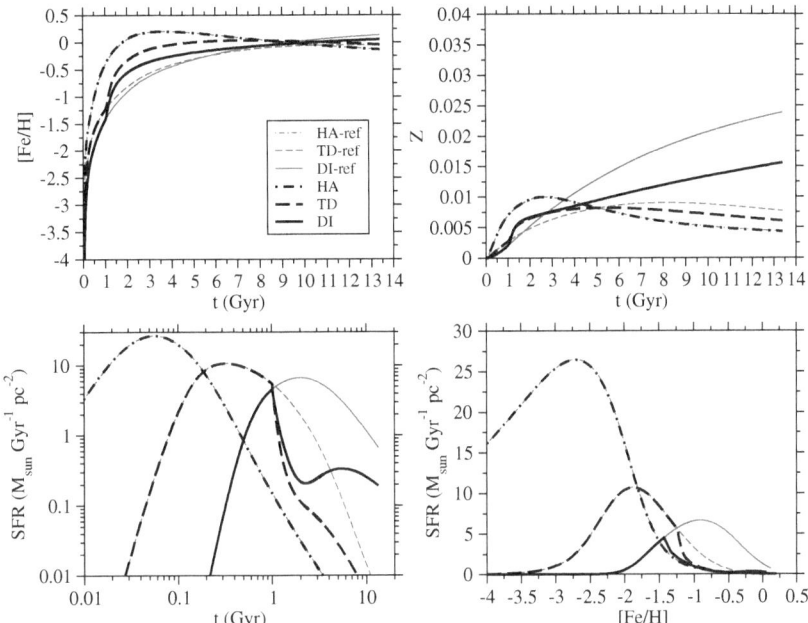

Fig. 4. Sample history of a collisional multizone multi-phase model with stripping at 1 Gyr lasting for 0.13 Gyr with subsequent refilling.

mass loss profile derived for cluster stripping by Quilis et al. [4] but here adapted to the case of impulsive mass loss. In our models, these two cases produce nearly indistinguishable development with time although they yield distinct final abundances. The timing of the first removal of gas, relative to the peak in the star formation rate, determines the subsequent metallicity evolution. This also affects the luminosity and population evolution of the galaxy after the gas is removed.

In a closed box, the star formation rate can only decrease with time. Thus the timing of a collision is irrelevant for the metallicity history of such a galaxy; the final abundance of metals will always increase relative to the moment of stripping. In contrast, in an open system, the global star formation rate can recover due to infall of externally or internally supplied diffuse gas and clouds. The metallicity history then becomes sensitive to the timing and frequency of such events. The final star formation rate may resemble an anemic systems.

A more complete analysis of the variety of models has been submitted to A&A (Valle, Shore, & Galli 2004).

References

1. Ferrini, F., Matteucci, F., Pardi, M. C., & Penco, U. 1992, ApJ, 387, 138
2. Valle, G., Ferrini, F., Galli, D., & Shore, S. N. 2002, ApJ, 566, 252
3. Comins, N.F. & Shore, S.N. 1990, A&A 237, 345
4. Quilis, V., Moore, B., & Bower, R. 2000, Science, 288, 1617

Chemical Abundances in the Local Group: Where Are We Going Next?

G. Gilmore

Institute of Astronomy, Madingley Road, Cambridge, UK

Abstract. Spectacular advances in the volume, the precision, and the accuracy of chemical abundance data for stars in many environments are now being delivered. Corresponding kinematic data are starting to appear, while the essential distances and calibrations await GAIA, still a decade away. Quantitative galaxy models with high spatial resolution are being developed. How do we use these data and models? Applications to calibrate the input functions to chemical evolution models are underway, such as supernova and AGB yields. The more basic question, what are the dominant processes in Galaxy formation and evolution, remains as a challenge: our goal is to identify the failings of the models, and make progress. Many specific areas of potential progress are now clear, ranging from the true age and abundance distribution in the Galactic Bulge, through the origin of the old thick disk, to the accretion history of the outer halo. Since the technology is new, most effort is still being devoted to the observationally easiest questions: all questions require massive surveys to address them. Many such surveys are underway: progress will be rapid, if well-focussed.

1 The Context: Abundances and Galaxy Formation

It is just a half-century ago that the concept of real abundance variations became well-established. It is less that 50 years since the famous B^{2}FH paper was published. Quantitative CCD-based spectroscopy is only some 20 years old, while quantitative multi-object spectroscopy has really begun only in the last decade. These rapid observational advances were enabled by impressive advances in instrumentation, combined with increasing software power and complexity. In parallel, significant advances in stellar atmospheric modelling, and the requisite atomic and molecular data, have allowed analyses of superb precision for large numbers of stars.

Apart from the rather self-referential application to deducing supernova and AGB yields and the convective physics of late stages of stellar evolution, what is all this for? Are abundance studies helping us to understand galaxies? The answer here is very clear: abundances provide the only useful clock which can be analysed to determine the relative importances of *in situ* star formation and assembly of (partially) stellar systems during galaxy formation. Though that knowledge is not yet available to us. The evolutionary process is illustrated in figure 1, the Hubble-Toomre diagram, which reminds us that major mergers happen, and may be the only evolutionary path to some parts of the Hubble tuning-fork. But are significant late mergers as important for the most common type of galaxies, the spirals, as current models would have us believe?

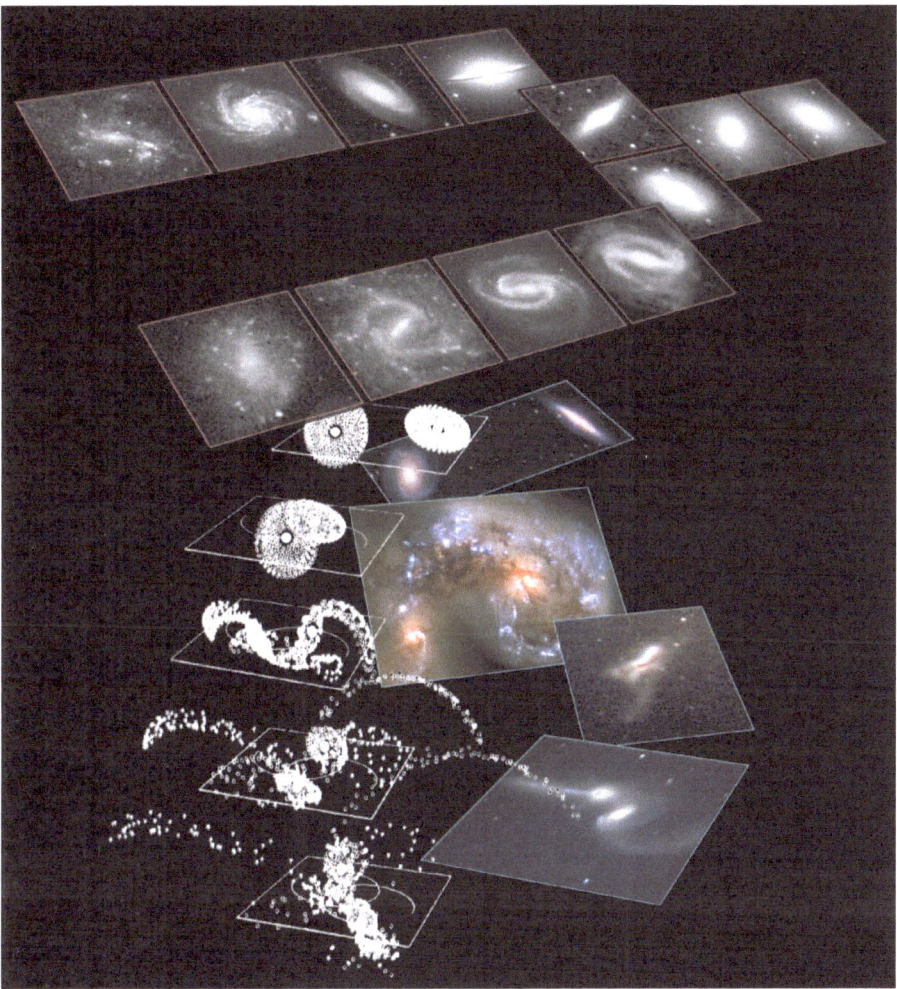

Fig. 1. The Hubble-Toomre diagram, relating the present morphologies of (large) galaxies and their (major) merger histories. Quantification of at least one merger and assembly history, for the Milky Way, is required to test the generality of these evolutionary paths.

The modern model of galaxy formation and evolution is consequent on gravitational dominance by a unknown mass, CDM, starting from an extremely simple scale-free initial condition. It is quite remarkable how successful this model is at reproducing large scale structure in the Universe, given how little physics is included. Fortunately, it is less successful on galaxy scales, so we may hope to learn something new: it is only by detailed analysis of the places where models fail that science progresses.

It is also salutary to note figure 2, which reminds us that agreement and correctness are not always linked. [This figure is from the on-line dBase of particle properties http://pdg.lbl.gov.] Systematic errors always exist, and may be much larger in amplitude than expected. In general, deducing from uncertain data that a model is acceptable is not useful scientific progress. One learns from the failure of models, not from their successes.

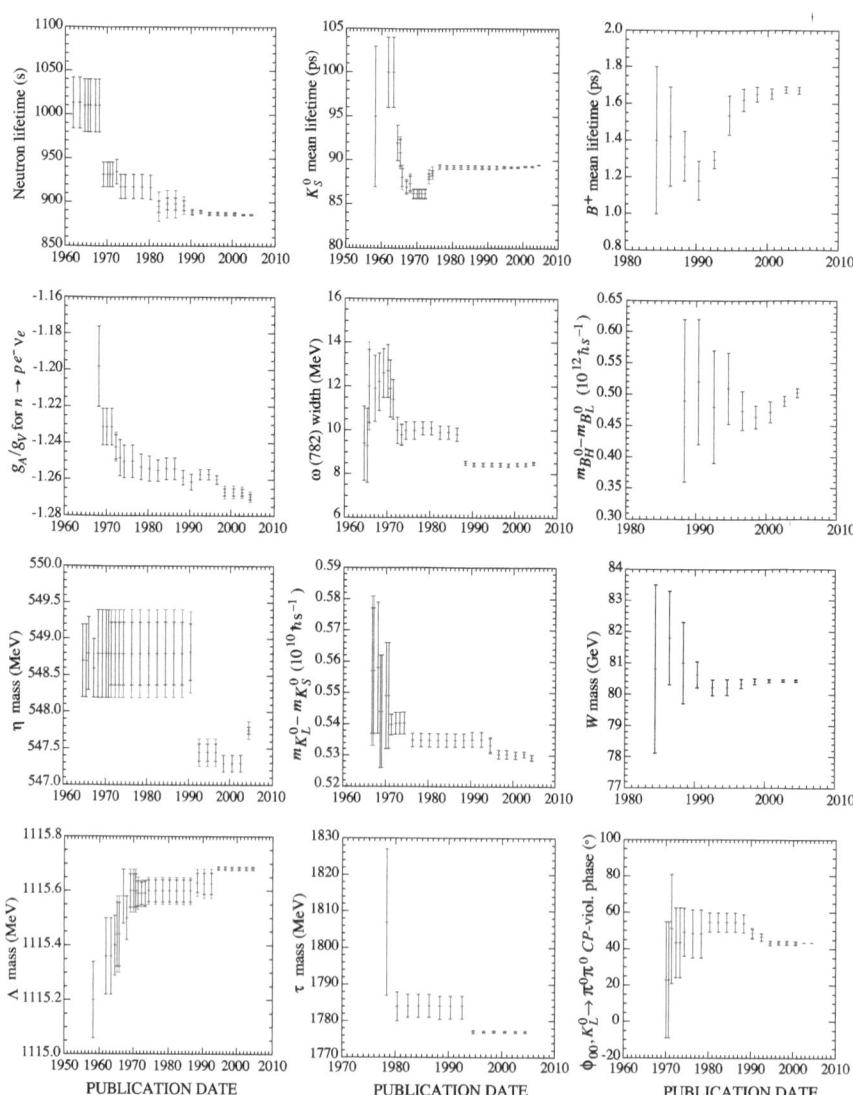

Fig. 2. An historical perspective of values of a few particle properties tabulated in this *Review* as a function of date of publication of the *Review*. A full error bar indicates the quoted error; a thick-lined portion indicates the same but without the "scale factor."

2   The Galaxy: Still Growing, But How?

Stars are known in the Galaxy with abundances ranging over some 5dex in [Fe/H] and 10+Gyr range in age, with at best a very weak correlation between age and metallicity. The overwhelming majority of stars are in a narrow abundance range within a factor of two of solar, and most star-forming gas reached that abundance very long ago. Most stars at all abundances have a relatively narrow range of element ratios, allowing deduction of local enrichment histories, the evolution of the stellar IMF, and so on. Several studies presented here illustrate that parametric evolutionary models can be derived to reproduce observed distribution functions in specific cases. One can imagine a convergence to some agreed model outcome by successive approximation, as data and model improve, much as is happening for supernova yield modelling. But are these approaches appropriate for Galaxy models? Or should we be more concerned with generic properties than with specifics? A viable galaxy model must be general, while at the same time explain the specific: there are some fundamental aspects of observation which tend to be given too little weight, but which are not simply consistent with common models.

Let us consider just one example, the Galactic Bulge. It is "well known" that "bulges" are old and metal rich. Yet, as figure 3 recalls, the galactic central regions host a large reservoir of molecular gas, fueling continuing star formation. Current high-mass star (and cluster) formation is evident in the central regions. The current rate of central star formation is indeed sufficient to build a bulge over a Hubble time. So what is happening here? Is the bulge old? This is one situation where a comprehensive survey is needed, to quantify the actual age and abundance distribution functions, and replace assumption with evidence.

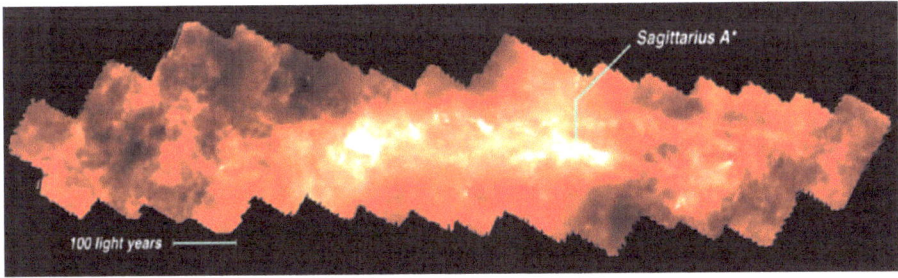

Fig. 3. Scuba map of the inner Galaxy: the central regions of 'old spheroids' host the majority reservoir of molecular gas for current and future star formation. This figure is from Pierce-Price et al. 2000.

It is currently very fashionable to identify new fragments of accretion events in the outer Galactic halo, each of which becomes evidence to support some concept of accretion onto the galaxy as an important process. Yet there has been strong evidence for many years, now made irrefutable by the abundance studies reported here, that accretion of small galaxies over the last substantial fraction of a Hubble time has not contributed significantly to the Galactic halo field star population. Stellar accretion certainly happens, as the Sagittarius dwarf illustrates; but gas accretion also happens, as figure 4 recalls. The scientifically interesting question is not to ask if stellar mergers happen – the Hubble Heritage WWW site is full of beautiful pictures of galaxy mergers: they happen (figure 1). It is also self-evident, from the galaxy metallicity-luminosity relation, that one can never make a large galaxy by merging small stellar systems: when the average star forms it 'knows' the potential well in which it will finally live. The science will come from quantification of the relative importances of gas and stellar accretion, and the relationship, if any of these two to dark matter substructures, where we already know gas accumulation prior to star formation must be the most significant process. This is where large abundance surveys, with accurate element ratio determinations, will advance the field.

3   The Future: Surveys Are Probing Galaxy Formation

The future is a survey. Surveys, allowed by the new multi-object instrumentation (thank you, instrument builders), are rapidly expanding our Galactic horizons. Large-area precision optical and near-IR photometric surveys are underway or planned (2MASS, DENIS, SDSS, VST, EIS, VISTA, MEGACAM, WIRCAM, ...) allowing efficient target pre-selection. Some large-area low-dispersion spectroscopic surveys exist (Hamburg-ESO,...), many more targeted multi-slit or multi-fibre projects are underway, some reported here. In the near future the ex-

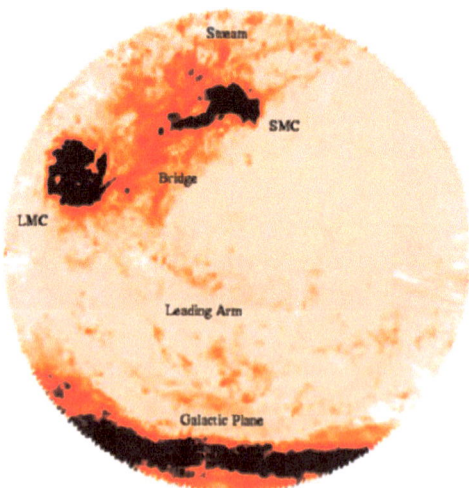

Fig. 4. The Magellanic stream, reminding us that continuing inflow of partially enriched gas is a significant factor in Galactic evolution.

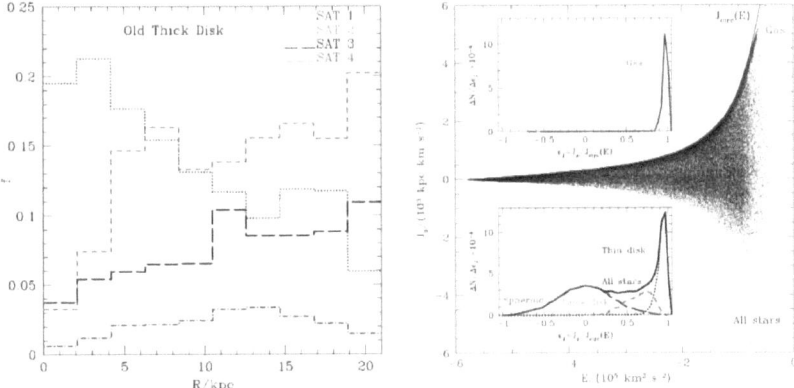

Fig. 5. Quantitative tests of numerical galaxy models are now becoming feasible. It is interesting to note how little apparent substructure is evident in the best conserved quantity, angular momentum, and how complex is the evolutionary history of a 'stellar population', such as the thick disk, in this model. This figure is from Abadi et al. 2003.

tensive RAVE survey (www.rave-survey.aip.de) using the 6-deg field UK Schmidt Telescope will have substantially improved study of the 9-13mag sky, complementing recent heroic star-by-star efforts (eg the Geneva-Copenhagen survey, Nordstrom et al. 2004). RAVE has in its first few months derived accurate kinematics and stellar abundances for 50,000 stars, illustrating just how large is the increase in data currently being implemented. Starting soon, the SEGUE extension of the Sloan survey to (anti-centre) galactic structure, targeting 200,000 stars, will revolutionise knowledge of the outer disk and halo. At the AAT the new AAOmega facility has the capability for a similar study in the south. Similar scale efforts are planned for the new wide-field Chinese survey telescope. We can ensure that the 8-m telescopes are not left behind, by pushing FLAMES to its limits, as has been done so impressively with UVES over its life to date.

In the next very few years stellar abundances and kinematics will be available for as many stars as redshifts are now available for galaxies. This abundance of information can, provided we approach the analysis and interpretation with due imagination, advance the astrophysics of galaxy formation as much as Cosmology has advanced over the last few decades. No doubt our image of galaxy evolution will be similarly revolutionised.

References

1. Abadi, M., Navarro, J., Steinmetz, M. & Eke, V. 2003 ApJ 597 21
2. Nordstrom, B., Mayor, M., Andersen, J. et al. 2004 A&A 418 989
3. Pierce-Price, D., et al. 2000 ApJ 545 L121

Author Index

ESO ASTROPHYSICS SYMPOSIA
European Southern Observatory

Series Editor: Bruno Leibundgut